D1064556

OIL AND GAS DICTIONARY

OIL AND GAS DICTIONARY

Editor
Paul Stevens

NICHOLS PUBLISHING: NEW YORK

First published in the United States of America by
NICHOLS PUBLISHING
Post Office Box 96
New York, NY 10024

Nichols Publishing is an imprint of GP Publishing Inc.

Library of Congress Cataloging-in-Publication Data

Stevens, Paul, 1947
Oil and Gas Dictionary/edited by Paul Stevens.
p. cm.
Bibliography: p.
ISBN 0-89397-325-4: $78.50
1. Petroleum industry and trade—Dictionaries.
2. Gas industry—Dictionaries. I. Title.
HD9560.5.S755 1988
333.8'23'0321-dc19 88-19626 CIP

Printed in Great Britain by
Richard Clay Ltd, Bungay, Suffolk

Contents

For Cassie

Acknowledgements

In a work of this nature, the editor inevitably accumulates a great many debts. The most obvious debt is to the other contributors who cheerfully accepted my weedling to extract yet further entries above and beyond those originally agreed. Their names are listed elsewhere, and I hope they find the finalized version within acceptable bounds.

At the start of the project, a major input was made by Jeanne Connelly who, as the Macmillan Research Fellow, helped enormously with the creation of the keyword index trawling through the various books and trade press for sources of jargon. She also acted as whipper-in on the contributors, and generally through her efficiency and hard work lulled the editor into the sense that the dictionary would write itself. When this proved to be an illusion, thanks must go to Rosemary Foster and the team at Macmillan who provided comfort and encouragement despite a succession of missed deadlines. Liz Blakeway of the Economics Department at Surrey University entered most of the text into the computer, and when the first draft of the manuscript was finished Dennis Lee of BP's Research Centre at Sunbury arranged to have the technical entries checked for howlers. Guy Pulham also assisted with the creation of the chronology.

When I had finally finished the manuscript, my detestation of it was so great that Macmillan wisely provided me with the excellent services of Ann Ralph to edit my editing and to ensure the cross-referencing was adequate and accurate.

Ideally I would love to be able to blame any errors on those listed above for not spotting them earlier, but alas I am obliged to accept the usual responsibility of the editor for any errors which may have crept in and remained undetected.

My final thanks are due to my wife Cassie to whom the book is dedicated. Her patience and sympathy when faced with the phrase 'when the dictionary is finished' for over four years were decidedly above and beyond the call of duty. It was, however, much appreciated.

Introduction

There is an old saying which states that 'if you are not confused you don't know what is going on'. During the 1950s and 1960s, the world was generally content to leave the business of oil and gas to the oilmen. Although the oil companies were regarded by many with suspicion, in general it was felt that they were best left to get on with the job of market-markers in the sense of getting energy from underground to the consumer in a useable form. Attempts to gain access to this world were invariably met by a barrier of ignorance which for many was simply too high to be scaled. The number of academic economists in the world with any interest or knowledge of the international oil and gas industry could probably be counted on one hand. Industrial country governments were in general woefully ignorant of the way the international industry operated, and it is probably true to say that knowledge about the oil and gas industry was greater in the Third World. Outside the industry very few knew what was going on because few thought it important enough to attempt to scale the barriers of ignorance.

This basic complacency was shattered in 1973 when the price of oil quadrupled within three months, and the prospects of politically motivated interruptions to oil supplies loomed large on the horizon. All of a sudden, energy in general and oil and gas in particular entered centre stage. Not only were oil and gas crucial as energy inputs per se, but they also began to impinge on other areas such as banking and international financial matters in a way that was previously unthought of. This clamour of attention on the industry has not gone away. The Second Oil Shock of 1979–80 reinforced the central position of oil and gas in the world's economic scene. If interest was high when the price of oil was high, the level of interest did not fall as the price of oil fell. The 1980s saw more than ever uncertainty, and interest was generated by the fall in oil price together with its likely implications.

All of a sudden after 1973, government officials, businessmen, bankers and a host of others involved in the world economic scene began to take a serious interest in the activities of the oil and gas industry. However, they immediately came up against the curse of communication — jargon. This can be categorized into three types: technical jargon (mainly engineering and scientific terminology); economic and financial jargon ; and industry-specific jargon used by scientists, engineers and economists. These three categories are covered in this book, but in different ways.

Since the industry began in 1859 it has tended to attract what might be described politely as 'colourful' characters. In what was (and is) a rough and tough industry there developed the inevitable camaraderie between those involved whether they were competing or colluding. Inevitably to secure their identity as different from mere mortals they developed their own technical language which frequently bore little or no relationship to what was being described. Hence, although a term might describe an engineering concept, an engineer outside the industry would be baffled by this term. In addition, there was the straightforward jargon of the engineering and natural sciences which, although being accessible to fellow engineers or scientists in other lines of work outside the oil and gas industry, leaves the 'arts' person bewildered.

A literature on the subject of technical jargon has developed, and many dictionaries and handbooks have appeared which have tried to explain to the non-technician the intricacies of the jargon. The problem is that much of the explanation is too technical, and it provides far more than the non-technician usually requires. For example, someone, who wishes to discover that a cat cracker is not an animal-hater but part of the refining process, does not necessarily want an explanation with which he could then obtain a degree in chemistry or chemical engineering. Many of the attempts at exposition assume so much prior knowledge on the part of the reader that they rapidly become unintelligible.

With this in mind it was decided to include technical entries written by non-technicians. The logic behind this is that frequently the best teacher is one who has learnt the subject recently, since they are aware of the problems and pitfalls that face the newcomer. Many of these technical entries are defined twice. Once individually where a normally brief explanation provides some guidance to the curious, and once again in a wider context and in greater detail for the reader who wishes for a little more depth. However, the oil engineer or scientist will learn nothing from such entries and will possibly be irritated by their simplicity. They have been explicitly designed for economists, financiers, etc. and therefore where possible the economic and financial implications of the technical term are emphasized.

The second category of jargon is the economic and financial terms. Unlike the technical terms these have remained relatively unadulterated. The rough and the tough were engineers not economists. Thus as defined here they will to the professional economist seem at best simple, although where possible their relevance to the oil and gas industry has been stressed. However, hopefully to the engineer they will shed light on the jargon used by the financial people to explain to the engineer why they cannot have what they want. To the best of my knowledge, there is no such dictionary accessible to engineers and other technical people; admittedly there are dictionaries of economics, finance and accountancy, but these dictionaries are general works and frequently (like the technical dictionaries mentioned above) are aimed at someone with prior knowledge. Also they do not make explicit the link to the oil and gas industry. This the book attempts to do.

Finally we come to the industry jargon. This is the jargon which the professional outside of the industry would find baffling. In this book it is explained for the non-industry specialist. As such it might be expected to provide little light for the insider. However, my experience of the industry suggests that increasing specialization within the companies at all levels means that people in one section know less and less of what goes on (and the language used) in another section. Hopefully therefore even some insiders will also be illuminated.

The above outline is one way in which the entries could be categorized. Another method of categorization which is relevant when the book is used is the distinction between fact and opinion. Many of the entries are simple facts and (allowing for Karl Popper's definition of a scientific law as a hypothesis yet to be disproved) incontrovertible. Thus it is a matter of fact that in October 1973 a number of Gulf countries announced an increase in the price of oil. Hopefully there are many such 'facts' in this book, and equally hopefully none of them are wrong. However, also the book contains other information of an analytical and explanatory nature. For example, *why* a number of Gulf countries increased the price of oil in October 1973. It is at this point that the dictionary takes on an encyclopaedic role. Indeed I would like to argue that this 'why' information is more interesting and important than the 'what' information of the facts.

There is a problem however: unlike the precision possible in engineering and the natural sciences, economics, politics and the other social 'sciences' are far from precise. The making of jokes about economists failing to agree have taken on the characteristics of a cottage industry. Thus the explanations contained in this book must not be seen as the last word on the subject. Indeed if my fellow energy economists read the book and are not thoroughly irritated by much of the content because they disagree then in part I have failed in the process of editing. Where possible I have indicated that a particular explanation or piece of analysis is controversial, but perhaps I have not always made this point strongly enough. The reader therefore needs to treat what is with a healthy scepticism. He or she should also be warned that the oil and gas industry is traditionally a graveyard for consensus views. Once they emerge they (it would appear almost invariably) are rapidly shown to be wrong. This, of course, is unsatisfactory for the non-expert reader. Unfortunately, this is in the nature of the beast, and the real world does not and cannot operate on certainty. This tends to be even truer in the international oil and gas industry. In a sense for the complete beginner it is a salutory lesson on the nature of industry. The only advantage of the book in this context is that it attempts to provide an analytical framework based upon (mainly) economic theory in which to ask the 'why' questions. Thus it has been said that the acute observer of the industry is not more knowledgeable than the others, he just has his ignorance better organized. Hopefully the reader of this book will equally find some sense of order in which at least to contemplate the uncertainties.

A

absorptive capacity. The ability of the oil-producing countries to spend internally or absorb their oil revenues. The concept emerged following the FIRST OIL SHOCK when oil revenues increased sharply and was linked to the RECYCLING problem. The implication was that oil producers with a low absorptive capacity would be unable to spend their revenues. This would reduce world levels of spending, thereby leading to a recession. In the event most countries designated as having a low absorptive capacity exhibited a remarkable ability to spend money, although how useful this spending was for the country's development remains highly questionable. The producers usually identified as low absorbers are Saudi Arabia, Kuwait, the United Arab Emirates, Libya and Qatar. The expression is often used to designate a split of interest within the oil producers. Hence low absorbers are assumed to have a lower need for immediate revenues than high absorbers and in theory might be willing to accept lower prices or production volumes.

ACC. *See* ANNUAL CAPITAL CHARGE.

accept or reject decision. The decision as to whether or not to go ahead with a specific project as opposed to a situation of whether to go ahead with one of several different project options. *See also* INVESTMENT APPRAISAL.

accommodating transactions. *See* BALANCE OF PAYMENTS.

accommodation platform (flotel). A PLATFORM or SEMI-SUBMERSIBLE RIG whose function is to act as a hotel for OFFSHORE personnel. The accommodation platform is often attached to a PRODUCTION PLATFORM.

accounting rate of return. One possible criterion by which to evaluate the economic viability of a project or to evaluate the performance of a company. It may be measured by the ratio of income to the BOOK VALUE of the asset or by the ratio of income to the cost of the investment. The main weakness of the measure is that it ignores the time value of money (*see* TIME PREFERENCE). *See also* INVESTMENT APPRAISAL.

accumulator. A vessel for the temporary storage of oil or gas. It is used to ensure a continuous flow operation if there are capacity imbalances in the system.

acetylenes. *See* ALKYNES.

Achnacarry Agreement ('as is' agreement, pool association). An agreement negotiated in September 1928 at Achnacarry Castle in Scotland between Royal Dutch Shell, the Standard Oil Company of New Jersey and Anglo–Persian Oil Company, and later joined by most of the other large oil companies. Preceded by an intensive price war between the MAJORS (*see* CRUDE OIL PRICES), the agreement was a form of loose CARTEL intended to restore order in the international oil market by, among other things, acceptance and maintenance of market shares and the sharing of facilities to avoid over-capacity. The agreement also tried to minimize price competition by strengthening the GULF PLUS basing point pricing system. In essence, the agreement was designed to restrain competition which followed from the growing excess capacity to produce crude oil. The agreement provided broad principles on which further detailed agreements could be made. By the early 1940s the US ANTI-TRUST LEGISLATION effectively precluded the US companies from adhering to the agreement. Details of the actual agreement did not become public until US Senate hearings forced disclosure in the early 1950s. Sometimes the Achnacarry Agreement is compared with the OPEC LONDON AGREEMENT of March 1983, because both were seen as attempts by the main actors to cope with excess producing capacity.

acid gas (hydrogen sulphide). A gas that can cause considerable problems of corrosion and therefore has to be scrubbed (*see* SCRUBBING) or removed from a gas flow as soon as possible.

acidization. *See* ACID TREATMENT.

acid sludge. *See* SLUDGE.

acid treatment (acidization). A method of improving the flow of oil in certain types of RESERVOIR in order to maintain oil production. *See also* PRODUCTION, NATURAL DRIVE MECHANISMS.

acreage. The commonly used term for a land area which has been granted for exploration purposes.

acyclic. A chemical term describing the arrangement of atoms within an organic compound. In acyclic compounds the carbon atoms are linked in an open chain, as distinct from a ring or CYCLIC structure. *Compare* AROMATICS. *See also* CHEMISTRY.

adaptive forecasting. Forecasts that adapt to the latest data available to the forecasters, generally by giving most weight to the most recent information. Exponentially weighted moving averages, for example, use a system of geometrically declining weights to give less and less weight to data the further back in the past they are.

administered price. *See* CRUDE OIL PRICES.

ad valorem tax. A percentage INDIRECT TAX based on the value of a good rather than on the weight or some other measure of the good. Ad valorem tax is usually compared with a specific tax, which is expressed as a fixed sum of money per unit of the good. During the early 1970s many of the consumer governments' indirect taxes on petroleum products were specific. Thus when the price of crude oil increased in 1973, the share of indirect taxes in the final price to consumers fell, thereby cushioning the consumer to some extent against the rising oil price. This was generally the result of a deliberate policy by the consumer governments who were concerned with inflation and the competitiveness of their exports.

Advanced Petroleum Revenue Tax. *See* PETROLEUM REVENUE TAX.

advanced recovery methods. The secondary recovery and tertiary recovery (i.e. enhanced oil recovery) methods of increasing the amount of oil recovered from a RESERVOIR when primary recovery (i.e. natural drive mechanisms) is causing a sharp decline in production due to falling pressure. *See also* PRODUCTION, NATURAL DRIVE MECHANISMS; PRODUCTION, ENHANCED OIL RECOVERY; PRODUCTION, SECONDARY RECOVERY MECHANISMS.

affiliate. A company owned by another company.

AFM prices (Aussen Handelsverband für Mineralöl ev prices). Prices of oil based upon actual transactions in the market in the previous week; these are similar to ROTTERDAM PRICES. Started in the early 1960s by a group of German importers, they were intended to overcome the uncertainties associated with other reported spot prices. *See also* SPOT MARKET.

AFRA (average freight rate assessment). A monthly assessment of the freight rates actually paid in the preceding month for the carriage of oil in chartered tankers. It is published by the London Tanker Brokers Association in order to provide the oil industry with an independent evaluation of the freight cost element in transactions with and between the oil companies. The period covered by each assessment is from the 16th day of one month to the 15th day of the next. Although expressed in terms of WORLDSCALE rates, it can be converted into dollars per ton by reference to the tables of base rate values available to subscribers to AFRA.

Since its inception in 1954, AFRA has been frequently revised to take account of the differences in rates enjoyed by vessels in different size categories. The present scheme provides AFRA rates for vessels in the following size ranges (thousand dwt): 16.5–24.999 (general-purpose); 25–44.999 (medium-range); 45–79.999 (large-range 1); 80–159 (large-range 2); 160–320 (very large crude carrier, VLCC); 320 and over (ultra-large crude carrier, ULCC). AFRA is a weighted rate for each size category calculated over vessels on long-term CHARTER (i.e. over one year), short-term charter (i.e. less than one year) and a single voyage, on the basis of a notional average voyage for the world fleet. The main difference between AFRA rates and the market rates is the inclusion of time charter rates currently operating for vessels in the particular time period. Hence AFRA rates tend to follow a smoother pattern than market rates.

From an economic point of view AFRA rates are of little value in assessing the current marginal cost of oil transport. They are largely of accounting and legal interest, having been widely used in transfer pricing (*see* TRANSFER PRICE) within oil companies and in tax disputes between companies and producing countries.

Agip. *See* ENI.

air pollution. The introduction into the atmosphere of substances that could be harmful to the environment. The two commonest air pollutants are carbon monoxide, which results from the operation of motor vehicles, and sulphur dioxide, which results from the burning of fossil fuels containing sulphur. Both carbon monoxide and sulphur dioxide have economic implications. In the case of sulphur dioxide its production is a function of the extent of sulphur in the fuel, hence low-sulphur fuels command a market premium (*see* SULPHUR PREMIUM) because it is expensive to scrub (*see* SCRUBBING) the emissions. Measures taken to control sulphur dioxide emissions that were once acceptable are now under challenge as concern grows over acid rain. Pressure is growing to reduce the emission of carbon monoxide and other substances — most notably lead — from internal combustion engines. This has implications on the cost of the fuel (e.g., LEAD-FREE MOTOR SPIRIT as well as the cost of the engines and exhaust systems). *See also* QUALITY, CRUDE OIL; QUALITY, DIESEL OIL; QUALITY, DISTILLATE AND RESIDUAL FUELS.

alcohols. There are various types of alcohol. However, in the context of this book, the one commanding most interest is ETHANOL (ethyl alcohol), which is derived from the fermentation of plant products such as sugar-cane or grain. Ethanol can be used as a direct substitute for motor spirit, but it requires a specially designed engine. Alternatively, with minimal change to the engine, a percentage of the motor spirit can be replaced by ethanol.

After the rise in crude prices during the 1970s many governments, most notably in Brazil and the USA, introduced programmes designed to substitute alcohol for motor spirit. Although the economics of many of these programmes was very uncertain, once the projects were begun governments were forced to continue with them. This is either because the car stock had been redesigned to run on ethanol and/or because elements within the agricultural sector have a strong vested interest in the continuation of the programmes. Thus, irrespective of crude oil prices, the programmes develop a momentum of their own.

aliphatic. A description of organic compounds with open chains of carbon atoms (e.g., ALKANES, ALKENES, ALKYNES), as compared with the closed chains of the AROMATIC compounds. *See also* CHEMISTRY.

alkanes (paraffins, saturated hydrocarbons). One of the main classes of HYDROCARBON compounds; the other two main classes are unsaturated hydrocarbons (i.e. ALKENES and ALKYNES) and AROMATICS. *See also* CHEMISTRY.

alkenes (olefins). A group of unsaturated HYDROCARBON compounds characterized by one or more double carbon–carbon bonds. This group provides the major petrochemical raw materials. *Compare* ALKANES; ALKYNES; AROMATICS. *See also* CHEMISTRY; PETROCHEMICALS.

alkylation. A REFINING process by which the OCTANE NUMBER (or ANTI-KNOCK RATING) of the lighter components is improved in order that they may be used in the manufacture of motor spirit. *See also* MOTOR SPIRIT UPGRADING.

alkynes (acetylenes). A group of unsaturated HYDROCARBONS containing triple carbon–carbon bonds. Because alkynes are highly reactive, combining easily with other substances, they play a key role in the production of PETROCHEMICALS. *See also* CHEMISTRY.

American Petroleum Institute. *See* API.

American Society for Testing and Materials. *See* ASTM.

American tanker rate schedule (ATRS). *See* WORLDSCALE.

AMI. Area of mutual interest (*see* JOINT VENTURES).

amortization. The extinguishing of a debt by means of a SINKING FUND or writing down the value of assets over a period as in DEPRECIATION.

amyl nitrate. An additive that theoretically may be used to increase the CETANE NUMBER of diesel; in practice it is little used. *See also* QUALITY, DIESEL OIL.

anchor string. *See* DRILLING.

Anglo-American oil agreement. An inter-government agreement signed in 1944 that was intended to provide US companies with access in the Middle East and to cope with the anticipated future glut of oil. This was to be carried out via an International Petroleum Commission which would recommend action to both UK and US governments. However, the oil companies did not welcome the agreement with the result that it did not come into operation.

Anglo-Iranian Oil Company. *See* BP.

Anglo-Persian Oil Company. *See* BP.

aniline point. A property of DIESEL OIL relevant to the calculation of the DIESEL INDEX, which is used to predict its combustion properties. *See also* QUALITY, DIESEL OIL.

Annex A. A document provided by an oil company to the UK government that gives an overall description of the oil-field project. Its official title is the Field Development and Production Programme, which accurately describes its function. *See also* ANNEX B.

Annex B. A more detailed version of ANNEX A. It includes such details as manpower resources and anticipated well performance. Its details have to be agreed by the UK Secretary of State for Energy before production can begin.

annual capacity. *See* CAPACITY.

annual capital charge (ACC). A criterion used to evaluate a project in DISCOUNTED CASH FLOW methodology. It assesses whether or not the net CASH FLOWS are sufficient to provide an adequate rate of return on the initial capital invested in the project and to allow recovery of the full capital cost.

annuity. A constant sum of money per period to be paid over a number of future time periods. Many financial analysis textbooks include tables showing the PRESENT VALUE of annuities at different rates of discount and for different periods of time. *See also* PERPETUITY.

annulus. The space between the drilling string, or pipe, and the wall of the hole being drilled. *See also* DRILLING.

anticipations data. Survey data obtained by asking households or companies about their future plans on, for example, spending. Such data can be useful adjuncts to FORECASTING.

anticlinal trap. A geological term for the commonest form of oil-bearing rock formation. *See also* GEOLOGY.

anti-knock rating. A product quality specification of motor spirit that defines the resistance of a fuel to KNOCK (or spontaneous ignition). It is expressed in terms of OCTANE RATING. *See also* QUALITY, MOTOR SPIRIT.

anti-oxidant inhibitors. Compounds added to motor spirit in order to prevent the formation of gum during storage. *See also* QUALITY, MOTOR SPIRIT.

anti-trust legislation. US legislation designed to control MONOPOLY power. The original act — Sherman Anti-trust Act — was designed specifically to break up large amalgamations of firms, of which the STANDARD OIL trust was a classic example. Subsequently it was very strictly applied. US oil companies have to be extremely careful to avoid contravention, even to the point of seeking special permission for employees to attend conferences at which representatives of other oil companies may be present. During the early 1970s, when joint approaches were required by the companies over the negotiations leading to the TEHERAN AGREEMENT and the TRIPOLI AGREEMENT or in the LIBYAN PRODUCERS AGREEMENT, special permission had to be obtained by the US oil companies to set aside the anti-trust legislation. *See also* ACHNACARRY AGREEMENT; MONOPOLY LEGISLATION.

API (American Petroleum Institute). A US institute that was created in 1919. It acts as a national trade association and as an information centre. In addition, it is responsible for publishing weekly statistics on the US petroleum industry. *See also* API GRAVITY.

API gravity. An arbitrary scale for the measurement of crude oil gravity that was devised by the American Petroleum Institute (API). API values are expressed in degrees API (°API) and are inversely related to specific gravity (*see* QUALITY, CRUDE OIL). The higher the API value, the lighter the crude, with most crudes falling within the range 30–45°API). In general the higher the API value the more valuable the crude. This is simply because the higher the API number the lighter the crude and therefore the more LIGHT ENDS or products (worth more than heavy ends or products) will be produced from a simple refining process.

appraisal drilling. *See* APPRAISAL WELL.

appraisal well (outstep well). A well that is drilled to provide information about the size and type of a find after oil has first been found. The main purpose of an appraisal well is to help to establish whether a field is commercial (*see* COMMERCIAL FIELD). Appraisal wells also provide initial information to help locate the DEVELOPMENT WELLS.

appreciation. An increase in the value of a ASSET over a period of time. A natural resource such as oil and gas held in the ground, for instance, may appreciate in value, as may a piece of property or a currency. *Compare* DEPRECIATION.

APQ (average programmed quantity). The formula that was used by the IRANIAN CONSORTIUM to determine production levels and their distribution between the partners. Each October the participants in the Consortium were told the expected production capacity for the following year. The participants then each nominated what they would like to lift (*see* LIFTING). Based upon their equity share this would produce a notional output. The participants were then ranked with the highest notional output at the top, and their equity share was cumulated until 70 percent was reached (or exceeded), at which point that output became the APQ which would be produced. Those companies in the list above the APQ company would receive less oil than they wanted, and those below would have the option on more crude than they wanted. The latter companies (the underlifters, *see* OFFTAKE NOMINATIONS) could if they wished use their unwanted portion of production to supply the companies who were left short of crude. The APQ system was a example of how the MAJORS managed to control the oil industry in the 1950s and 1960s. The starting point of the exercise was capacity, and since the APQ was always below capacity this exerted a downward pressure on the development of that capacity. This dampening effect was reinforced by the fact that the three main Aramco partners — EXXON, SOCAL and TEXACO — invariably put in relatively low notional output levels (derived from low nominations), which helped to pull down the APQ. The 70 percent cutoff meant that the European and US groups had the power to block each other since neither alone could reach the 70 percent equity required.

APRT. Advanced Petroleum Revenue Tax (*see* PETROLEUM REVENUE TAX).

aquifer. A pool or layer of underground water mixed into porous rook. *See also* GEOLOGY.

AR. Average revenue (*see* REVENUE).

ARA. The commonly used abbreviation that denotes the vast amount of refining and storage capacity located in and around Amsterdam–Rotterdam–Antwerp.

Arabian–American Oil Company. *See* ARAMCO.

Arabian Gulf. *See* GULF, THE.

Arabian light. *See* MARKER CRUDE.

Arab oil embargo. In the spring of 1973, King Faisal of Saudi Arabia promised Egypt that, in the event of an Arab–Israeli war, an oil weapon would be used by the Saudis. On 17 October 1973, 11 days after the start of the October War (sometimes called the Yom Kippur War) the Saudi-sponsored plan was revealed. Countries were classified as: 'most favoured' (which included the UK, France and other Arab and Islamic countries), who were to get their full current demand requirement; 'preferred', which was an earlier version of 'most favoured'; 'neutral', who were to receive no special treatment; 'embargoed' (which included the USA and The Netherlands), who were to receive no oil from the Arab oil producers. However, it was realized that, because of the interlinked nature of the international oil industry, the embargoed countries could simply import from non-embargoed countries. It was therefore announced that the Arab oil-producing countries would cut production by 5 percent each month (based on September 1973 production) until the Israelis withdrew from Arab territories and Palestinian rights were restored. In November 1973, the cutback was raised to 25 percent. The logic behind these production cutbacks was that other countries would be too concerned about their own supplies to redirect supplies to the embargoed countries. The measures were relaxed towards the end of December 1973 and effectively disappeared by March 1974.

With the benefit of hindsight it can be seen that the embargo was ineffective, both in terms of reduced crude oil availability to the USA and The Netherlands, and in terms of pressurizing either government. Despite this, however, the embargo was very important. Following the sharp rise in the price that was instituted the day before, it injected a note of panic into the market. In particular it raised to a serious level the prospects of further supply disruptions for purely political reasons. The subsequent scramble for supplies did much to cause a strengthening of the oil market, which in turn encouraged OPEC to complete the FIRST OIL SHOCK by increasing prices further at the Teheran meeting in December 1973.

Aramco (Arabian–American Oil Company). The main oil operating company in Saudi Arabia. Created in 1948, it consisted of four of the US MAJORS — EXXON, SOCAL, TEXACO and MOBIL — and was effectively taken over by the Saudi Arabian government in 1976, although the former Aramco partners retain very close links.

Aramco advantage. During the SECOND OIL SHOCK Saudi Arabia tried to prevent a general rise in oil prices, although at times the policy was confused and confusing. It tried to do this by increasing its own production and by refusing to charge the premium on official prices that most other producers were charging. Thus those who could gain access to this crude, such as the former partners to Aramco, were effectively gaining cheap crude when the international price was rising. The Aramco advantage existed to one degree or another until October 1981 when Saudi Arabia finally brought its prices into line. *See also* CRUDE OIL PRICES.

arbitrage. The opportunity for arbitrage arises when the price of the same commodity (or yield of similar assets) is different in different markets. Arbitrageurs would then buy such a commodity in the lower price market and sell it in the higher price market in pursuit of a safe profit. As a result, prices in different markets will tend to move towards a unique level, as buying pressure drives the price up in lower-price markets and selling pressure drives the price down in higher-price markets. Arbitrage occurs most frequently in financial markets, such as foreign currency and security markets. In commodity markets, differences in price ratios are often more relevant than differences in price levels to indicate the existence of profitable arbitrage opportunities; thus, goods would be bought in markets where they are relatively inexpensive and sold in markets where they are relatively expensive (taking into account trading costs). In this case the arbitrage forces tend to equalize price ratios across markets.

arc elasticity. *See* ELASTICITY.

area of mutual interest agreement. *See* JOINT VENTURES.

area rental. A cash sum paid in instalments throughout the period of a production licence. An area rental is required in both Norway and the UK. An alternative is a fixed signature bonus (*see* FIXED BONUSES), which is equivalent to the area rental, but is paid in a lump sum at the commencement of the licence; The Netherlands employs both systems. In general the amount of area rental paid depends upon the degree of interest in the ACREAGE. The more interest from the companies the more they would be willing to pay. However, in some cases where the government desperately wishes to find oil, area rentals are often reduced, or even dispensed with altogether, in return for a higher commitment on exploration expenditure.

arm's length. A description of goods traded outside of the affiliates of a MULTINATIONAL corporation (i.e. trade between independent companies). Arm's length crude oil has been especially important in the development of the international oil market. During the 1940s and 1950s the quantity of such crude was extremely small. However, during the 1960s and after, new sources of crude oil became available to companies who lacked sufficient refinery capacity to process the oil and who urgently required the revenue from the crude to recover their investment in the development of crude-producing capacity. This led to a gradual expansion of the arm's length market which, because of the general excess crude capacity in existence, tended to generate price competition, causing an erosion of crude oil prices. Since the break up of the vertically integrated structure (*see* VERTICAL INTEGRATION) in 1979–80, the majority of international crude oil traded is arm's length crude.

aromatics. One of the main group of HYDROCARBONS, the other two being saturated hydrocarbons (*see* ALKANES) and unsaturated hydrocarbons (*see* ALKENES, ALKYNES). *See also* CHEMISTRY.

aromatics content. A product quality specification for aviation fuels, limiting the amount of AROMATICS present, since these burn with a smoky and hence undesirable flame. *See also* QUALITY, AVIATION FUELS.

ASCOOP Agreement. An agreement (oil and gas cooperative association) signed in 1965 between Algeria and France to govern fiscal relations between the Algerian government (represented by SONATRACH) and the French oil companies.

ash content test. A measure of the residual waste of fuel following burning.

'as is' agreement. *See* ACHNACARRY AGREEMENT.

asphalt. Bitumen (*see* FINISHED PRODUCTS, BITUMEN).

asphaltenes content. A quality characteristic of crude oil that indicates its yield of residue (from which heavy products are made). *See also* QUALITY, CRUDE OIL.

asphaltic crude. *See* HEAVY CRUDE OIL.

assay. *See* QUALITY.

assets. An item of value, forming part of the wealth of the owner. Assets may be 'real' (e.g., an oil or gas reservoir, buildings and machinery) or 'financial' (e.g., money, BONDS and EQUITIES). In accounting terms, everything owned by a company that has a monetary value is classified as an asset. On the balance sheet these assets can be categorized under four headings:

(a) *current assets.* Items that are either cash or can easily be converted into cash (e.g., bills receivable, stocks and work in progress and marketable securities).

(b) *trade investments.* Investments in subsidiary or associated companies.

(c) *fixed assets.* Items that are usually valued at cost less DEPRECIATION (e.g., land, buildings, plant and machinery).

(d) *intangible assets.* Items that have no tangible value (e.g., goodwill and patents).

A collection of SECURITIES is known as a portfolio, although this latter term is also applied to any collection of assets. For example, production of oil and gas involves exchanging one asset (e.g., underground oil and/or gas) for another asset (i.e. cash paid for the oil and/or gas). This implies that the money received for producing and selling the oil or gas is not in fact income, but merely a reshuffling of the country's asset portfolio. This view has led to suggestions that oil and gas revenue (and revenue from other EXHAUSTIBLE RESOURCES) should be excluded from measures of national output (*see* DOMESTIC INCOME).

assignments. The selling of part or all of a company's rights to another company when it obtains a concession/licence to explore for, develop or produce oil or gas. This often happens when the company finds itself short of cash or when production exceeds the handling or marketing ability of the company. In most cases the company must acquire government approval before assigning or sub-leasing any rights of the initial licence. A method used to avoid non-assignments in the UK is a CARVED-OUT OVER-RIDING ROYALTY, whereby a right to receive a proportion of oil produced is sold for cash which can then be employed in exploration.

associated gas. Gas that is present as a by-product in crude oil. When the oil reaches the surface it is mixed with the gas. Before the crude oil can be pumped for transportation this gas must be separated from the oil. In many cases because oil-fields tend to be located far from the market for energy and because of the problems of transporting the associated gas it has been burnt at the oil-field (*see* FLARING).

Flaring has been a cause of concern to producer governments who have regarded the gas as a valuable national resource which they considered was being wasted. In most cases prior to the rise in energy prices during the 1970s (*see* FIRST OIL SHOCK), it was not economic to sell the gas, although at some cost it could have been reintroducing into the oil-field to help maintain the reservoir pressure (*see* GAS INJECTION). As oil-producing governments began to take control of the production operations during the early 1970s, one of their first moves was to reduce gas flaring. This was initially done by reinjecting the gas into the reservoir, but gradually gathering systems were developed to deliver the gas for use either as an energy source or as a petrochemical FEEDSTOCK. Use of the gas, however, carried important implications for the crude oil production decision: as the countries began to depend on the associated gas this placed a constraint upon how far they could reduce oil production. This was particularly significant for OPEC following the production cutbacks which were part of the 1983 LONDON AGREEMENT. If oil production fell too far then a shortage of associated gas meant gas-processing equipment would be running below capacity and there would, in effect, be an energy shortage in the gas-using sectors such as electricity generation. The only solution, apart from increasing oil production, was to try to discover and develop sources of non-associated gas which could be used as a 'stored' substitute. This, however, took both resources and time. *See also* NATURAL GAS.

associated liquids. Liquid hydrocarbons found in association with NATURAL GAS.

Association of British Independent Oil Exploration Companies. *See* BRINDEX.

ASTM (American Society for Testing and Materials). The US organization responsible for the issuing of many of the standard methods of assay used in the petroleum industry. It also provides specifications for products that are used internationally.

ATK. *See* AVIATION TURBINE KEROSINE.

atmospheric distillation. A basic REFINING process in which crude oil is heated in order to break it down into a number of intermediate components (*see* CUTS), from which finished products can be made. *See also* REFINING, PRIMARY PROCESSING.

atmospheric residue. The residue remaining after CRUDE OIL DISTILLATION. It consists basically of a HEAVY FUEL OIL type of material. *See also* REFINING, PRIMARY PROCESSING.

ATRS.. American tanker rate schedule (*see* WORLDSCALE).

attapulgite (salt gel). A clay used in salt water drilling muds (*see* DRILLING FLUID) to improve the mud's carrying capacity by its ability to hold solids in suspension. This prevents cuttings produced by the drilling from settling at the bottom of the WELL BORE.

auction system of licensing. There are many types of auction or competitive bidding systems for the allocation of oil licences.
(a) The bonus bidding (or lease sale) system involves a company, or more than one

company offering a joint bid, entering a competitive auction, bidding monetary payments (i.e. the BONUS PAYMENTS) either paid in a lump sum or in instalments for the lease. This is the most common form of auction employed in the USA.
(b) A variation of the bonus bidding system is royalty bidding (or oil pledging), whereby companies compete by offering to the licensing authority a proportion of the oil to be produced. This system is equivalent to a tax on production.
(c) Net profit bidding (or percentage profit rate bidding) involves a percentage of net annual returns being paid to the licensor. This differs from the bonus bidding system in that costs are considered.
(d) Rent bidding takes place based on a rental rate for the lease payable until the lease expires.

Other forms of auction systems include the method of bidding according to research and development expenditure. In its most simple form the company willing to spend the most on research wins the licence.

The service lease system requires oil companies to bid for the right to provide specific services, such as exploration drilling, thus the licensing authority is contracting out various functions.

Bidding on the basis of work programmes may occur whereby the technical and financial ability of an oil company is determined by the submission to the licensor of a detailed work plan. This is likely to be used in conjunction with other licensing systems.

The carried interest (or shared interest) system requires oil companies to agree to 'carry' the licensor's interest through the exploration and appraisal stages. If oil reserves are discovered the licensor may then exercise the right to participate and has only to contribute some proportion of past costs incurred by the oil company. It may be the case that these costs are, in part, paid in the equivalent value of oil produced. In Norway the interests of the state oil company — Statoil — have been carried by foreign oil companies.

Frequently auctions take the form of sealed-bid auctions (or blind auctions) in which the bidders must enter a confidential bid by a specific date. It has been argued that the great advantage of such a system of granting licences is that it enables the government to capture effectively any ECONOMIC RENT that may be associated with oil or gas production provided that the bidding is competitive. Furthermore, supporters of such a view argue it is a more effective method of rent capture than taxing subsequent profits since the latter can introduce disincentive effects.

Aussen Handelsverband für Mineralöl ev. *See* AFM PRICES.

autonomous transactions. *See* BALANCE OF PAYMENTS.

avcat. *See* AVIATION CARRIER TURBINE FUEL.

average freight rate assessment. *See* AFRA.

average programmed quantity. *See* APQ.

average total costs. *See* COSTS.

avgas. Aviation gasoline (*see* AVIATION SPIRIT).

aviation carrier turbine fuel (avcat, JP5). A KEROSINE-type jet fuel used mainly by naval aircraft. *See also* FINISHED PRODUCTS, AVIATION FUELS.

aviation fuels. *See also* FINISHED PRODUCTS, AVIATION FUELS.

aviation gasoline. *See also* AVIATION SPIRIT.

aviation spirit (aviation gasoline, avgas). A fuel consumed by piston-engined aircraft. *See also* FINISHED PRODUCTS, AVIATION FUELS.

aviation turbine kerosine (ATK, avtur, Jet A). The principal aviation fuel, consumed by most commercial jets. *See also* FINISHED PRODUCTS, AVIATION FUELS.

aviation turbine gasoline (JP4 and Jet B). A wider CUT than AVIATION TURBINE KEROSINE which therefore gives it a lower FLASH POINT. It is less widely used than aviation turbine kerosine. *See also* FINISHED PRODUCTS, AVIATION FUELS.

avtag. *See* AVIATION TURBINE GASOLINE.

avtur. *See* AVIATION TURBINE KEROSINE.

B

backhaul. An arrangement by which a cargo is carried for all or part of a return journey of a tanker. More frequently, however, a return journey is usually made in BALLAST. Backhaul is in contrast to crosshaul in which vessels carrying similar crude or products sail the same route, but in opposite directions. In such a situation it would seem to be logical that the cargoes should be exchanged rather than transported. This, however, is not always possible.

back-to-back agreement. A situation that occurs essentially in the context of middle men. Once a sale has been contracted the seller then makes a specific purchase to meet that sale. This means the seller does not have to carry stocks and effectively minimizes his risk in the transaction.

back-up. The process of holding stationary one section of drill pipe while another section is added or removed. The back-up line is the wire rope used to hold the TONGS attached to the drill section during a back-up operation. The dead end of the back-up line is attached to a back-up post. The frequence of a back-up operation and the speed with which it is performed are important determinants of drilling time and therefore costs. *See also* DRILLING.

back-up line. *See* BACK-UP.

back-up post. *See* BACK-UP.

bailer. A drilling tool used for removing cuttings from the hole. *See* DRILLING.

balance of payments. A representation of a country's accounts with the rest of the world. Normally it is presented in a standard format containing the following items:

(a) *current account.* (i) Visibles/balance of trade measures the import and export of physical goods. (ii) Invisibles includes the import and export of services such as transportation and financial services together with the flows of remittances, interest on loans and dividends on investments. On current account a transaction that causes an inflow of foreign currency (i.e. an export) has a positive sign whereas one that causes an outflow of foreign currency (i.e. an import) has a negative sign.
(b) *capital account, long-term and short-term.* This means the inflow/outflow of financial funds with a distinction being made between private- and government-sector flows. The sign follows the convention of the current account. An inflow of foreign exchange (i.e. a capital import) is positive whereas an outflow (i.e. a capital export) is negative.
(c) *reserve movements.* Because the accounts are effectively a double-entry system of book-keeping the inflows and outflows of foreign currency must balance. Hence any imbalance in current and capital accounts must be accounted for by a change in the foreign exchange reserves. Thus if the inflows exceed the outflows foreign exchange reserves rise. However, rather confusingly, this is given a negative sign in order to balance the surplus, because it can in fact be regarded as an outflow of domestic currency. Similarly, if outflows exceed inflows the deficit must be paid for by drawing down the reserves. The sign here is positive, representing an inflow of domestic currency.
(d) *balancing item (errors and omissions).* Because of the double-entry nature of the accounts they must ultimately balance. However, normally due to mistakes in the collection of the statistics, inflows do not match outflows. These mistakes are corrected by inclusion of the balancing item

which is simply an arithmetic residual.

Since the balance of payments must balance it is necessary to define what constitutes a balance of payment surplus or deficit. This can be most clearly established by distinguishing between autonomous and accommodating transactions within the accounts. Autonomous transactions are those transactions which are carried out independently of the state of the balance of payments. Thus they are motivated by a desire for profit or satisfaction, or to meet a contractual obligation. Autonomous transactions include therefore all current account and long-term capital account transactions. By contrast, accommodating transactions are those carried out in response to the state of the balance of payments usually in order to secure the ultimate balance, These are transactions on short-term capital and reserve movements, although the former may well include elements of autonomous transactions. Thus conventionally balance of payments surpluses/deficits refer to the balance on current and capital accounts.

Oil and gas enter the balance of payments in two areas. In one area the export of oil and gas as physical goods enters the balance of trade on current account, as would any imported equipment used in production. Similarly, any services involved, ranging from insurance to transportation, would appear in invisibles. The other area involves financial flows either in the form of foreign capital invested in the sector or in the investment overseas of revenues from the sales of oil and gas. These flows would appear in the capital account, whereas the costs or profits/revenues from the flows would appear in invisibles.

balance of trade. *See* BALANCE OF PAYMENTS.

balancing item (errors and omissions). The item in the BALANCE OF PAYMENTS accounts that allows for errors in the collection of the statistics and forces the accounts to balance.

ballast. In oil tankers, seawater carried in the permanent ballast tanks in order to provide stability in all weather conditions, immersion of the propeller on ocean journeys and reduction of vibration. The need for large amounts of ballast — up to 50 percent deadweight tonnage — on return journeys when the tanker is empty creates potential environmental hazards arising from the disposal of oily water and the venting of gas. A UN agreement (MARPOL '73/78) required signatories to implement the complete segregation of ballast tanks for all new tankers over 70 000 dwt and for all existing tankers over 40 000 dwt as from June 1981. Such a system not only reduces pollution of the environment, but also has the advantage of reducing delays in loading and discharge at the port.

bareboat charter (demise charter). An agreement to hire or charter a tanker without crew, stores or fuel for a specified period. The charterer — usually an oil company — assumes responsibility for all operating and labour costs, and merely rents the vessel without restrictions on use. He is responsible for keeping the hull and machinery in good repair and is obliged either to return the vessel in the same condition as when chartered, ordinary wear and tear excluded, or to pay an agreed sum to the owners in the event of loss. A bareboat charter is attractive to charterers in periods of rapidly rising spot freight rates as it provides stability in transport costs. However, in the depressed tanker market following the FIRST OIL SHOCK, with low-cost tanker capacity available in the spot or voyage market, bareboat chartering has declined in importance. *Compare* TIME CHARTER; VOYAGE CHARTER.

barefoot completion. The simplest and easiest method of completing a productive well. The final string of the CASING is cemented to the CAP ROCK immediately above the reservoir. The oil then flows up the casing to a simple wellhead. *See also* DRILLING.

barge. A non-self-propelled vessel that can be used as a support facility in oil operations. The most common use of a barge, however, is to transport products in a river or coastal context. The size of barges can vary enormously; some can carry up to 5000 tonnes.

barge prices. *See* CARGO PRICES.

barium sulphate. *See* BARYTES.

barrel. The basic unit of volume used in relation to crude oil. It is defined as 42 US gallons. It is a remnant from the early days of the industry before the development of bulk handling. *See also* APPENDIX, TABLE 2.

barrels per day. *See* BD.

barrels per stream day. *See* STREAM DAY.

barter deals (counter-trade deals). The sale of crude, products or gas in return for other goods or services rather than for cash. Such transactions occur in situations either in which the buyer is short of foreign exchange or in which the seller wishes to disguise the fact that he is cutting prices. Barter deals are less popular with the seller than straight cash purchases. This is because they limit flexibility. A worldwide expansion in such arrangements is usually a sign of a weak market from the viewpoint of the seller.

barytes (barium sulphate). A compound that is added to DRILLING FLUID to increase its weight.

base load demand. *See* PEAK LOAD.

basic petrochemicals. The relative simple organic chemical compounds (e.g., ETHYLENE and BENZENE) from which other more complex PETROCHEMICALS are derived. For those countries that have tried to diversify away from crude oil and gas exports into petrochemicals the development of capacity to produce such basic petrochemicals was the first stage. During the 1970s concern grew among the traditional petrochemical producers — the OECD countries — about the potential competition from this newly developed capacity of the oil and gas producers. This would result in the areas that had traditionally produced petrochemicals having to import the basic petrochemicals and to develop them further. However, the prospect of developing an import dependence on what was seen by OECD as politically insecure sources was viewed by some as a powerful counter-argument.

basing point pricing (delivered pricing). A pricing system in which prices are calculated from one or more 'base' locations to all customers. To the base price are added standard freight charges, which vary according to the distance of the customer from the nearest base. Customers are obliged to pay delivered prices, all transport being arranged by the suppliers. From the point of view of the seller such a scheme offers a reduction in direct price competition. In addition, the seller may obtain sales outside of its own market area without the need to reduce prices in all markets by accepting lower 'netback' prices — a form of price discrimination between markets. Basing point pricing was practised in the international oil industry until about 1948. *See also* GULF PLUS.

Basrah Petroleum Company. *See* IPC.

batching. A process by which different crude oils or products are transported through the same pipeline. The different throughputs can normally be separated by controlling the velocity of flow. Where this is not feasible, the separation is achieved physically by the insertion of a batching pig or sphere.

Batching enables greater capacity use of pipelines, which is crucial to the economic viability of a pipeline, since such a high proportion of the costs are fixed costs. Thus below-capacity operation spreads the fixed costs over a smaller throughput thereby sharply increasing average fixed costs.

batching pig. *See* BATCHING.

batching sphere. *See* BATCHING.

BD (B/D, barrels per day). A measure of production or capacity commonly used in the oil industry. For crude oil the day is a calendar day; for refineries it may be a calendar day or a STREAM DAY. When applied to a refinery it normally refers to the capacity of the PRIMARY DISTILLATION UNIT.

bed. A term used in geology to describe a layer of sediment or sedimentary rock of considerable thickness and uniform composition and texture. *See also* GEOLOGY.

benchmark crude price. *See* MARKER CRUDE.

bentonite. A magnesium aluminium silicate added to drilling fluid. Bentonite is also used in refining as a treating agent (to remove undesirable constituents) and as an additive to grease.

benzene. The simplest member of the ARO-

MATICS group of hydrocarbon compounds. It is used as a petrochemical raw material. *See also* CHEMISTRY; PETROCHEMICALS.

benzine. *See* PETROLEUM SPIRIT.

BHA. *See* BOTTOM HOLE ASSEMBLY.

BHDP. *See* BOTTOM HOLE DIFFERENTIAL PRESSURE.

bill of lading. A document issued by a ship owner or agent that acknowledges receipt of goods to be shipped as specified in a pre-arranged contract — a contract of affreightment. In the document is recorded, among other things, the name of the shipper, the ship's name, a full description of cargo, the port of shipment, the port of discharge, details of freight payment and optionally the name of the consignee. Bills of lading are governed by the Bills of Lading Act 1855, and the Carriage of Goods by Sea Act 1924 and 1971. A bill of lading binds the shipper to deliver the cargo in a pre-specified condition to the consignee at the port of discharge, for a stipulated freight.

Bills of Lading Act. *See* BILL OF LADING.

biogas (biological gas). Gas that is produced from the fermentation of animal and vegetable residues. Many view this as a fuel with great relevance for rural areas in THIRD WORLD countries. Other energy forms can have very high transport/transmission costs which make their supply to such areas prohibitively expensive especially when the poverty of such areas is remembered. Biogas can be produced *in situ* by using locally available feedstock in a biogas digester. However, these digesters, which are effectively concrete tanks, can also be expensive to build given the poverty levels. There is also the problem of purchasing a gas-utilizing appliance. As a result of these factors, the initial enthusiasm that greeted the development of biogas on an organized scale has been somewhat muted.

biogas digester. *See* BIOGAS.

biological gas. *See* BIOGAS.

bit. The basic cutting tool used in DRILLING for oil. The resistance to wear of the bit is a crucial determinant in the cost of drilling. The harder the bit the less often must the DRILLING STRING be withdrawn, which is a time-consuming operation. Suitably hard materials for the cutting edge have been known for some time, but the problem lay in a means of firmly attaching the cutting edges to the bit. Recent developments in this area have produced much better bits.

bit breaker. A metal plate on the DRILLING TABLE that holds the BIT while it is being unscrewed from the DRILL COLLAR. *See also* DRILLING.

bitumen (asphalt). A heavy, normally solid, petroleum product used mainly in road surfacing. *See also* FINISHED PRODUCTS, BITUMEN.

black products. The heavier and darker products such as heavy diesel oil and fuel oil; used in comparison with 'white products' from the lighter end of the barrel such as motor spirit and kerosine. The distinction is most relevant with respect to transport and storage, since a black product carrier must be cleaned before it can be used for another product.

blanketing. The replacement of air in or around equipment with an inert gas to reduce the risk of fire hazard or explosion. Blanketing is also used to prevent oxidation.

blind auction. *See* SEALED-BID AUCTION.

blocks. The block and tackle system used in a DERRICK to raise and lower the DRILLING STRING. The system usually consists of a fixed system of pulleys — the crown block — to which is attached a moving pulley system — the travelling block). *See also* DRILLING.

blown oil. Oil in which the viscosity has been increased by blowing it with air at a higher temperature. This is usually applied to BITUMEN.

blow-out. An eruption of oil or gas from a well due to a failure of the containment system. The result is extremely serious since the escape of the oil and/or gas (usually under very high pressure) is uncontrolled, thus highly volatile liquids and/or gas are being sprayed into the environment. This can lead to explosions and fires, as well as pollution of the environment. *See also* BLOW-OUT PREVENTER.

blow-out preventer (BOP). A device used to control hydraulically the pressure in the well in order to prevent a BLOW-OUT.

BNOC (British National Oil Corporation). The UK state oil company. BNOC, which was established in 1976, was responsible for the participation of the UK government in the North Sea oil-fields derived from the 1974 White Paper (Cmnd 5696). It was also responsible for taking the UK government's royalty oil and for advising the Minister of Energy.

The major reasons put forward for the creation of BNOC were that it would permit control over the disposal of North Sea oil and be an effective instrument by which a national oil policy would be implemented. The establishment of BNOC by a Labour government was overwhelmingly a political act in order that the government would be seen to be protecting a national asset, by controlling the activities of foreign oil companies and by putting forward the country's interest within the industry.

At the same time the existence of BNOC added an additional organization, with its own criteria and preferences, active in the oil policy process. Hence it would be possible for BNOC to engage in tactical and strategic bargaining in an attempt to steer, or at least influence, government policy.

The status of BNOC in the offshore industry was unclear, and its terms of reference were vague. It was uncertain if it was a competitor in the industry or a neutral partner of the oil companies that listened and advised the government. Because of the presence of BNOC in North Sea consortia it would have direct access to considerable information. Thus there would be an imbalance of information between BNOC and the government, providing BNOC with the potential for influencing oil policy, thus possibly lessening the influence of the Department of Energy.

In 1982 the Conservative government privatized the exploration and production sectors of BNOC (with the formation of Britoil). However, the Conservative government recognized the political value of BNOC, both in its ability to control disposal and to act as an intermediary between government and industry on policy issues. On occasions since the beginning of 1983 the government used BNOC in order to intervene in the oil market and to attempt to maintain oil prices because of the importance of tax revenues.

During the period 1982–85 BNOC played a crucial role in the international oil market because of an institutional quirk. BNOC received the PARTICIPATION OIL and ROYALTY CRUDE OIL. Although BNOC theoretically had the powers to develop its activities DOWNSTREAM, it had not done so. Thus BNOC had to dispose of the crude oil. This was normally achieved by BNOC selling the crude oil back to the other oil companies at prices that were publicly known. However, if the potential buyer did not like the price demanded by BNOC one of two implications followed. BNOC, without refining capacity and with very limited storage, had either to reduce its asking price — tantamount to a visible price cut — or to sell on the SPOT MARKET, thereby causing spot prices to fall. BNOC had to do one or the other since it kept receiving the oil whether it wanted it or not and had to sell it.

The result was that either the price of North Sea crude fell officially (usually in a glare of publicity) or it fell in the market place. Normally Norway followed the official price cuts. This then provoked Nigeria, which produces crudes similar to the North Sea, into cutting its prices in order to protect its volume. A Nigerian price cut immediately threatened any OPEC agreement. In effect this institutional oddity provided an automatic 'self-destruct mechanism' to any OPEC agreement. OPEC members and observers generally did not understand this mechanism with the result that the actions of BNOC were seen as part of a plot by the UK government to destroy OPEC. This was reinforced by the oil companies' practice of SPINNING in response to the UK government's fiscal regime.

The abolition of BNOC, announced in March 1985, seems to have been a result of the continual criticism of the Corporation being employed by the Conservative government to intervene in the market. More importantly, the losses made by BNOC since the autumn of 1984, due to BNOC having to sell large quantities of oil at the spot price having previously bought it at the higher official price, caused the government considerable embarrassment.

B/O carrier. Bulk/oil carrier (*see* COMBINED CARRIER).

boiling point. The temperature at which a liquid vaporizes. Since it is related to gravity — the heavier the product, the higher the boiling point — it is frequently used as an alternative description of product type, particularly in the context of REFINING, with heavy products such as fuel oil, for example, being referred to as 'high-boiling-point materials'.

boiling range. A single pure substance has a given BOILING POINT at a given pressure. However, a mixture of substances, of which crude oil is an example, has a range of temperatures at which boiling/DISTILLATION takes place. Thus at lower temperatures the LIGHT END of the barrel boil first. This range of temperatures is the boiling range and varies between types of crude. *See also* REFINING, PRIMARY PROCESSING.

boil off. *See* LNG.

bonds. Marketable fixed-interest securities such as government GILT-EDGED stock and debentures, which pay the holder a certain sum per period until their redemption date. Preference shares in a company also often carry a fixed rate of interest, although they do not invariably do so. Since bonds carry fixed-interest payments, changes in the price at which the bonds are bought and sold implicitly change the rate of interest to be earned on the bond since there is a fixed return on a variable outlay.

bonus bidding (lease sale). Monetary payments (lump sum or instalments) offered in a competitive bid for the lease in an auction for acreage. Bonus bidding is most commonly used in the USA. *See also* AUCTION SYSTEM OF LICENSING; BONUS PAYMENTS.

bonus payments. Payments made by a company to the owner of the acreage, which is normally a government. These can be signature bonuses paid when the concession or contract is signed or production bonuses paid at pre-set levels of production usually on a sliding scale. Thus the higher the production level the higher the bonus.

book value. The value of an ASSET in a company's books of accounts — normally the written-down (after DEPRECIATION) value of those assets at a point of time. Book value is not generally the same as market value since the price at which the company's assets could be sold in the market will reflect their perceived future earnings potential rather than the sum to which they happen to have been depreciated. A variant is updated book value, which is the book value adjusted for inflation to bring the book value more in line with the present-day replacement cost.

booster platform. A PLATFORM attached to an underwater pipeline in order to boost the pumping process.

BOP. *See* BLOW-OUT PREVENTER.

border price. *See* SHADOW PRICE.

bottled gas. Normally PROPANE or BUTANE gas (or a mixture) for domestic use.

bottom hole assembly (BHA). The lower end of the drill string comprising the drill BIT, the DRILL COLLAR and the heavyweight DRILL PIPE. *See also* DRILLING.

bottom hole differential pressure (BHDP). The difference between the RESERVOIR PRESSURE and the pressure at the bottom of the producing well. This is a crucial figure in the economics of production since the higher the BHDP, other things being equal, the faster the rate of flow to the well bore. *See also* PRODUCTION.

bottoms. (1) The heavy end of the feed to a distillation process. (2) The sludge collected at the bottom of tanks; sometimes called bottoms settlings. This comprises a mixture of oil and water, and sometimes waxes, asphalt and mud. If allowed to collect it can significantly reduce the storage capacity of the tank

bottoms settling. *See* BOTTOMS.

bottoms settling and water. *See* BS&W.

BP (British Petroleum). A MAJOR. It began life in 1909 as the Anglo-Persian Oil Company based upon the oil discoveries made in Persia by D'Arcy. In 1914 the UK government took a controlling interest in the company for £2.2 million following the conversion of the ships of the Royal Navy to oil-fired boilers. The name was subsequently changed to Anglo-Iranian Oil. The present name was adopted in 1954. In 1987, the UK government sold its share in BP as part of the Conservative government's general programme of privatization.

BPSD. Barrels per stream day (*see* STREAM DAY).

branched-chain paraffins. *See* ISO-PARAFFINS.

branded distributor. A wholesaler of motor spirit who, although he uses his supplier's brand name, provides his own facilities and finance. *See also* BRANDED TANK TRUCK DISTRIBUTOR.

branded tank truck distributor. Similar to a BRANDED DISTRIBUTOR, but without permanent storage.

breakout. (1) The unscrewing of one section of a DRILL PIPE from another. (2) Promotions within the drilling team.

bridging crude. *See* BUY-BACK CRUDE.

bright stock. A LUBRICATING OIL of high viscosity.

BRINDEX (Association of British Independent Oil Exploration Companies). A lobby group made up largely of the smaller, independent oil companies active in the North Sea.

bringing in. The act of starting production from a well. Once the DRILLING STRING is removed the back pressure on the reservoir is reduced by removing the mud (*see* DRILLING FLUID) until the oil begins to flow.

British National Oil Corporation. *See* BNOC.

British Petroleum. *See* BP.

Britoil. A company created in 1982 by the UK government. It was given the government's participation shares in North Sea oilfields previously held by BNOC to allow their sale to the public. In 1988 it was taken over by BP.

Brookhaven Energy Reference System. A model of the role of energy in an economy that was developed by the Brookhaven National Laboratory in New York. *See also* ENERGY ANALYSIS.

British thermal unit. *See* BTU.

Brown Book. The commonly used name for the annual publication entitled *Development of the Oil and Gas Resources of the United Kingdom* produced by the UK Department of Energy. It is a combination of a report on progress and a statistical abstract.

BS&W (bottoms settlings and water). Contaminants in crude oil that can be separated off by gravity or centrifugal force. *See also* BOTTOMS.

BSD. Barrels per stream day (*see* STREAM DAY).

BSW (bunkers, stores and water). Items other than oil that are carried by oil tankers, representing a reduction in carrying capacity

Btu (British thermal unit). A measure of heat energy; the amount of energy needed to raise the temperature of one pound (1 lb) of water from 39.2 to 40.2°F at sea level. *Compare* JOULE. *See also* APPENDIX.

BTX (benzene, toluene and the xylenes). The main AROMATICS used as petrochemical FEEDSTOCK. *See also* PETROCHEMICALS.

bubble cap. A covered hole in a tray of a FRACTIONATING COLUMN. Its purpose is to disperse the vapour in small bubbles beneath the surface.

bubble point. *See* GAS SATURATION PRESSURE.

buffer stock. A stock of a commodity that is built up, usually via an international agreement, in order to minimize price fluctuations by trying to smooth out variations in supply (and demand). In times of excess surpluses are bought, and buffer stock is released at times of shortage. It is possible to see the relevance of buffer stocks to agricultural produce where supply is subject to random influences such as weather and/or disease. However, for such commodities as oil its relevance is limited, although the concept forms the basis of the holding of STRATEGIC RESERVES.

bulk consignee. A US term referring to a gasoline (motor spirit) wholesaler who operates on commission from a supplier.

bulk/oil carrier. *See* COMBINED CARRIER.

bulk-supply tariff. The price negotiated for the large-scale supply of certain energy forms. Gas supply utilities normally have fixed-supply tariffs. However, if there is a large potential gas consumer then this consumer would be in a strong bargaining position and may be able to force down the price. This would be particularly true if consumption were to be in off-peak periods.

bundwall. A safety earthwork or wall surrounding a tank containing liquid. In the event of tank rupture or leak the contents would be contained by the bundwall.

bunker C. A heavy RESIDUAL FUEL OIL used by ships, industry and for large-scale heating. *See also* FINISHED PRODUCTS, FUEL OIL.

bunkering. Oil or coal taken on board ship as fuel is known as bunkers; the term is also used of fuel for aircraft. In the case of ships, the fuel used predominantly is heavy residual fuel oil (e.g., BUNKER C). This can be burnt both in steam turbines and large diesel engines. A smaller quantity of MARINE DIESEL fuel is also used. Approximately 5 percent (133 million tons out of 2775 million tons) of all liquid fuel products were consumed as bunkers in 1982, of which 32 million tons were used in aircraft, whereas 101 million tons were used in ships' engines. Bunkers' share of total liquid production has declined with the development of short-haul trades in recent years. Normally bunkers exclude fuels used by the military. In addition, domestic consumption tends to exclude bunkers since almost by definition they are consumed offshore.

bunkers. *See* BUNKERING.

bunkers, store and water. *See* BSW.

bunter. A geographical term indicating the lowest series of Triassic formation. It includes sandstones that may contain hydrocarbon traps. *See also* GEOLOGY.

Bureau of Mines. Despite its name, this US government office (attached to the Department of the Interior) has always had an involvement in the US oil industry, largely as a source of information (from its own researchers) and statistics. In earlier periods the Bureau's estimates of demand were an important input into the US oil control policies.

burning oil. *See* KEROSINE.

bury barge. A BARGE used to dig a trench for underwater pipes.

business cycle. The periodic fluctuations in the level of economic activity when measured by macroeconomic variables, one of the most common being output. It represents the fluctuations of such a variable around its long-term trend. Taking GROSS DOMESTIC PRODUCT (GDP) as an example, a full business cycle can be described as starting from the point where actual GDP becomes higher than the level indicated by its long-term trend, reaches a maximum level, declines to a minimum level below the long-term trend and rises eventually to reach again its long-term level. In this case, the business cycle is characterized by an expansionary phase (GDP above its trend) followed by a contractionary phase (GDP below its trend).

The pattern of fluctuations can be described by two dimensions: (a) the amplitude of the cycle, which is the difference in the level of the variable (e.g., GDP) between the highest and the lowest values; and (b) the period of the cycle, which is the time necessary to complete a full cycle. In terms of oil and gas, insofar as the business cycle reflects the level of economic activity, it would be expected that demand for oil and gas (and energy in general) would follow the cycle. This has caused and will continue to cause problems in analyzing demand for oil and gas. Both the FIRST OIL SHOCK and the SECOND OIL SHOCK were followed by economic recession. This made it very difficult, if not impossible, to disentangle the income effect (*see* SUBSTITUTION EFFECT) from the PRICE EFFECT to explain the subsequent fall in oil and gas demand. This was particularly important in the weak oil market that developed after 1981. The general consensus was that the fall in oil demand was caused by the recession. Thus OPEC, in anticipation of better days to come when the contractionary phase gave way to the expansionary phase, felt able to make short-term sacrifices by cutting production. However, as time progressed, it became apparent that the recession was less important as an explanation of falling demand. This made it much harder for OPEC to accept the sacrifices which now appeared to be necessary for very much longer than originally envisaged.

butadiene. An ALKENE, derived from BUTANE and containing two double bonds, that is used in the manufacture of synthetic rubber (i.e. STYRENE–BUTADIENE RUBBER).

butane. An ALKANE that is a gas at atmospheric pressure and normal temperature, but is easily liquefied by pressure for handling and use (*See* LPG). *See also* FINISHED PRODUCTS.

buy-back crude. When PARTICIPATION by the Middle East producing governments in their concessionary companies was first being discussed in the late 1960s one of the problems was disposal of crude oil. Whatever EQUITY share would be agreed would entitle the government to that percentage share of the production. This would create two problems. Initially it would mean that the concessionary companies would lose access to the government's share, thereby running the risk of having to disappoint customers. It also meant that governments with limited experience would have to market a lot of crude oil. To overcome these difficulties the concepts of bridging and phase-in crude were developed. In effect this enabled the companies to buy back the government's crude entitlement. The bridging crude would be sold at market price because it allowed the companies time to fulfil their contract obligations. The phase-in crude would be sold to the companies at a small discount because it allowed the governments time to develop marketing networks. Since then buy-back crude in general refers to crude oil owned by the government by virtue of its equity share which is sold to the company partners for disposal.

buy-back price. The price of BUY-BACK CRUDE.

by-pass. A pipe connected around a valve or other control mechanism that permits continuous flow while maintenance or repair is carried out on the valve or mechanism.

by-product. A substance obtained incidentally as a result of the production of another product. It is the 'incidental' production that makes it different from a joint product. A by-product, which is a common phenomenon in refining and petrochemicals, may have an economic value, but it may also have an economic cost, especially if it has no use and/or is toxic.

C

C_3/C_4. Liquefied petroleum gases (i.e. propane, C_3H_8, and butane, C_4H_{10}).

C_5s+ (pentanes plus). Low-boiling-point liquid HYDROCARBONS in the approximate range C_5 to C_{10}. *See also* CHEMISTRY.

cable drilling (cable tool drilling). An early form of DRILLING powered by steam.

cable tool drilling. *See* CABLE DRILLING.

cabotage. An expression used in maritime trading to indicate coastal trade. In Europe it tends to mean transport operations within one country.

calcium petroleum sulphonates. Petroleum-based products that are used as detergents.

calorie. A measure of heat energy. It is the amount of energy required to raise one gram of water by one degree centigrade at sea level. *Compare* BTU. *See also* APPENDIX.

calorific value. The amount of heat in a given volume of fuel. It is usually measured either in British thermal units (BTU) or CALORIES. If BTU is used, it is linked to another imperial measure (e.g., BTU per pound for solid fuels or BTU per cubic foot for a gas). If calories (usually kilocalories or Calories) are used, the link is to another metric measure (e.g., kilocalories per tonne for solid fuels or kilocalories per cubic metre for gases).

There exists a gross value that includes the latent heat of condensation of the water vapour which is produced by burning. If the water is removed the net value is measured. Given that water is more likely to remain as vapour the net value is a better guide to the useful heat obtained. It is the difference in the calorific values of fuels that to a large extent determine their relative prices for a given measure, although other factors such as ease of handling and the degree of 'dirtiness' can also influence relative prices. *See also* QUALITY, DISTILLATE AND RESIDUAL FUELS; APPENDIX.

Caltex. *See* TEXACO; SOCAL.

Table C.1

Tonnage (SCNT)	Vessel type	Transit toll ($ per SCNT) Laden	Ballast
First 5000	Tankers & bulk carriers	4.75	3.80
Next 15 000	Tankers & bulk carriers	2.60	2.08
Next 65 000	Tankers	1.25	1.00
Rest of tonnage	Tankers & bulk carriers	1.45	1.16

canal tolls. Levies imposed by canal owners on transiting vessels. In the case of the Suez Canal tolls vary according to size and type of vessel. The rates, expressed in SDRs per Suez Canal net ton (SCNT), as they applied on 1 January 1986 to tankers and bulk carriers, are given in the Table C.1.

Table C.2

	June 1981	June 1983	June 1985
	(thousand US$)		
Panama	45.5	49.4	49.2
Suez	77.0	67.0	65.0

Rates for the other important canal — the Panama Canal — are somewhat lower than those for the Suez Canal, although the gap has closed substantially in recent years. The total fee for a 60 000 dwt tanker (laden voyage) using each canal has varied as shown in Table C.2.

capacity. The ability to produce, process and transport a stream of oil or gas. In terms of processing and transport, capacity refers to the maximum amount that the equipment can handle each day the plant operates. In this case, the annual capacity is determined by how many days the plant is capable of operating and is not experiencing DOWNTIME for repair or maintenance. Production capacity is, however, more complex, and understanding may be assisted by reference to capacity over three different time periods.

(a) In the MARKET PERIOD it is assumed that everything is fixed except the rate of flow. Production capacity is determined by the maximum amount that can be extracted and handled, including transportation, and is therefore determined by the weakest link in the chain. This rate-limiting stage can vary according to the time of year, specifically with respect to the ability to load (e.g., in winter bad weather can inhibit the ability of a tanker to load). Market period capacity is subject to two constraints. The first concerns the state of the equipment. If capacity remains unused then its theoretical maximum is likely to decline in practice as equipment deteriorates. The rate of this decline is determined by the extent of maintenance. The relationship between maintenance and decline is something of an unknown quantity, and thus erosion of capacity from this cause is a matter of guesswork, albeit educated. The second constraint on market period capacity concerns RATE SENSITIVITY. If a field is rate-sensitive then the faster the flow the less oil or gas is eventually recovered. Thus current high production can be at the expense of future production. This has given rise to the concept of maximum sustainable capacity, which is the maximum flow that can be sustained without invoking excessive rate sensitivity. This is part of what is known as GOOD OIL-FIELD PRACTICE. Maximum sustainable capacity also takes into account downtime for maintenance and repair.

(b) In the SHORT TERM it is possible to vary elements in the production chain (e.g., by drilling more wells or by installing more equipment). The constraining factor now becomes the amount of RECOVERABLE RESERVES. As in the market period, short-term capacity is also influenced by rate sensitivity since this influences the amount of reserves that can be recovered.

(c) In the LONG TERM all factors in the production chain become variable. Most obviously the amount of recoverable reserves can be increased by the process of exploration and development.

The history of the international oil industry has always been characterized and influenced by the existence of excess capacity in all three time periods. There are a number of reasons for the existence of this excess capacity.

(a) There is the random element in discovering oil, although it is important to remember that capacity is developed as a result of an investment decision and is not simply fortuitous. However, companies cannot go out to discover a certain size of resource base; they find whatever they find. The history of the industry is littered with examples of large destabilizing finds, ranging from EAST TEXAS to the Middle Eastern oil-fields.

(b) Companies have developed capacity to ensure security of supply for their refinery operations. High fixed costs mean that refineries operating below capacity face exponentially rising costs. Thus it has made sense to have access to greater crude oil production capacity than refinery capacity in case a particular source is lost to the company.

(c) Seasonal fluctuations in the demand for oil have meant that the need for a peak capacity, in order to meet winter demands in the northern hemisphere, has lead to an overcapacity in the summer. It is cheaper to store crude below ground than above ground.

(d) Probably the main reason for the existence of excess crude-producing capacity has been the frequently successful attempts by those controlling the capacity to restrict its use. This carries two implications. Restricted supply keeps the price above its marginal cost (*see* MARGINAL), thereby making it profitable to develop capacity and produce it. As this new capacity emerges, those who have been restraining existing capacity must restrain it even further if prices are not to be eroded. If they are successful this calls forth even more capacity. Linked to this is the fact that restricted use of the existing capacity forces refining companies that are short of crude oil to develop greater crude-producing capacity, particularly if it is cheaper per barrel to develop capacity than to buy crude in the market place.

Because the existence of excess capacity in the industry is endemic, the history of the industry can be written in terms of attempts to control the excess capacity and protect prices. *See also* DESIGN CAPACITY.

capacity utilization ratio. The ratio of actual output to full CAPACITY output. It is in effect a measure of excess capacity.

capital. In layman's language capital is usually thought of in financial terms. In economics it has a different meaning which has two components: (a) physical goods used in production which have themselves been produced (e.g., drilling rigs, pipelines, machines, etc.); (b) circulating or working capital, which consists of stocks of raw material (e.g., crude oil) and semi-finished goods. Capital is a stock concept, thus at any point in time there is fixed quantity of capital. The stock level is changed by a flow of investment. This flow may be positive if new investment is undertaken, or it may be negative if the capital stock is used up and not replaced.

capital account. That part of the BALANCE OF PAYMENTS which measures financial flows between a country and the rest of the world; normally broken up into short-term and long-term and government and private. It excludes the costs (interest) and benefits (dividends) associated with such flows which appear in invisibles (*see* INVISIBLE TRADE).

capital budgeting. The process of controlling the expenditure of CAPITAL over time. In effect it represents a company's investment plan. *See also* INVESTMENT APPRAISAL.

capital charges. That part of a company's accounts which shows interest paid on capital, DEPRECIATION and/or loan repayment.

capital consumption. *See* DEPRECIATION.

capital cost. *See* DISCOUNT RATE.

capital rationing. *See* EXTERNAL CAPITAL RATIONING; INTERNAL CAPITAL RATIONING.

cap rock. An impervious (*see* IMPERMEABLE) layer of clay or rock that overlies a RESERVOIR ROCK thereby preventing the petroleum from leaking to the surface. At the risk of oversimplification, the integrity of the cap rock determines whether a well is dry or not. *See also* GEOLOGY.

carbon black. *See* CONVERSION OIL.

carbon residue. A quality characteristic of crude oil that indicates its yield of residue (*see* RESIDUAL FUEL OIL) from which heavy products are made. *See also* QUALITY, CRUDE OIL.

carburetted water gas. A fuel that is made by enriching WATER GAS with a hydrocarbon oil.

cargo prices. Prices quoted for crude oil and for products in quantities of more than 5000 tons. This contrasts with barge prices, which are quoted for loads of less than 5000 tons, the latter figure being the maximum size of a BARGE. Since a 'cargo' represents a bulk buy, cargo prices are usually slightly below barge prices.

carriage, insurance and freight. *See* CIF.

Carriage of Goods by Sea Act. *See* BILL OF LADING.

carried interest system. A system in which a company pays for all, or part, of a partner's costs during exploration, appraisal and possibly development. During these stages the company risks that no oil or gas will be found. If a commercial find is made then the carried partner has to repay its share of the expenses incurred. The carried interest system has been commonly used in SERVICE CONTRACTS and JOINT VENTURE arrangements. *See also* AUCTION SYSTEM OF LICENSING.

carried partner. *See* CARRIED INTEREST SYSTEM.

cartel. A group of separate firms that have agreed to try to act as a collective MONOPOLY. In so far as the cartel is the sole supplier, the group faces the market's downward-sloping DEMAND CURVE. The firms agree to restrict output (e.g., through production quotas) below what it would have been, thus raising the price. As a result, profits are higher than they would be in the face of greater competition and the entry of more firms into the industry. The success of the cartel in achieving higher prices acts as an incentive to member firms to cheat secretly on the deal and to produce and sell more. Consequently, cartel agreements are inherently unstable, and the members are likely to need to 'police' the agreement and to be able to employ sanctions to ensure compliance.

The legal status of cartel agreements varies according to the type of product and according to the countries involved. Some agreements are deemed to be unlawful because of their restraint on trade, whereas others are given the force of a legally binding contract. Over the years there has been considerable debate as to whether or not OPEC has behaved as a cartel.

carved-out over-riding royalty. A mechanism that can be used to raise capital for oil and gas exploration/development activities. The seller of the 'right' receives cash, whereas the buyer receives a right to obtain a percentage of any oil or gas produced. It has been used in situations where the government granting the exploration/development licence forbids the licensee from assigning some or all of his rights to a third party.

cash flow. The inflows and outflows of cash over the lifetime of a project. The gross cash flow is the gross profit (after payment of fixed interest) plus DEPRECIATION. It therefore represents the cash that is available to the company for investment, DIVIDENDS and tax payments. Net cash flow is retained earnings and depreciation, normally after tax. Net cash flow can be thought of more simply as all revenues less all costs per time period. A positive net cash flow is equivalent to a project's profit. Normally at the start of a capital-intensive project such as in the oil and gas industry the net cash flow is negative, reflecting the cash outflow and investment. *See also* DISCOUNTED CASH FLOW.

casing. A steel tube inserted after a hole has been drilled in order to consolidate the walls. *See also* DRILLING.

casing hanger. Equipment installed at the surface of a well nearing production from which the final section of steel pipe (i.e. the CASING) which lines the hole is suspended. *See also* DRILLING.

casing head. A part of the surface equipment used for suspending sections of steel pipe (i.e. the CASING) which lines the hole during DRILLING operations.

casing head gas. *See* NATURAL GAS.

casing head gasoline. *See* NATURAL GASOLINE.

casing perforation. Holes in the last string of CASING — the production string — to allow the oil or gas to flow into the production tube. The holes are normally produced by a series of charges fitted to a PERFORATING GUN. *See also* PRODUCTION, NATURAL RECOVERY MECHANISMS.

casing seat. The lowest point in a well at which the CASING is set.

catalyst. A substance that aids or speeds up a chemical reaction, but it does not itself enter the overall reaction and remains unchanged. *See also* CATALYTIC CRACKING.

catalytic converter. Equipment that is used to carry out CATALYTIC CRACKING.

catalytic cracking (cat cracking). A refining process that converts heavy, relatively low-value compounds (e.g., FUEL OIL) into lighter, more valuable ones (e.g., MOTOR SPIRIT). The process consists of passing the heavy product over a CATALYST, which is normally heated, in order to break down (i.e. crack) the heavy hydrocarbon molecules into lighter molecules. Its significance is that it increases the value of the product barrel. *See also* MOTOR SPIRIT UPGRADING; REFINING, SECONDARY PROCESSING.

catalytic polymerization. One of the less common processes for improving the quality of motor spirit components. *See also* MOTOR SPIRIT UPGRADING.

catalytic reforming. One of the most commonly used processes for improving the quality of motor spirit by increasing its OCTANE RATING or ANTI-KNOCK RATING. *See also* MOTOR SPIRIT UPGRADING.

catchpot. A vessel in a pipeline used to remove liquid droplets or solid particles from a gas stream.

cat cracking. *See* CATALYTIC CRACKING.

CBA. *See* COST–BENEFIT ANALYSIS.

CDU. *See* CRUDE DISTILLATION UNIT.

ceiling price. A maximum price, invariably imposed by government. An example of its use was in the USA following the FIRST OIL SHOCK when crude oil ceiling prices were defined as the posted price (*see* POSTED PRICE, CRUDE) plus $1.35 per barrel.

cementing. The pumping of cement through the CASING in a well and up into the annular space (*see* ANNULUS) between the casing and the well bore to hold the casing in place.

centistoke. *See* VISCOSITY.

centralizer. A device clamped on to the CASING to hold it away from the well bore wall prior to CEMENTING.

cetane index. A mathematical relationship based on certain properties of DIESEL OIL from which its combustion properties can be estimated. *See also* CETANE NUMBER; QUALITY, DIESEL OIL.

cetane number. A numerical description of the length of time taken for a DIESEL OIL to ignite. *See also* QUALITY, DIESEL OIL.

ceteris paribus (other things being equal). *See* PARTIAL EQUILIBRIUM ANALYSIS.

CFP (Compagnie Française des Petroles). An oil company that has been dubbed as the French MAJOR. CFP was created in 1924 and concluded an agreement with the French government which gave it control of French international oil interests including its interest in the forerunner of the IPC. Its operating name with the consumer is Total.

charters. Contracts for the carriage of oil by sea. There are four principal types of charter: voyage charter, time charter, bareboat charter and those arranged on a contract or lump-sum basis.

In a voyage charter a ship is chartered for a one-way voyage between specified ports for a designated cargo paid for by the charterer at a rate per ton of cargo. The ship owner pays all costs incurred in the voyage. Occasionally this type of charter is extended to cover a series of consecutive voyages.

A time charter is an agreement to hire a vessel for a specified period of time or for specified journeys. In this case the rate of hire is expressed in terms of dollars per deadweight ton per month. The charterer bears the cost of fuel (bunkers) and stores required for the period.

A bareboat charter is a long-term agreement under which the charterer assumes complete responsibility for the vessel and has to hire its crew, buy stores, pay for insurance, etc. This type of contract was attractive to the larger oil companies as a means of obtaining reliable freight transport services.

A lump-sum charter is an agreement on the part of the owner to ship a given quantity of a specified cargo for a stated rate per ton of cargo; the actual vessel to be used need not be specified. This type of charter is popular in the coal trade and other bulk trades. The freedom to choose an appropriate vessel is advantageous to the owner, but against this must be weighed the possible extra costs of any disputes that may arise.

The pattern of chartering among oil companies has altered drastically since the mid-1970s. At that time most used a mixture of owned and long-term chartered tankers to cover their predictable business, and used the voyage charter to cover their unforeseen requirements and for speculative reasons. Since then, because of increased uncertainty, the use of time charters has declined in significance, and the relative importance of the voyage charter has increased. *See also* AFRA.

chemical tankers. *See* TANKER TYPES.

chemistry. The smallest chemically distinct particle of an element is an atom, and it is the joining together of individual atoms of different elements to form a molecule that results in a compound being formed. Of the many compounds that make up petroleum, the main constituent element is carbon. The chemistry of petroleum lies within the field of organic chemistry, which is the study of carbon-containing compounds. Carbon is unique among the 106 chemicals elements because of the way in which it forms compounds, with the result that the number of carbon compounds exceeds that of all the other elements put together. Carbon is able to join (i.e. form a bond) with other carbon atoms, resulting in the formation of carbon chains and carbon rings. Carbon is also able to bond to atoms of other elements.

Bonding is best illustrated by the use of ball and stick models, which represent the three-dimensional arrangement of the atoms in a molecule. In this convention, a carbon atom can be represented as a ball with four sticks symmetrically protruding from it thus:

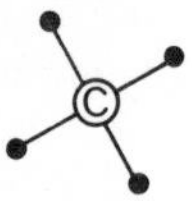

Each stick can join onto (i.e. form a bond with) a similar stick belonging to another atom, which may be either an atom of carbon or of another element (shown here as hydrogen). This can be represented two dimensionally as:

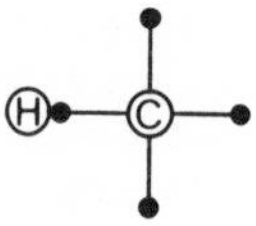

Petroleum is not a single substance; it is a mixture of many different organic compounds. In all the compounds found in petroleum the main element combined with carbon is hydrogen, either on its own or together with other elements, principally nitrogen, oxygen or sulphur. Compounds containing only carbon and hydrogen are called hydrocarbons, but in the context of petroleum the term is often extended to include compounds in which carbon and hydrogen are the major, although not the exclusive, elements.

The three main classes of hydrocarbon compounds relevant to petroleum and its products are saturated hydrocarbons, unsaturated hydrocarbons and aromatics. The classification is carried out on the basis of similarities of structure. All members of a class of compounds have similar chemical properties, contain the same elements and can be represented by a single general formula.

saturated hydrocarbons (paraffins or alkanes). When all four bonds of a carbon atom are attached to other carbon atoms or to other elements a compound is said to be saturated. The simplest saturated hydrocarbon compound is METHANE, which has one carbon atom with four hydrogen atoms bonded to it. Methane can be represented as follows:

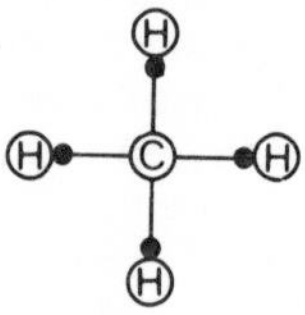

Each successive compound in the saturated hydrocarbon series contains one more carbon atom and two more hydrogen atoms. Thus METHANE is CH_4, ETHANE is C_2H_6 and PROPANE is C_3H_8. For hydrocarbon compounds containing four or more carbon atoms these atoms can be arranged not only in straight chains, but also in branched chains. The straight-chain hydrocarbons are known as normal paraffins (n-paraffins), whereas the branched-chain ones are called iso-paraffins (i-paraffins). The differences in the arrangement of the carbon atoms results in distinct compounds with different physical and chemical properties. Compounds that have the same chemical composition, but different structural arrangements of atoms within the molecule are said to be isomers. The existence of isomers accounts for much of the scale and complexity of organic chemistry, since the number of possible isomers of a compound increases exponentially with the number of carbon atoms present (e.g., butane, with four carbon atoms, has only two possible isomers, whereas decane, with 10 carbon atoms, has 75, and the hydrocarbon comprising 15 carbon atoms has over 4000).

In addition to straight-chain and branched-chain paraffins (collectively described as acyclic compounds), a third paraffin subgroup exists known as cycloparaffins; cycloparaffins are commonly called naphthenes in the petroleum industry. In these compounds the carbon atoms form a ring structure, which results in less hydrogen atoms being accommodated. Cycloparaffins (cycloalkanes) have the general formula C_nH_{2n}. Their names are derived by adding the prefix cyclo- to the corresponding acyclic compound with the same number of carbon atoms, thus propane (C_3H_8) cyclopropane (C_3H_6)

As a group, paraffins do not react very readily with other substances to form new compounds (as reflected in their name derived from the Latin *parvum affinis* meaning little activity). Modern refining processes, however, render conversion to other compounds possible when necessary (*see* MOTOR SPIRIT UPGRADING; REFINING, SECONDARY PROCESSING).

unsaturated hydrocarbons. Hydrocarbons are said to be unsaturated when they contain less than the maximum potential number of hydrogen atoms per carbon atom. This occurs when multiple, rather than single, bonds are formed between carbon atoms, so that, for example, the arrangement of atoms is

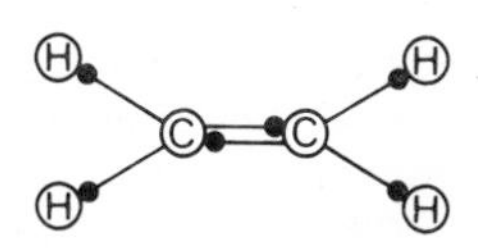

There are two groups of compounds within the class of unsaturated hydrocarbons: (a) alkenes (olefins) are compounds containing double carbon–carbon bonds; and (b) alkynes (acetylenes) are compounds containing triple carbon–carbon bonds.

A major distinction between saturated and unsaturated hydrocarbons is that the latter are chemically very reactive, combining readily with other substances to form further compounds. This is because multiple bonds can be regarded as representing unused potential for additional atoms to be added, and in many of the chemical reactions involving unsaturated hydrocarbons these 'spare' bonds are taken up by other atoms. It is for this reason that unsaturated hydrocarbons are the principal source of most PETROCHEMICALS.

aromatic hydrocarbons. Aromatic compounds are more difficult to characterize than the saturated and unsaturated hydrocarbons. Although such compounds are highly unsaturated (*see above*) with only one hydrogen atom for each carbon atom (formula: C_nH_n), they are chemically stable or unreactive, and hence are more like the saturated hydrocarbons.

The common structural feature is illustrated in benzene (C_6H_6), the simplest member of the group, which has a cyclic hexagonal structure of carbon atoms with alternating single and double carbon–carbon bonds:

Benzene reacts mainly by substitution (i.e. by replacing one or more of the hydrogen atoms in the benzene ring with an alternative atom or group of atoms, called a substituent). This type of reaction is the starting point of a wide range of products. When two substituents are attached to a benzene ring, their relative positions must be designated. This is done by the use of the prefixes *ortho-* (*o*-, 1,2-) (adjacent positions), *meta-* (*m*-, 1,3-) (one carbon atom removed from each other) and *para-* (*p*-, 1,4-) (two carbon atoms removed from each other), thus in the case of xylenes:

ortho- *meta-* *para-*

The main aromatics used as petrochemical FEEDSTOCKS are BENZENE, TOLUENE and XYLENE (often referred to collectively as BTX).

As a general rule, the number of carbon atoms in a product gives an indication of its physical state (i.e. whether it is a solid, liquid or gas): hydrocarbons with up to four carbon atoms are gases at normal temperature and pressure; those with five to 20 carbon atoms are liquids; those with more than 20 carbon atoms are solids. It is fairly common in the oil and gas industry to use the number of carbon atoms as a shorthand means of referring to products or groups of products, namely C_n, where *n* is the number of carbon atoms. For example, C_3/C_4 refers to LPG (liquefied petroleum gases, or propane and butane); C_5s+ (pentanes plus) refers to the lower-boiling-point liquid hydrocarbons in the approximate C_5–C_{10} range.

In refining, distinctions between chemical types become of significance only in the later stages. Basic crude distillation (*see* REFINING, PRIMARY PROCESSING) fractionates hydrocarbons into groups according to the number of carbon atoms they possess (e.g., motor spirit includes hydrocarbons ranging from C_6 to C_{11}). It does not, however, differentiate between chemical types (i.e. normal or iso-paraffins, saturated or unsaturated hydrocarbons) contained within each STRAIGHT-RUN fraction. The secondary refining stage (upgrading or conversion processes) alters chemical types in order to produce finished products of the required quality and value (*see* MOTOR SPIRIT UPGRADING; REFINING, SECONDARY PROCESSES). For example, catalytic reforming acts by converting the straight-chain paraffins of straight-run gasoline, which have low octane numbers, to branched-chain paraffins and aromatics of high octane number that are required for finished motor spirit. The various cracking processes achieve an increase in finished product value by breaking up the large molecules of heavier (lower-value) products to small molecules of lighter (higher-value) products.

nomenclature of hydrocarbons. With over a million known organic compounds, considerable problems are involved in assigning names to individual compounds. The complexity is increased by the existence of two systems of nomenclature. This results in the same compound, or group of compounds, having two different names. The common system (C) of naming organic compounds is the better known in relation to the simpler hydrocarbons, but is the less rigorous of the two systems. The IUPAC system (I) (devised by the International Union of Pure and Applied Chemistry) provides a basis for unequivocally assigning a name to any individual hydrocarbon, by numbering the carbon chain and identifying the precise location of branched chains. Although such a system is essential for naming isomers of the higher compounds, which are beyond the scope of conventional prefixes, the IUPAC names are rarely used for the simpler compounds (e.g., iso-butane is rarely referred to as 2-methylpropane). The more important guidelines of nomenclature under both systems are summarized below.

Suffixes
-ane denotes a saturated hydrocarbon (I)
-ene denotes an unsaturated hydrocarbon with a double bond (I)
-ylene denotes an unsaturated hydrocarbon with a double bond (C)
-yne denotes an unsaturated hydrocarbon with a triple bond (I)
-acetylene denotes an unsaturated hydrocarbon with a triple bond (C)

Prefixes
normal- (n-) denotes a straight-chain saturated hydrocarbon (C,I)
iso- denotes the first branched-chain isomer of a saturated hydrocarbon (C)
neo- denotes the second branched-chain isomer of a saturated hydrocarbon (C)
-,- denotes a branched-chain isomer of an unsaturated hydrocarbon (C)
1-, 2-, etc. denotes a branched chain isomer of an unsaturated hydrocarbon (I)

Chevron. *See* SOCAL.

choke. A wellhead control to regulate the flow of oil. *See also* PRODUCTION.

christmas tree. The complete assembly of valves and connections at the WELLHEAD.

Church Committee. The US Senate Multinational Corporations Subcommittee (chaired by Senator Church) during 1974. Like many of its predecessors it was a source of previously confidential information on the operations of the multinational oil companies. Without this information much less would be understood today.

cif (carriage, insurance and freight). Effectively the cost of transportation. Normally imports are measured cif, whereas exports are measured fob (i.e free on board, or net of transportation costs). In the oil and gas industry, because reserves tend to be located away from markets, cif has always played an important role in the economics of the industry. The development of larger tankers has significantly reduced the level of cif on crude oil transport. In the 1960s the relative decline in crude cif compared with product cif encouraged the already developed trend to locate refineries on the markets away from the oil-fields (*see* REFINERY LOCATION). In gas transportation other than by pipeline it has always been the very high cif element that has been an important factor in inhibiting liquefied natural gas trade.

cif trading. *See* COMMODITY MARKET.

circulation. The process of continuously pumping drilling fluid through the drill string and up the annulus while DRILLING.

clean-oil vessels. A ship for carrying clean products (*see* BLACK PRODUCTS).

clean products. *See* BLACK PRODUCTS.

cloud point. The temperature at which certain components in DIESEL OIL start to crystallize out and affect engine performance. *See also* QUALITY, DIESEL OIL.

club agreements. Agreements between companies concerning market behaviour. They are in effect a form of CARTEL.

Club of Rome. The organization (formed in 1968) responsible for the publication in 1972 of the *Limits to Growth*. This was an important document in the sense that it fed a popular belief held during the 1970s that oil and gas supplies would run out. This gave rise to a perception at a relatively popular level of ever-rising energy prices.

Club of Zurich. Five of the European non-MAJORS — Elf-Aquitaine, Veba, CFP, Petrofina, ENI — in the mid-1970s began meeting in Zurich in order to make policy recommendations to the EEC. The group, sometimes referred to as 'The Five', often tried to provide a counterweight to the power of the majors.

CMEA. *See* COMECON.

coal gas. A gas produced by heating coal. It has about one-half the CALORIFIC VALUE of NATURAL GAS.

coal liquefaction. Coal has all the CRUDE OIL CHARACTERISTICS except one: oil is liquid, whereas coal is a solid. Considerable effort has gone into the development of processes to convert coal into a liquid (*see* HYDROLIQUEFACTION) for two reasons: (a) to enable coal to be used as oil in those countries that have plentiful coal, but whose access to oil is difficult; (b) to give coal the enormous economic advantages accruing to oil because of its liquid characteristics (i.e. handling advantages and ECONOMIES OF SCALE). One of the commonest methods is the FISCHER–TROPSCH PROCESS, developed in 1925, in which the coal is gasified, and then the gas is condensed to a liquid under high pressure using a CATALYST. Research and development into other methods of hydroliquefaction received a significant boost during the perceived oil crisis of the 1970s. Two approaches present distinct possibilities. The first approach is to crush the coal and create a SLURRY by mixing it with water. Although this technique allows liquid handling, the presence of water reduces the CALORIFIC VALUE of the fuel. An variation on this theme is to form a slurry of coal in METHANOL, which increases the calorific value of the fuel. The second approach involves the use of biochemical methods. These are only in their infancy, but they could provide major technological breakthroughs in the future.

Cobb–Douglas production function. *See* PRODUCTION FUNCTION.

cobweb model. One of the earliest and simplest examples of dynamic analysis in economics. It characterizes a situation in which price changes do not lead to shifts in the DEMAND CURVE and SUPPLY CURVE. The model assumes that a finite period of time is necessary for supply to adjust to price changes. The name cobweb model is used because if the movements of prices and quantities are traced on a conventional demand–supply diagram the resulting pattern looks like a cobweb (*see* Fig. C1). The spot tanker market is an example in the energy field in which the cobweb model can be applied because changes in the spot price of tanker services are not accompanied immediately by a change in the quantity of tanker services supplied. Some delay is likely to be experienced in bringing in idle tankers in order to satisfy increased demand if the spot price should rise; similarly a delay in withdrawing tanker services is likely should the spot price fall. The spot price behaves in a predictable manner oscillating around a stable equilibrium price. Suppliers will keep making incorrect guesses as to the right quantity to supply because of the time lag in matching supply to the perceived price. The model may either be stable, in which case the actual price will converge towards the equilibrium price, or it may be unstable or explosive and the actual price will diverge away from the equilibrium price. The nature of the oscillation depends on the relationship between the slopes of the demand and supply curves. The model described here has severe restrictions which limit its usefulness; however, it can be a useful aid in thinking about the causes of price changes.

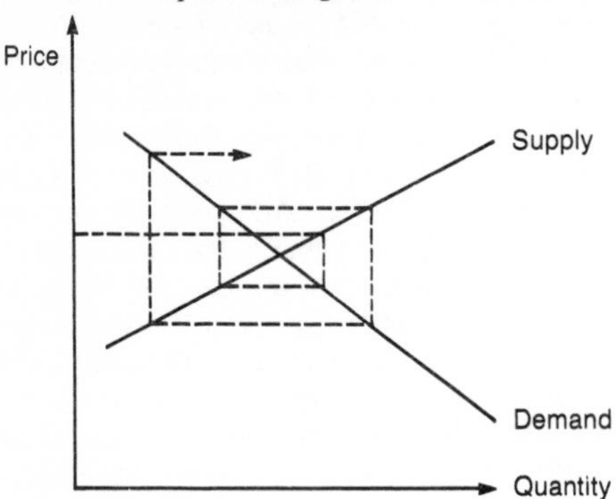

Fig. C.1

coincident indicator. *See* LEADING INDICATORS.

coking. The production of PETROLEUM COKE by heating a petroleum FEEDSTOCK and then holding it in the coking drum for an extended period. An indication of the potential coke yield of the feedstock is given by the conradson carbon residue content.

collateral. ASSETS against which loans are made.

collective heating. *See* DISTRICT HEATING.

collusion. Overt cooperation between companies to control the market normally experienced in an oligopolistic (*see* OLIGOPOLY) industry structure. It is the activity of CARTELS.

combined carrier. A vessel designed to carry either liquid or dry cargoes. Combined carriers are generally classified as ore/bulk/oil (OBO), ore/oil (OO) or bulk/oil (BO) carriers according to the degree of flexibility offered to the user. The economic advantage of a combined carrier is due to its ability to minimize BALLAST operations by being able to take on cargoes for the return trip. Against this, however, is the fact that the capital cost is about 15–20 percent greater than that of an equivalent-sized oil tanker. In depressed markets a combined carrier suffers from a further disadvantage due to the difficulty of finding a suitable combination of cargoes for the outward and return journeys.

Comecon (Council for Mutual Economic Assistance, CMEA). An organization set up in 1949 in order to develop the economies of the member countries on a complementary basis in order to achieve self-sufficiency. The members of Comecon are the USSR, East Germany (German Democratic Republic), Czechoslovakia, Hungary, Poland, Bulgaria, Romania, Cuba, Vietnam and Mongolia.

coming out of hole. The withdrawal of the DRILLING STRING. *See also* DRILLING.

commercial energy. Energy that moves in corporately controlled markets and whose production requires a gross input of foreign exchange. Examples of commercial energy include oil, gas, coal, electricity and the EXOTICS. The expression is used to differentiate from TRADITIONAL ENERGY when energy issues in developing countries are being discussed.

commercial field. An oil- or gas-field that produces an 'acceptable' rate of return. Its main significance lies in JOINT VENTURES and SERVICE CONTRACTS where a company is exploring at its own risk or using a CARRIED INTEREST SYSTEM. Once oil has been discovered a decision has to be made as to whether or not the field is commercial. This decision carries legal implications since if the field is declared commercial the government then incurs certain obligations. For example, in joint venture agreements, after the field has been declared commercial, the government must begin to repay its share of the exploration expenses and invest its share of the development costs. In some of the earlier agreements the commercial criterion was couched in technical terms. However, this proved to be too rigid, and subsequent agreements moved towards the use of the RATE OF RETURN criterion.

commodity market. A market in which the title to commodities is bought and sold. Initially it was based on warehouse auctions where the commodity could be physically inspected. However, the development of efficient grading systems and modern communications has led to the development of cif trading whereby someone can buy the commodity unseen in the country of origin for delivery cif to a specified port. The market normally involves trading for future delivery in an effort to allow the users of the commodity to hedge against price fluctuations. Oil and gas and their products have never been traded in such markets until recently with the rise of markets containing elements familiar to commodity exchange. At present it is looking increasingly as though oil and its products will be bought and sold like any other commodity.

common carrier pipeline. An oil or gas pipeline where rights of access are available to suppliers other than the owner of the pipeline. In the absence of such rights a degree of monopoly power is conferred on the pipeline owner because of the high capital cost and locational advantages of the pipeline. To limit this power, various countries require pipeline owners in specified circumstances to allow their line to function as common carriers or, as in the UK, provide for rights of access to publicly owned pipelines.

common property resource. A natural resource that is not owned by a single owner, but to which a number of people have access rights. Oil and gas under the LAW OF CAPTURE are an example. The problem with such resources is that in the absence of regulation it is likely the resource will be overused.

Compagnie Française de Petroles. *See* CFP.

comparative advantage. Cost advantages in the production of a specific good. This has a specific technical meaning in international trade theory. An example is the argument that the oil-producing countries have a comparative advantage in refining and petrochemicals.

compensated demand curve. *See* DEMAND.

compensation. The amount paid by governments to companies in order to settle a NATIONALIZATION. This is an extremely contentious area; often the governments refuse to even use the word in agreements. For example, in the GENERAL AGREEMENT ON PARTICIPATION the term 'consideration' is used instead of compensation. Dispute arises because the companies normally wish to be paid the maximum, whereas governments normally wish to pay the minimum. The use of 'normally' here is important since often negotiations over compensation occur in a wider context which may require either party to concede on the nationalization issue for the sake of gains elsewhere. In general the company seeks to gain going concern value of the company by use of DISCOUNTED CASH FLOW (DCF) methodology. Thus expected revenues and costs provide a NET CASH FLOW which by use of the chosen discount rate is discounted to a present value (usually on the day of seizure) which forms the basis of compensation. In effect the companies are trying to arrive at a figure that a potential buyer of the company would be willing to pay. There are two major drawbacks to this approach.

The first problem is that the company is including in its assets the underground reserves which constitute the basis of its revenue-earning potential. However, under most legal systems these reserves are already the property of the state. Therefore to pay compensation on such a basis would mean the state is having to buy what it already owns.

A second problem is that the DCF methodology cannot produce objective results. It involves forecasting future cash flows, and there is no forecasting method that can produce irrefutable results apart from an after-the-fact record of what actually happened. In theory, uncertainties over the cash flow are accounted for by adjusting the discount rate for that risk. However, the risk adjustment itself is a matter of subjective judgement. In the situation where DCF is normally used — to determine price between buyers and sellers or to decide within companies on investment decisions — such subjectivity in the DCF result does not invalidate the method. This is because the buyer, seller or decision-maker is free to accept or reject the different DCF results. Furthermore if the result is accepted it is backed by money — 'putting your money where your mouth is'. Indeed, it is the speculative nature of DCF which gives the methodology its value because it forces the decision-maker to consider the underlying assumptions. However, in so far as compensation — especially if being determined in an arbitration situation — requires an objective valuation DCF cannot be used.

Thus apart from the fact that a DCF approach to compensation is unacceptable to governments, it is also inappropriate. In practice in most (if not all) nationalizations of oil companies, NET BOOK VALUE has been used as the starting point for negotiations. This is the written-down value (i.e after DEPRECIATION) of the company's physical ASSETS above the ground. How much above or below this figure the compensation is finally agreed depends largely on how important the goodwill of the company is to the government. If the government anticipates that it will need the services of the company in the future, then it may add 'sweeteners' to the net book value figure. For example, in the General Agreement on Participation signed in October 1972, because the governments felt they would need the companies' technical expertise updated BOOK VALUE was used, although this was later changed to straight net book value.

compensation principle. The basis for a decision criterion often used in COST–BENEFIT ANALYSIS. If those who stand to gain from a changed situation (e.g., a project to build a new airport) could hypothetically compensate those who stand to lose, and still themselves retain a surplus, then the change is said to pass the compensation test. In the terminology of WELFARE ECONOMICS, such a change would constitute a 'potential Pareto improvement' (this would become an actual Pareto improvement — nobody being worse off and at least one person being better off — only if the compensation were actually to be paid). The gains and losses are normally measured in terms of people's WILLINGNESS TO PAY to experience a gain or to avoid a loss. The compensation principle is sometimes associated with the names of two British economists, Hicks and Kaldor. *See also* PARETO OPTIMAL.

compensators. Hydraulically operated equipment designed to offset the upward or downward motion of a floating rig or drill ship while drilling is in progress.

competitive bidding. *See* AUCTION SYSTEM OF LICENSING.

competition. At its simplest, the competing of individuals or institutions for something (e.g., resources, markets, etc.). In economics it is generally used as shorthand for perfect competition which is a market in which there are a large number of buyers and sellers and price is determined by the free interaction of supply and demand.

complementarity. A situation in which a good is usually bought in conjunction with a different good (e.g., motor spirit and cars). Thus if the price of one changes (e.g., motor spirit price increases) the demand for the other good changes (i.e. car demand falls). In terms of oil and gas this is an important economic concept since fuels require appliances for their use. *See also* CROSS-PRICE ELASTICITY OF DEMAND; DEMAND.

completion. The installation of permanent production facilities over a well which is about to come into production. *See also* DRILLING.

completion and performance guarantee. A contract or a clause within a contract that imposes financial penalties on a contractor if work is completed late or unsatisfactorily.

compounding. A financial process whereby interest is calculated on the basis of the original sum plus the accumulated interest. This can lead to very rapid increases in the sum of money realized. For example, at an annual rate of 10 percent a sum of money will double in 7 years 3 months and treble in 11 years 7 months.

compressor. Equipment used to compress gas either for reinjection into a well (*see* GAS INJECTION) or for pumping through a pipeline.

compulsory stocks. Stocks that oil companies are forced by legislation to hold. The government sets these stock levels based on essentially strategic considerations (*see* STRATEGIC RESERVES) although INTERNATIONAL ENERGY AGENCY (IEA) members have to meet the IEA agreed quantities and would use the compulsory stocks as part of the IEA commitment.

concentration ratios. Measures that are aimed at revealing how many firms are active in a market. At the simplest level a concentration ratio may be the percentage of total sales emanating from the three largest companies in the industry. At a more sophisticated level it may involve measures such as the GINI COEFFICIENT or the HERFINDHAL INDEX. Its use is generally to establish the extent of competition in an industry. Concentration ratios cannot, however, pick up COLLUSION which may arise because of the presence of a CARTEL or because of INTERLOCKING DIRECTORS.

concessionaire. *See* CONCESSIONS.

concessions. Contractual arrangements between oil-producing countries and oil companies — the concessionaires — whereby the concessionaire has certain rights over petroleum reserves in a designated area for a prescribed period of time. Concession agreements have varied between countries and over time, but they are generally characterized by their longevity — early agreements in the Middle East were over 80 years, later ones around 45 years — and their size (the four main concessions in Iran, Iraq, Kuwait and Saudi Arabia covered 88 percent of countries). Although early agreements had no RELINQUISHMENT (or surrender) provisions, later concessions provided for areas to be handed back to host governments after a certain period, in order that the area might be offered to other companies. Other common features of concessions were the managerial freedom given to companies and the financial provisions. The early method of payment to the producer government was a ROYALTY made at a fixed rate per ton. After the late 1940s, renegotiations altered the method to a system based on a tax of 50 percent of profits attributed to oil production. More recent trends have been for the licensing of tracts for shorter periods (*see* LICENSING SYSTEMS).

The history of the granting of each concession is usually a long story and often an intriguing and exciting one. However, a common feature during the inter-war years was the intervention by the governments of the oil companies (i.e. the UK, France and the USA) in order to try to secure interests for their companies in the concessions, especially in the Middle East. The result was a pattern of joint ownership and control that was to be of crucial importance in the economics of the industry after World War II because of its implications for the development of HORIZONTAL INTEGRATION. This practice continued for some time, in particular in the 1954 settlement following the IRANIAN NATIONALIZATION of 1951, which led to the creation of the IRANIAN CONSORTIUM. During the 1950s and 1960s much of the history of the international oil industry revolved around attempts by the governments to renegotiate the terms of the concessions in order to improve their terms. Although much was negotiated, the main sticking point on which the companies refused to make any changes was the issue of managerial freedom. It was this that led to the gradual takeover by governments during the early 1970s by NATIONALIZATION and PARTICIPATION.

conch system. *See* LNG/LPG CARRIERS.

concrete platform. A PRODUCTION PLATFORM made of reinforced concrete rather than of steel.

condensates. (1) Liquid hydrocarbons recovered from non-associated gas-fields (*see* NON-ASSOCIATED GAS). They now tend to be used as refinery and petrochemical FEEDSTOCKS. (2) LPG, which is gaseous at normal temperatures, but can be liquefied by pressure or refrigeration. This use of the term can be misleading since although LPG chemically is composed of C_3 and C_4 hydrocarbons condensates are composed of C_4 and higher hydrocarbon numbers.

condenser. Equipment that is used to convert vapour at high temperature to liquid by cooling. Condensers are used in distillation (*see* REFINING, PRIMARY PROCESSES).

conference system. A situation in which an association of ship-owners engaged in a particular trade agree to limit competition between themselves. It may also include pooling agreements whereby earnings on a particular route are shared between the owners. It is effectively a variant of a CARTEL.

conglomerate. A business organization that consists of a holding company and a group of subsidiary companies engaged in different activities. *See also* CONGLOMERATE INTEGRATION.

conglomerate integration. A form of diversified expansion; when a firm joins with another firm via merger or takeover or sets up a new firm in an activity unrelated to its current activities. A good example of this was following the FIRST OIL SHOCK when a number of oil companies began to diversify by purchasing firms involved in activities ranging from hotels to supermarkets. It is not clear how far the moves of the oil companies into coal, gas and nuclear activities are a form of conglomerate integration or HORIZONTAL INTEGRATION.

connate water (interstitial water). The water that remains in the pores of the RESERVOIR ROCK when the oil and gas migrates into this rock. *See also* GEOLOGY.

conradson carbon residue content. *See* COKING.

conservation. The more efficient use of a given RESOURCE, resulting in resource savings without sacrificing the output of goods and services or reducing social welfare. This objective may also be achieved by resource substitution. In terms of oil and gas, conservation can occur via one of two mechanisms.

The first mechanism involves changing behaviour such as lowering the thermostat, switching off lights, reducing car use for leisure, etc. However, it is a moot point as to whether this is genuine conservation as opposed to deprivation. Thus, although oil and gas inputs are reduced, so too (it might be argued) is social welfare since people are operating for example in a colder and darker environment. This conservation mechanism can operate very quickly since it relates to the utilization of existing energy-using appliances. Many governments ran publicity campaigns and legislative programmes after the 1973 oil price rise (*see* FIRST OIL SHOCK) in an effort to secure such behavioural changes.

The second mechanism is to change the stock of oil-and gas-using appliances, creating new equipment which produces the same output (heat, light or work) with less energy input. This approach can be a very slow one, and there are three components to the time lag: (a) it is necessary to undertake the research and development to produce the prototype energy-efficient appliance; (b) production lines must then be changed and rebuilt to manufacture the appliance or existing appliances must be retro-fitted; (c) most important is the time to turn-over the existing appliance stock since appliance owners do not immediately scrap their existing equipment simply because a more fuel-efficient version is available. This can take a long time; for example, 10–15 years are required to turnover the vehicle stock. If five years are added for research and development and retooling then the full impact of a rise in motor spirit prices can take up to 20 years to be felt fully. Many analysts have failed to take account of this time lag. Thus after 1981 (*see* SECOND OIL SHOCK) when oil demand began to decline and people debated how much was due to recession versus conservation, many did not realize that a good part of the fall was conservation stemming from 1973–74. This explains why so many were surprised when the fall in demand became an established trend.

These mechanisms for conservation raise crucial questions with respect to the future prospects for energy use. Will an energy-efficient appliance be used more intensively because it is now cheaper to run? Above all, to what extent will falling oil prices reduce the incentive to produce yet more efficient energy-using appliances? One point, however, is noteworthy. The engineers who design and make the appliances are trained to obtain the maximum output for the minimum input. This is not because they are good economists, but because their objective is efficiency. Thus even during the 1950s and 1960s, when real oil prices were falling, electricity generation was becoming more efficient (i.e. there was a higher generation of electricity for a given oil input). During the 1970s, these engineers along with the population in general were subjected to massive publicity campaigns to the effect that energy is scarce and will become even more so. These campaigns had three sources: (a) enviromentalist groups, who sought conservation for the sake of conservation; (b) consumer governments, who sought to reduce their oil import bill; (c) oil companies, who wished for a perception of an imminent energy crisis in order to pressurize governments either to deregulate their activities or to adopt more favourable fiscal regimes. Thus the engineers' urge for greater conversion efficiency was reinforced at every stage. Falling oil prices are unlikely to have much effect on these urges. Particularly because the energy shortage campaigns are unlikely to diminish since, in general, the motives behind the campaigns are unlikely to be dented by falling prices. This, of course, is not saying oil demand may not pick up, but if it does it will be as a result of fuel switching back to oil rather than a reversal in the trend to conservation.

Consortium. *See* IRANIAN CONSORTIUM.

consortium. An alliance or grouping of companies or institutions. It has no special meaning in economics.

constant prices. Prices that have been adjusted because of changes in the purchasing power of money (i.e. inflation). This is achieved by deflating by an appropriate index number of prices. Constant prices are sometimes called prices in real terms as opposed to prices in nominal terms which have been left unadjusted.

consumption. In a purely economic context, the expenditure in an economy on goods and services which are then used up relatively quickly. In terms of crude oil and gas, consumption refers to oil and gas actually processed and to products that are 'used up' in the sense of either being burnt or converted into something else. Consumption may be thought of as satisfied demand. The difference between consumption and production is accounted for by stock changes. This can create problems because if consumption figures and/or production figures are inaccurate then so too is the stock level, which in most cases is simply an arithmetic residual.

The level of oil and gas consumption is determined by a number of factors including the stock of oil- and gas-using appliances, the price of oil and gas and other fuels, the level of income, people's preferences, etc. *See also* DEMAND.

containment boom. A floating, flexible boom that is used to contain an oil spill at sea.

continental shelf. The edge of a continental mass which extends under the sea in comparatively shallow water (up to 200 metres). Its significance for oil and gas is that because the water is relatively shallow it is easier to drill, develop and produce oil- or gas-fields.

contract of affreightment. A contract to hire a ship, or part of a ship, for the carriage of cargo. The contract may take the form either of a charter party (*see* CHARTERS) or a BILL OF LADING.

controlled crude. In the USA after 1973 crude oil that was subject to a CEILING PRICE.

conventional oil recovery techniques. A term used to describe what were formerly known as primary (natural drive mechanisms) and secondary techniques for the production of oil, namely those relying on pressure from water or gas, whether naturally occurring or artificially introduced. *Compare* PRODUCTION, ENHANCED OIL RECOVERY. *See also* PRODUCTION, NATURAL RECOVERY MECHANISMS; PRODUCTION, SECONDARY RECOVERY MECHANISMS.

conversion efficiencies. In order to use energy a two-stage conversion is required; (a) the primary energy form must be converted into a useable form (e.g., crude oil must be refined into products, and gas may also need to undergo SCRUBBING and be stripped before use); (b) the useable form must be converted into useful energy (e.g., heat, light, work) by using it in an appliance such as a car or a boiler. During both conversions some energy is lost, and the size of that loss is given by the conversion efficiency. The greater the efficiency the less primary energy is needed to produce a given unit of useful energy. The level of conversion efficiency is therefore a function of CONSERVATION.

conversion oil. FEEDSTOCK used in the manufacture of carbon black, a form of finely divided carbon.

conversion process. A refining process that converts one type of component (usually the heavier, lower-value ones) into another (normally the lighter, more valuable ones). The most common type of conversion process is CRACKING. *See also* MOTOR SPIRIT UPGRADING; REFINING, SECONDARY PROCESSES.

copper chloride treatment. A refining process that removes the pungent smell of sulphur from lighter products. *See also* REFINING, SECONDARY PROCESSES.

copper strip corrosion test. A test carried out on finished MOTOR SPIRIT to establish that the sulphur content has been reduced to acceptable levels. *See also* QUALITY, MOTOR SPIRIT.

core. A cylindrical sample of material that is cut in rotary DRILLING using a special BIT. The samples or cores can then be used to help develop the geological picture of the formation being drilled. *See also* CORE BARREL.

core barrel. A tool that is used in DRILLING operations to take samples of rock formations (i.e. CORE).

coring. Taking samples of rock formations during DRILLING operations. *See also* CORE.

corporate debt. The value of a sum of money or property owed by a corporation to individuals or bodies outside the corporation.

Corporation Tax (CT). A tax that is charged on the profits made by companies operating in the UK. With respect to oil companies active in the British sector of the North Sea, however, there is a RING FENCE, which prohibits the deduction of losses made outside the North Sea from North Sea profits. Profits for CT purposes include the landed value of gross revenues less capital expenditure, less operating expenditure, less interest on debts, less accrued ROYALTIES and less a deduction for the amount of PETROLEUM REVENUE TAX that would have been paid in the absence of Advanced Petroleum Revenue Tax. CT losses may be carried forward and deducted from profits made in subsequent years. In the 1984 Budget the Chancellor of the Exchequer announced a reduction of the rate of CT from 52 to 50 percent in the 1983–84 tax year, to 45 percent in 1984–85, to 40 percent in 1985–86 and to 35 percent thereafter. Furthermore, the 100 percent first-year capital allowance in CT has been reduced (thus delaying the time when capital expenditure may be claimed against income) to 75 percent in 1984–85, 50 percent in 1985–86 and 25 percent thereafter.

corrosion. The progressive breakdown of a metal structure by chemical or electrolytic attack. It is a particular problem associated with high-sulphur crude oil.

cost and freight. A type of contract in which the seller provides the product and the vessel and delivers to the nominated discharge point. *Compare* CIF; FOB.

cost–benefit analysis (CBA). A method for appraising public investment projects. It differs from commercial project appraisal by including costs and benefits to all members of society (i.e. social costs and benefits) not just the costs and revenues of the agency directly involved in the project. The changes in welfare are measured on the basis of each individual's WILLINGNESS TO PAY for benefits or to avoid costs. Individual benefits and costs are then aggregated to deduce the resulting net change in society's 'welfare'. The decision criterion is usually based on a variant of the COMPENSATION PRINCIPLE: a project is worthwhile if the gainers from it could more than compensate the losers, even if they do not actually do so. An unweighted sum of individuals' gains and losses is often used, implicitly assuming that a unit of money is of equal value to everyone, regardless of their income level. In other cases distributional weights may be employed.

In CBA, it is first necessary to specify the relevant project(s) to be analyzed. Then the physical outcomes have to be specified: who and what will be affected and in what ways (and with what probabilities). Here the net is cast widely and takes account of EXTERNALITIES. In the case of an electric power station, for example, this would include the impact of air pollution. All outcomes have next to be valued. In cases where goods and services are marketed and where market prices reflect valuations of social costs and benefits as they would be in a 'first best' PARETO OPTIMAL resource allocation, these prices can be used. However, in some cases prices are 'distorted' (e.g., through IMPERFECT COMPETITION or EXTERNALITIES). In other cases there may be no market at all (e.g., there is no market for clean air or for time saved as a result of reduced road congestion, or for changes in the probability of dying, or for the use of unpriced recreational facilities, such as parks). Here SHADOW PRICES must be estimated in an attempt to correct for the distortions and to supply valuations for the non-market items.

Account has to be taken of the fact that society may place different relative values on costs and benefits occurring in different time periods. Before they can be made commensurable and so reduced to a PRESENT VALUE, society's TIME PREFERENCE rate has to be taken into account by applying an appropriate social DISCOUNT RATE to the streams of costs and benefits. This social discount rate may not be the same as the private market discount rate. CBA offers a set of techniques widely used by governments and international agencies, although not without controversy because of the value judgements and valuation problems involved.

cost curves. *See* COSTS.

cost of capital. The cost, measured as a rate of interest either paid on money borrowed or lost on internal funds used, of CAPITAL. It is used either directly in determining the DISCOUNT RATE or as a point of comparison with the INTERNAL RATE OF RETURN.

cost of oil. *See* CRUDE OIL PRICES.

cost plus. A method of price fixing whereby price is determined by costs plus a percentage increase. The problem with cost plus is that there is no incentive to keep costs low. However, it can be useful in situations where new ground is being broken and thus it is difficult to predict actual costs.

costs. Economists' definitions of costs are not necessarily the same as those of accountants. Economists are concerned with opportunity costs: the opportunity cost of a resource is the benefit foregone by using it. It is the value of a resource in the best of its alternative uses. The concept highlights the opportunities that one has to sacrifice when taking a decision to use a resource in a particular way. Clearly, if resources were not scarce then the opportunity cost of using them would be zero.

Private cost is a measure of the opportunity cost of the resources that a private agent (person or company) uses. Private cost can be further broken down into EXPLICIT COST, where, for example, an explicit money price has to be paid for an input, and implicit cost or imputed cost, where there is no explicit money cost or payment. An example of implicit costs is a factor input that a producer already owns and does not, therefore, have to pay for in order to buy or hire. Owner/managers of small enterprises are notorious for failing to subtract the implicit opportunity costs of their managerial skills, time and personal capital from revenues, when calculating the economic profits of their firm.

In contrast to private cost, social cost measures the opportunity cost of using a resource to society as a whole. If the private and social opportunity costs of using a resource diverge, then a private agent's optimizing decisions will not be based on the same considerations as would the decisions of society as a whole. Social costs are explicitly taken into account in COST–BENEFIT ANALYSIS.

A producer's cost function for a product relates each level of output per time period to the minimum cost of producing it. Cost functions can be defined for the SHORT TERM, when at least one input cannot be varied, and the LONG TERM, when all inputs are variable. A variety of cost curves can be defined, relating output per period to minimum cost, including total cost (TC), average cost (AC) and marginal cost. Total costs can be split into total fixed costs (TFC) or SUNK COSTS and total variable costs (TVC).

In the short term, average total costs (ATC) per period of output (Q) can be split into average fixed costs (AFC) and average variable costs (AVC):

$$ATC = TC/Q = (TFC/Q) + (TVC/Q) = AVC + AFC$$

AFC declines with output, whereas AVC will eventually rise because of DIMINISHING MARGINAL RETURNS to the fixed factor input.

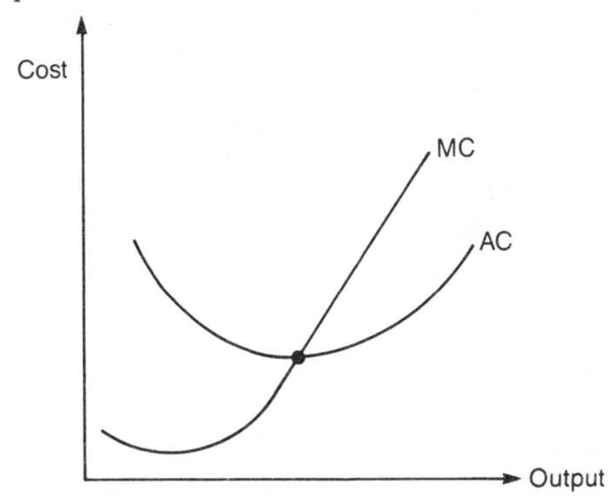

Fig. C.2

In the long run all costs are variable, and the shape of the ATC curve is determined by ECONOMIES OF SCALE and RETURNS TO SCALE in the production function. Marginal cost (MC) is defined as the change in total cost resulting from a small change in output: MC = small change in TC per small change in Q. In calculus terms MC = dTC/dQ. MC depends only on variable costs. The relationship between MC and AC is given by:

$$MC = AC + Q(\text{slope of AC})$$

Thus if AC is rising, MC lies above it, whereas if AC is falling MC lies below it; MC = AC when AC is neither rising nor falling (*see* Fig. C.2).

Profit-maximizing firms set marginal cost equal to marginal revenue (MR). However, for firms in perfect COMPETITION, MR equals price (P), so they will be effectively using MARGINAL COST PRICING. For firms in imperfect competition, who have a downward-sloping DEMAND CURVE, marginal revenue lies below average revenue (price) (*see* Fig. C.3)

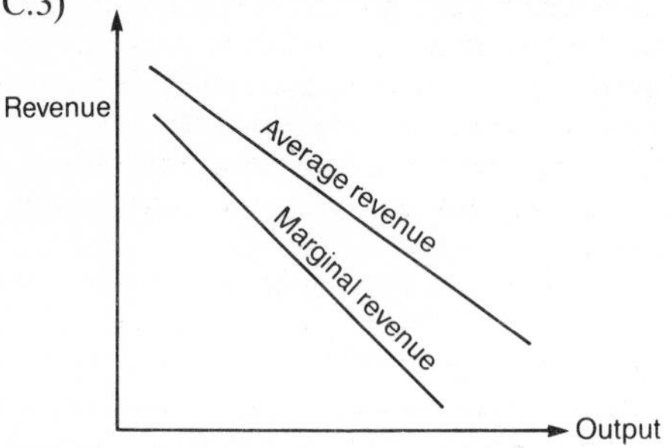

Fig. C.3

Council for Mutual Economic Assistance. *See* COMECON.

countertrade deals. *See* BARTER DEALS.

COW. Crude oil washing system (*see* TANKER TYPES).

cracker feedstock. The input into a CRACKING process. *See also* FEEDSTOCK.

cracking. A process whereby hydrocarbon molecules of high molecular weight (*see* CHEMISTRY) are broken down to lighter molecules. This is achieved by the application of high temperature usually in the presence of a CATALYST. Cracking results in the conversion of heavier products into more highly valued lighter products. THERMAL CRACKING was first used in 1913; CATALYTIC CRACKING appeared in the inter-war years and fluidized-bed catalytic cracking (FCC) in 1940. The process has become increasingly common, with important implications on the demand for heavier crudes. *See also* MOTOR SPIRIT UPGRADING; REFINING SECONDARY PROCESSES.

crane barge. A BARGE carrying a large crane (or cranes) that is used in OFFSHORE work.

credit worthiness. A reflection of the ability of a borrower to repay a loan together with interest. This ability obviously influences the willingness of the lender to lend. The term can be used in a rather vague way, but it can also be used very specifically when corporations are classified on a sliding scale. Increasingly political risk analysis tries to put countries on a quantifiable scale.

creep. A process whereby a pipeline expands and becomes longer due to high pressure or temperature of the contents and thus tends to shift position.

critical temperature. The maximum temperature at which a gas can be liquefied. Above the critical temperature the phase boundary disappears.

crosshaul. *See* BACKHAUL.

cross-price elasticity of demand. A measure of the change in DEMAND of one good in response to the change in price of another good. If the figure is negative then the goods are complements showing a relationship of COMPLEMENTARITY. For example it would be reasonable to assume that the cross-price elasticity of demand for fuel oil in relation to coal would be positive and greater than one. The actual coefficient is measured by the percentage change in demand for fuel oil divided by the percentage change in the price of coal. The higher the number the greater the degree of substitutability (*see* SUBSTITUTION EFFECT) or complementarity.

cross-section data. *See* TIME SERIES.

crown block. A part of the equipment used to hoist the pipes used during DRILLING. *See also* BLOCKS.

crude. *See* CRUDE OIL.

crude assay. A procedure for determining the DISTILLATION CURVE and quality characteristics of a crude oil. It provides in effect a product description. *See also* QUALITY, CRUDE OIL.

crude carrier. A tanker specifically designed to carry crude oil. *See also* TANKER TYPES.

crude distillation unit (CDU). The basic unit of a refinery in which the crude oil receives its first major processing via a distillation process. *See also* REFINING, PRIMARY PROCESSING.

crude long company. *See* CRUDE SHORT COMPANY.

crude oil (crude). The oil produced from an underground RESERVOIR after being freed of any gas that may have been dissolved in it under reservoir conditions (associated gas), but before any other operation has been performed (i.e. before REFINING). Because crude oil from different sources has different quality characteristics, it is a DIFFERENTIATED PRODUCT. *See also* QUALITY, CRUDE OIL.

crude oil characterizations. Crude oil has four characteristics which makes it unique as a product. The oil is a liquid, is hidden, is depletable and has unlimited potential as a raw material; no other material has all four characteristics. These features have significantly affected the economics of the oil industry since its beginning in 1859.

The fact that oil is a liquid has two implications. (a) Because it flows in three-dimensional space, it can attract enormous ECONOMIES OF SCALE (*see* POINT SIX RULE) which makes the industry very capital-intensive. This in turn implies low MARGINAL costs of production, which means that it is tempting to produce ever-increasing quantities provided that the marginal revenue exceeds the marginal cost. This, coupled with the existence of excess CAPACITY, has meant there is a natural tendancy for there to be a downward pressure on the price of oil. (b) Oil is a PUBLIC GOOD. If a reservoir straddles a boundary, one landowner could capture the oil of another landowner (*see* LAW OF CAPTURE) because oil, being a liquid, can flow across this dividing line. Even if capture were illegal it would not be possible to detect the syphoning off of the crude oil. Without state intervention, either via state ownership or through some other mechanism, this would lead to excessive oil production, which apart from anything else, would certainly damage the fields due to RATE SENSITIVITY.

The fact that oil is hidden underground carries two implications. (a) It creates uncertainty over ultimate availability because the only way to determine the existence of oil is to drill for it. (b) It means that any agreements between a company and a government to extract oil are inherently unstable for two reasons: (i) the agreement (CONCESSION or LICENCE) to split the proceeds must be made in ignorance of the size of the proceeds to be divided; (ii) any such agreement is the outcome of bargaining power between the two parties. While the oil is hidden, the bargaining position of the oil company is strong, but as the oil is revealed the bargaining position of the government is vastly improved, leading invariably to a demand for the renegotiation of the original agreement. Thus in the oil industry there is a tendency for continued 'warfare' between companies and governments.

The fact that oil is depletable adds to the uncertainty. Oil produced today cannot be produced tomorrow, which means the production decision must weigh the present value of the oil with its future value. Views of the future are always uncertain.

Finally, because of its chemical properties oil has almost unlimited possibilities as a raw material. Thus if it is used up in one use (e.g., burnt) its potential is lost. This raises questions as to the advisability of allowing a pricing mechanism to determine its allocation to one use in the short term which in the longer term may lead to the misallocation of resources.

crude oil costs. During the 1950s the view gained ground that the oil industry was one of decreasing costs (*see* INHERENT SURPLUS). The implications were that in order to take advantage of lower costs output must be expanded. Thus there is a tendency in the industry for the development of a MONOPOLY as one company becomes increasingly dominant. However, this view can be clearly seen to be incorrect, both at the level of the overall industry and when considering individual fields.

In theory, the tendency would be to produce first from the cheaper areas (e.g., the Middle East), moving on to higher-cost areas as the cheaper ones neared capacity, but in practice the situation is more complex and other factors must be considered. (a) Buyers perceive a potential problem associated with these low-cost areas, namely political instability, whereas many higher-cost areas carry implicitly lower costs because of their perceived advantages in relation to security of supply. (b) Many countries with large capacity potential for low-cost oil are also countries with a low ABSORPTIVE CAPACITY. Thus the point is eventually reached where the UTILITY of the marginal revenue from the extra capacity, if produced, would be close to zero. Indeed, if one were to take a very traditional view of the society, the utility of the extra revenue would be negative (i.e. not only would the extra revenue be worth little to the country it would actually be harmful).

In terms of the individual field there is a natural tendency for exploration, development and production costs to increase. The trend in the case of exploration costs is ambiguous. Clearly a larger find means a lower per barrel finding cost compared with a smaller field, suggesting a trend towards falling costs. Offsetting this, however, would be the tendency to explore in the 'easier', more accessible places first (e.g., onshore before offshore), suggesting a trend towards rising costs. Because of the random element in the discovery of fields, the past costs of finding oil indicate little about future costs. The trend when considering development costs is clear. In any field the easier PAY ZONES are developed first. Also, as development proceeds, the chances increase of drilling a well beyond the limits of the field. Both of these factors would cause development costs to rise as output expands. Production costs are similarly unambiguous as expressed in the production decline curve. As output proceeds the reservoir pressure falls, and maintenance of production requires secondary recovery methods and enhanced oil recovery which would increase costs (*see* PRODUCTION, ENHANCED OIL RECOVERY; PRODUCTION, SECONDARY RECOVERY MECHANISMS).

The increase in costs, both in terms of individual fields and the sum of fields, with increasing output, however, is not the same as increasing costs over time. With time, technological improvements may result in the whole cost curve being shifted downwards.

Clearly if crude prices were to fall, then at some point the higher-cost fields would become uneconomic and would be closed down thus reducing supply. This is, however, too simplistic; account has to be taken as to which costs are relevant, and a distinction needs to be drawn between fields that are already operating, fields that have been discovered, but have not yet been developed, and fields which have yet to be discovered.

The key element in cost is marginal cost, which is the cost of producing additional output. The reason for the emphasis on marginal cost is that it is avoidable (i.e. the producer has the power to decide whether or not to incur the cost). For a field already operating, the marginal cost is simply the variable costs of production. In general, because of the capital-intensive nature of oil production, variable costs are very low even in so-called high-cost producing areas such as the North Sea. Even the very-high-cost STRIPPER WELLS and STEAM DRIVE wells in the USA have marginal costs below $5–6 per barrel, although there is considerable variation between wells. On this basis, crude prices would have to fall a very long way before supply would be significantly affected. In reality, marginal costs are much less important in determining current supply than price expectations.

Because oil is depletable (*see* CRUDE OIL CHARACTERIZATIONS), the decision on whether to produce or not depends upon the producers' view of future prices. If prices are low, but still above marginal costs, production may be reduced if it is expected that future prices will rise: oil produced and sold at $10 per barrel now cannot be produced and sold at $20 per barrel in the future. Producers with very large reserves have a different attitude. If a producer has reserves that would last for a very long time (e.g., more than 100 years) then the PRESENT VALUE of the 100-year barrel would be close to zero irrespective of future prices because of the power of DISCOUNTING. This assumes, however, a positive DISCOUNT RATE, and it can be argued that many of the producers face a zero or even negative discount rate, either because of the problems associated with foreign investment (e.g., inflation, exchange rate fluctuations, the risk of expropriation) or because of undesirable effects on the local economy/society. Thus even for producers with large reserves price expectations could be the main determinants of current production levels.

In terms of the decision to develop an existing find or to explore for new discoveries (i.e determining long-term supplies), marginal cost becomes more important because it now includes capital outlays and is therefore much higher. For example, in the North Sea marginal development costs can rise to $15–18 per barrel of daily capacity. Thus, in theory, if the price falls below marginal development costs new fields are not brought onstream, thereby threatening a future supply shortage. Again price expectations tend to complicate the analysis. If prices fall below marginal development costs the perception of future shortage will grow, but such a perception will also cause expectations of future price rises. It is perfectly feasible that oil companies seeking to outguess the market should develop capacity that may be currently uneconomic, but will become economic in time. This implies that the future shortage may not occur. Of course, this too is realized and in turn may affect the decision.

crude oil distillation. The basic refining process, in which crude oil is heated in order to break it down into a number of fractions, or intermediate components, from which finished products can be made. *See also* REFINING, PRIMARY PROCESSING.

crude oil prices. Oil prices, like prices of other commodities, are determined ultimately by the interaction of SUPPLY and DEMAND. This approach is often criticized on the grounds that it ignores political factors. Such criticism is misplaced because, since supply and demand reflect both ability and willingness to pay, there is no reason not to incorporate political considerations into the analysis. Politics simply changes the shape and position of the supply and demand curves.

Because of the nature and structure of the oil industry, supply and demand do not interact in a competitive market to produce the immediate price response that would be suggested from an economist's use of PARTIAL EQUILIBRIUM ANALYSIS. The price of oil has generally been an administered price rather than being a MARKET PRICE. This administration of price has often been direct and explicit, whereas on other occasions it has been indirect and implicit. If to this is added the time lags inherent in the industry then it does appear frequently that pricing is little influenced by supply and demand interaction. However, given time, the administered price will respond to market forces; the amount of time is largely a function of the power of those administering (directly or indirectly) the price coupled with the extent to which the administered price is out of line with market reality.

The history of the oil industry has been characterized by two factors of key importance for pricing: (a) the existence of excess CAPACITY to produce oil and (b) a low MARGINAL cost of production. In such circumstances the economist would expect the market to exert a downward pressure on prices. For much of the industry's history this is precisely what has happened, and the history of pricing comprises essentially the efforts of the oil producers (companies and governments) to resist these pressures. If this is remembered then pricing history becomes clearer, as too do future price trends.

The Early Period 1859 to World War II

Between 1859 when the first well was drilled by Edwin Drake in Pennsylvania and World War II, international oil prices were determined by the US prices. This was because the USA was the world's main source of oil products. Up to 1928, the industry was characterized by a series of price wars between the MAJORS. These were the inevitable result of excess capacity, low marginal cost and an oligopolistic structure (*see* OLIGOPOLY). Thus if one company were to shave prices to gain market share, other companies would respond, thus creating a chain reaction. The result of these price wars was invariably that similar market shares were retained, but at lower prices. The industry was therefore characterized by uncertainty between the rival companies under the shadow of reduced profitability.

The inevitable solution to this is oligopolistic uncertainty cartelization (*see* CARTEL) whereby the rivals agree on a division of the market. This came about in 1928 in the ACHNACARRY AGREEMENT. In order for such a system to work it was necessary to protect the international price system, until then determined by US domestic prices, from the growing availability of crude oil, especially from the Middle East, which was significantly cheaper to produce than US crude.

This was achieved by the adoption of the US 'Gulf Plus' system which was a BASING POINT PRICING system. Under this system product prices worldwide were set by the US domestic price on the Gulf of Mexico plus the CIF cost (i.e. transportation costs) from the Gulf of Mexico irrespective of the product origin. Any actual price difference from a lower-cost source was absorbed by a PHANTOM FREIGHT RATE. Thus low-cost sources could be brought into the market without causing price competition. At the same time WINDFALL PROFITS were created for those with access to cheaper crude. It is important at this point to note that the prices were product prices and not crude prices. This was simply because refineries tended to be located predominantly on the fields (*see* REFINERY LOCATION), and consequently it was products that were traded and transported rather than crude oil. Furthermore, the crude oil was moving between AFFILIATES of the same company in a vertically integrated structure (*see* VERTICAL INTEGRATION) with the result that almost no crude was being sold in the open (ARM'S LENGTH) market.

The US domestic prices were determined by supply and demand. Because of the LAW OF CAPTURE underground oil was the property of whoever could pump it, with the result that it was in the interest of each individual to pump the maximum possible, which resulted in gross oversupply. To overcome this PRORATIONING was introduced which enabled regulatory bodies (e.g., TEXAS RAILROAD COMMISSION) to restrain supply. Restrained supply coupled with relatively high production costs in the USA kept domestic prices, and therefore international prices, higher than would have been the case in a competitive international market.

During World War II the USA introduced pricing controls in order to keep the domestic price down. In 1944–45, however, the international pricing system underwent a significant change. The UK government objected to the fact that Royal Navy ships were buying oil at, for example, Abadan in the Gulf at a price which assumed that the oil had been shipped from the Gulf of Mexico. Therefore a second basing point (at Abadan) was introduced. The resulting DUAL BASING POINT SYSTEM meant that now the lower-cost Middle Eastern oil had a natural market which was defined by western and eastern watersheds. For example, initially the western limit was the east coast of Italy. Here it was cheaper to buy from the Middle East than from the Gulf of Mexico, whereas the reverse was true on the west coast of Italy.

At the end of World War II, the USA removed price controls with the result that domestic prices rose sharply. This rise was mirrored internationally as the companies tried to maintain the link between US domestic and international prices. However, the link was far from perfect with the result that the Middle East's markets, as defined by the dual basing point system, expanded to the West and East.

Changes and Defence of the Price Structure 1945–69

The late 1940s saw another major change in the pricing system. In 1948, Venezuela insisted on, and got, a 50 percent tax imposed on profits attributable to crude oil production. This was in place of the traditional ROYALTY. However, in the absence of an international market for crude oil, because of the vertically integrated company structure (*see above*), there was no price from which to calculate profits. To overcome this posted prices were used, which reflected the market value of crude oil (*see* POSTED PRICES, CRUDE). This new system, which rapidly spread to other oil-producing countries, was acceptable to the US oil companies because they could offset tax payments to producer governments against their US tax liability. Other countries soon changed their tax systems to accommodate their companies in a similar way.

During the 1950s the oil companies still tried to maintain the link between US domestic prices and international posted prices. This culminated in 1957 when US domestic prices increased followed by the posted prices. The price increase, however, was simply not tenable in the international market given the pressure of market forces, despite the majors' ability to control supply via their horizontally integrated systems (*see* HORIZONTAL INTEGRATION). Furthermore, reductions in posted prices reduced the TAX-PAID COST of crude to the companies, thereby increasing their profitability. Thus, in 1959, and again in 1960, the oil companies without consulting the governments whose tax revenues would be affected cut the posted prices. This led in September 1960 to the creation of OPEC with the specific intention of restoring the cuts in posted prices, and hence the tax revenues of oil-producing governments. As a result the companies were unable to cut the posted prices further.

Meanwhile the majors' grip over the excess capacity to produce crude oil was beginning to weaken as new sources of crude outside the horizontally integrated control structure began to emerge in world markets. The move was started with the USSR beginning to export crude oil in order to obtain hard currency, and was reinforced as newly discovered and developed sources in Algeria, Libya, Nigeria, the United Arab Emirates, etc. came onstream. This was as a result of the perceived profitability of oil operations outside of the USA coupled with the restraints imposed by the crude long majors (*see* CRUDE SHORT COMPANIES) on existing capacity. Although much of this new crude oil had been intended for the US market, in 1959 more stringent import controls were imposed by the USA which meant many of the US companies were effectively 'stranded' abroad with crude oil.

The growing supply meant that, despite rising demand, crude oil realized prices (market prices on arm's length deals) began to decline during the 1960s. This, in turn, encouraged more consumers to switch to oil. In addition, towards the end of the decade companies began to pump oil in anticipation of their losing control of the production to the producing governments (*see* NATIONALIZATION).

Because oil is a depletable resource, oil produced and sold today is lost forever. Therefore, the decision as to whether or not to produce the barrel today involves comparing current prices with expected future prices. Taking into account the time value of money (*see* TIME PREFERENCE) to make the expected future price comparable to the current price it is necessary to compute via DISCOUNTING the present value of the expected future price. This is done by means of the following formula, which is simply the reverse of the formula to compute future value in a situation of compound interest (*see* DISCOUNTED CASH FLOW):

$$Pv = \frac{Fv}{(1+r)^n}$$

where PV is the present value of the crude less costs, FV is the future value of the crude less costs, *r* is the discount rate and *N* is number of years. As is apparent from the formula, a reduction in the future expected value reduces the present value; a rise in the discount rate will have a similar effect. On this basis the production decision becomes straightforward. If the current price less costs is above the present value more money will be made by producing the crude now; thus supply will be increased. Alternatively, if the current price less costs is below the present value then more money will be made by postponing the production; thus supply decreases.

Using this analysis the increase in supply of the late 1960s caused an erosion of the price structure. Two points are relevant. (a) The companies' belief that the producing governments would take over the oil operations gave the companies a very high discount rate (i.e. the present value of a barrel produced in five years' time is zero because it would be the government and not the company that would sell the barrel). (b) During the 1960s the perception grew in the industry that oil prices would fall further. Thus expected future prices were falling which (as with higher discount rates) reduces the present value below current price encouraging further production.

The First Oil Shock 1970–74

The price of Arabian light (*see* MARKER CRUDE) rose from $1.80 per barrel in 1969 to $11.65 per barrel by 1 January 1974. The obvious question is why had the downward pressure on prices, which had typified the industry in the 1960s, been reversed so dramatically Several explanations will be considered below, but before that it is necessary to relate what happened.

During the winter of 1969 demand was higher than expected, particularly in western Europe, and supply was suppressed due to a shortage of tankers. Thus prices rose, as did profits particularly for those companies with access to Mediterranean crude because of the shorter distances over which the oil had to be transported. This led the Libyans to demand an increase in posted prices to secure an increase in the government TAKE. Ever since Exxon had set the first Libyan posted price in August 1961 (which other companies operating in Libya used as a marker crude) the Libyan government had argued that the price was too low. These demands coincided in September 1969 with a revolutionary take over in which King Idris was removed from power by a group of army officers. In January 1970 the new government demanded from the 21 oil companies operating in Libya higher prices. The companies refused on the grounds that only they had the right to decide posted prices. However, by September the companies had agreed to new price terms. This in effect was the start of the FIRST OIL SHOCK.

Several factors explain why the companies acceded to the Libyan government's demand. (a) There were a large number of companies operating in Libya for whom Libyan oil was their main (if not only) source of eastern hemisphere crude. Having sunk considerable capital investment in Libya they were seeking to recover as much of this as rapidly as possible. They were therefore vulnerable individually to pressure from the Libyan government in a way that the major companies with excess capacity elsewhere would not have been. (b) Libya was not yet dependent on revenues from crude oil and was therefore able to impose production limits on the companies. This aggravated the shortage of Mediterranean crudes, which was further reduced when in May 1970 Syria (negotiating over increased transit fees) stopped the flow of Saudi crude to the Mediterranean via TAPLINE, following a breach of the line. The result was the strengthening of Mediterranean crude prices. (c) The major oil companies were not adverse to a price increase in Libya as this would constrain the ability of the independent companies from indulging in price competition in Europe.

The increase in Libyan prices provoked an immediate response from the other OPEC members who also demanded posted price increases in order to increase their revenues. In December 1970, OPEC met in Caracas and decided upon a strategy of regional negotiations with the companies. The 24 companies who feared the divide and rule tactics so successfully employed by Libya formed the LONDON POLICY GROUP to insist on unified global negotiations. OPEC and the London Policy Group were in deadlock.

Although the US government initially supported the London Policy Group's joint approach, in late January 1971 it changed its policy and recommended that the companies agreed to regional negotiations. One school of thought argues that the US government acted sensibly on the grounds that unified negotiations with such a diverse group as OPEC would have tended to produce a result close to the more extreme demands of the producers. Others argue that the US government was actively seeking an increase in oil prices in pursuit of its own interests and was deliberately seeking to weaken the companies' bargaining hand. There are strong *a priori* grounds for the latter argument. During the 1960s, the US domestic industry had experienced a significant decline. In particular the reserve base had been declining. This was partly due to the natural depletion process, but was also due to the lack of exploration effort by the oil companies. The result was the prospect of increasing oil imports which carried serious implications for the US BALANCE OF PAYMENTS which were already ravaged by the Vietnam War, and for US import dependence. In addition, because of regulation and protection the price of oil to US domestic consumers was significantly higher than in Europe and Japan, which was reducing US export competitiveness. A substantial increase in the international price would be paid for largely by the European and Japanese consumers and would provide a much needed boost to exploration and development in the USA.

Whatever the reason for the change in US policy, the result was regional negotiations which produced the TEHERAN AGREEMENT in February 1971 for the Gulf producers and the TRIPOLI AGREEMENT in April 1971 for the Mediterranean suppliers. Both were five-year agreements providing for a price increase and price escalation clauses.

Following the Teheran and Tripoli Agreements the market continued to firm. World economic conditions were in a state of boom, and several countries introduced PRODUCTION CEILINGS for conservation reasons. In August 1971 the US dollar moved to a FLOATING EXCHANGE RATE, which effectively reduced oil prices until OPEC succeeded in negotiating a supplementary agreement to account for the DEVALUATION of the dollar: GENEVA I in January 1972 and GENEVA II in June 1973. The effect of all of these factors was that realized prices (i.e. MARKET PRICES) for crude oil began to increase and by 1973, in some cases, exceeded the posted prices. This led to calls for a renegotiation despite the Teheran and Tripoli Agreements. Negotiations with the companies were held in early October, but the companies stalled. On 16 October the six Gulf ministers met in Kuwait and announced unilaterally an increase in the price of Arabian Light by 70 percent to $5.119 per barrel; the other OPEC members rapidly followed. Many regard this as the start of the First Oil Shock, but in reality it was the logical culmination of events begun by Libya in January 1970 (*see above*). The significance of the 16 October was that the administered price of crude oil was now administered solely by the producing countries.

In addition to the price increase, the ministers also announced the ARAB OIL EMBARGO, which was to have a market impact out of all proportion to its effectiveness. The USA and The Netherlands were the countries on which the embargo was imposed. However, the international interconnections of the industry meant that simply determining an embargoed destination was not enough. For example, although Arabian oil could not go to New York, it could go to Genoa, and where it went after Genoa no one need know. Therefore it was felt necessary to reduce supply availability (by 5 percent per month, using September 1973 as a base) on the grounds that if non-embargoed countries were concerned about their own oil supplies they would be less inclined to assist the embargoed countries. Subsequent studies have shown that the embargo was ineffective in that neither the USA nor the Netherlands suffered from a shortage of oil.

The supply reduction was due to the embargo plus an expansion in demand as people sought crude stocks. This meant that realized prices in the market continued to rise. In effect the market was in a situation of panic. In December 1973 an auction of Iranian crude fetched $17.04 per barrel for a significant volume of oil, and apparently bids in excess of $20 per barrel had been made, but were dismissed as not serious. At the end of December, OPEC met in Teheran, and the main item on the agenda was further price increases. There was considerable unease, especially from Saudi Arabia, about further price increases, but rather than risk openly splitting with OPEC Saudi Arabia agreed, and the price rose to $11.651 per barrel. The First Oil Shock had happened.

The causes of the First Oil Shock are still a matter of controversy. One argument starts from the premise that the price rises stemmed from a genuine supply shortage. At the time, and subsequently, this lead to people viewing the shortage as a result of a CARTEL action by OPEC. Such a view, however, is untenable since there was no production limiting/sharing scheme within OPEC. However, there are three possible explanations for a supply reduction that are not mutually exclusive. It has already been shown that a fall in the discount rate and/or a rise in expected future prices increase the value of the future barrel, thereby reducing current supply. Both occurred during this period.

The first half of the 1970s saw the producer governments taking over at least at a *de jure* level oil production operations by means of a process of NATIONALIZATION and PARTICIPATION. It is argued by some that the governments had a lower discount rate than the oil companies (i.e. the oil companies placed a lower value on future barrels than did the governments). In theory this view is acceptable, but in practice the ECONOMIC THEORY OF POLITICS casts some doubt on the long-sightedness of governments. The early 1970s did see the rapid growth of the view that future oil prices would rise. This view stemmed from three sources. (a) There was the growth of the environmental lobby, drawing its inspiration from such documents as the report of the CLUB OF ROME on *Limits to Growth*. (b) The oil companies began a concerted campaign to generate fears of impending shortage. This was generally designed to encourage governments to be more generous in their fiscal treatment of the industry and specifically aimed at persuading the US government to deregulate the US industry. (c) There were fears of politically induced shortages, leading to higher prices following the Arab oil embargo.

A final explanation for lower supply concerns the revenue requirements of the oil producers. Many oil producers had become TARGET REVENUE EARNERS reluctant to accumulate foreign assets. Thus once they had secured sufficient revenue for their domestic needs they would produce no more. If prices were to rise then the same revenue could be earned with lower production levels. In ef-

bending.

All explanations are plausible, but they suffer from one major weakness. There was no real shortage of oil, although for November and December 1973 this could be debated. Thus a pure supply–demand approach to the First Oil Shock needs at least a supplementary explanation. This can be provided by the argument that both the buyers and the sellers of crude oil wanted higher prices. The motives of the sellers for higher prices are self-evident, whereas the motives of the buyers require some explanation since it is usually assumed that buyers wish to pay the minimum. The apparent contradiction can be resolved when it is remembered that the buyers were the oil companies. As a result of the ownership changes of the early 1970s (*see above*) the oil companies had lost a significant proportion of their owned crude. Because of the advantages of having owned crude (*see* VERTICAL INTEGRATION), they were thus seeking new sources. However, these new sources were in high-cost provinces such as the North Sea and North Slope Alaska. Such oil needed higher prices in order to be viable. Thus higher crude prices meant that the oil companies' crude-producing divisions became more profitable, and it was these divisions that had always been the profit centres of the companies as a result of the manipulation of the TRANSFER PRICES. As for the refining operations, although they faced much higher costs as a result of the rise in crude prices, these costs could easily be passed on to the consumer, and they were. Thus the profitability of the oil companies is increased as a result of the price increases. Furthermore, the US government also privately welcomed higher prices (*see above*), although it is probable that both the governments and the companies were somewhat alarmed at the extent of the rise in December 1973. Thus the First Oil Shock was something of an aberration and outside of the normal tendency of the industry.

The Aftermath of the First Oil Shock 1974–78
Once the First Oil Shock was over, although not the aftermath of it, the natural downward pressure reasserted itself, but from a higher base. At this stage it is necessary to explain why prices did not fall faster in the mid-1970s, particularly because one of the key mechanisms — joint control of supply by the companies (*see* HORIZONTAL INTEGRATION) — that had kept the excess capacity in check in the 1950s and 1960s had been swept away due to the change of ownership.

The controlling mechanism of VERTICAL INTEGRATION remained in place. This controlled the excess capacity by encouraging companies to lift only sufficient for their own refining operations, thereby limiting the size of the ARM'S LENGTH crude market. Some explanation is needed as to why vertical integration remained when the *de jure* production decision had been taken over by the producing governments. The answer lies in the reluctance of the producing governments to market their own oil. This was partly because of their inexperience, partly because the market was weak and partly because government officials were frightened of being accused of corruption if they became involved in what was a very esoteric exercise.

After the First Oil Shock a new mechanism to balance supply emerged: the willingness of Saudi Arabia to act as a SWING PRODUCER. Thus if excess supply began to put pressure on the pricing structure Saudi Arabia would cut back its production to relieve the pressure.

During the period 1974–78 considerable attention was given by observers of the industry to the role of OPEC in the pricing of crude. OPEC would meet and at the end of the meeting would announce the new marker price. Occasionally a small increase would be announced, but more often the price was frozen. The only break in this pattern came at the December 1976 meeting in Doha when Saudi Arabia and the United Arab Emirates split from the rest of OPEC with only a 5 percent increase compared with the 10 percent increase introduced by the majority. This led to the view that OPEC was determining prices and, in that sense, was behaving like a CARTEL.

There is another explanation for the way in which the price was administered. It can be argued that Saudi Arabia went into the meeting having decided on the price of the marker (which was Saudi Arabian light crude). The source of this price was a complex interaction between views of the current and future oil markets (derived from the ARAMCO partners) and Saudi Arabia's view of the world and its political objectives. Having made this decision the rest of OPEC talked, but ultimately the price that was announced was a Saudi decision. The only qualification for this was that sometimes the Saudis would enter the meeting with a narrow price range and allow the meeting to debate which end of the range would be chosen.

In theory a formula existed that could be used to arrive at DIFFERENTIAL PRICES once the market price had been set. Thus differences in API GRAVITY and SULPHUR CONTENT meant applying linear formulations. In practice other producers of crude simply charged whatever the market would bear given the characteristics of the crude. If the price was set too high the companies refused to lift, as they were allowed to do by their contracts. These companies would theoretically be short of crude, but they could then appeal to Saudi Arabia to make up any shortfall from Saudi's excess capacity. The beauty of this system is that it was seldom invoked; the threat was sufficient to control the pricing of other crudes. If this view is accepted then the implication is that the role of OPEC was much less important than much of the literature suggests.

After 1973, a combination of controlling mechanisms managed to curtail the excess supply such that although prices did erode in real terms it was a slow decline rather than a collapse.

The Second Oil Shock 1978–81

In the case of the SECOND OIL SHOCK the events are agreed, but their interpretation is controversial. During 1978, unrest in Iran against the regime of the Shah was growing. In October 1978 the Iranian oil workers went on strike, and Iranian production fell from about 5.5 million barrels per day to almost zero. Other OPEC producers expanded their output, although Saudi Arabia's expansion was extremely erratic. This was partly because the expansion ran into technical problems, but also because of a political decision following the peace treaty between the Egypt and the Israelis.

The result in the market was one of panic because stocks had been allowed to fall to very low levels. Spot crude oil virtually disappeared from the market, and it was spot product prices that led the upward movements in the price structure. Hence the market saw a scramble for products and crude oil by both companies and governments. A number of the oil producers began to impose premiums on their official selling prices. Furthermore, as prices in the spot market became ever higher several producers, by invoking force majeure clauses, began to take over crude oil marketing which had previously been carried out by the companies. OPEC met periodically during the crisis, but merely approved of what members had already done independently. Meanwhile, Saudi Arabia was trying to stem the rise in prices, fearing the long-term effects of such rises on future oil demand. Finally, in October 1981 OPEC managed to agree unanimously on a marker price of $34 per barrel.

It is extremely difficult to interpret and explain what actually happened. This is because the price rise was generated by perceptions and expectations of the various participants, and these views were being formed in a situation of inadequate and often inaccurate information coupled with a general atmosphere of panic. In such a situation the application of rational analysis is at best difficult.

It is possible to explain one key element. The shortfall of crude, although much less than thought at the time, immediately caused a spot market response, and unfulfilled demand from the TERM MARKET spilled into the spot market causing prices to rise. The take-over of crude marketing by the governments reduced term market availability channelling yet more demand into the spot market despite the fact that spot supplies also increased. Thus in the spot market demand and supply were rising, but because of the Iranian shortfall demand was always above supply which pushed prices ever higher. Inevitably the administered price in the term market responded, although in an individual and *ad hoc* manner. The result when the situation settled following OPEC's meeting of October 1981 was a price structure that bore no relationship to anything.

Almost immediately the familiar downward pressure on prices reasserted itself, but from a higher base. Demand for oil was falling; everyone agreed this fall in demand was due to a combination of RECESSION, FUEL SWITCHING and CONSERVATION, but there was disagreement about their relative importance. In the early stages, the consensus placed the major share of the responsibility on recession. On the supply side, the outbreak of the Gulf War in 1980 had removed some crude from the international market, but this had been more than offset as non-OPEC production began to expand rapidly as new sources of crude (called forth by the

1973–74 price increase) came onstream.

The Aftermath of the Second Oil Shock 1981–85

One of the important consequences of the Second Oil Shock was the great weakening, if not destruction, of one of the market control mechanisms — vertically integrated structure — as producer governments took over the marketing role from the companies. This left only the Saudi swing role to control the market's excess supply. A second consequence of the 1979-80 shock was the discrediting of long-term contracts, and hence a move towards short-term contracts and the spot market which meant the price structure now responded much faster to market conditions than before.

To this situation of limited supply control and price responsiveness was added a new element which arose from the peculiar position of the BNOC. BNOC was obliged to take the UK government's royalty/participation crude from the lifting companies. Because BNOC had neither refinery capacity nor much storage capacity it sold back the crude to the companies. If the companies thought the price was too high and refused to take the oil, BNOC either had to announce an official reduction in price or had to sell on the spot market forcing down the spot prices. This created competitive pressures on the African producers (e.g., Nigeria) whose crude oil quality was similar to that of North Sea crude. Thus what was in effect an automatic self-destruct mechanism was introduced into the market.

The Third Oil Shock 1986

As the oversupply began to increase the downward pressure on prices, Saudi Arabia reduced its production in an effort to protect the price structure. This reflected an important change in Saudi policy. The swing role provided the only control mechanism, and it had proved to be insufficient with the result that prices continued to weaken. Some additional support mechanism was needed. This emerged in March 1982 when OPEC, for the first time since the mid-1960s, introduced a production-sharing scheme. The agreement, however, was ignored, and the price weakness was aggravated. In March 1983 OPEC tried again at a marathon meeting which produced the LONDON AGREEMENT. This agreement coupled production control with a $5 per barrel cut in the oil price.

With the signing of the London Agreement OPEC undertook a series of holding actions — further cuts in both production levels and prices — in the belief that as the world moved out of recession, increased oil demand would come to the rescue. However, it was becoming increasingly clear that the fall in oil demand had far more to do with fuel switching and conservation than recession, and therefore would be unlikely to increase in the near future.

OPEC production control was suffering from two problems. The first was another effect of the Second Oil Shock, namely the loss of information on supply and demand. Thus when OPEC set its production level it simply did not know whether it was too high, too low or just right. The second problem was that the temptation for individual countries to cheat on their production quotas was irresistable. The pressures came from a combination of the need for revenues and the need for the ASSOCIATED GAS. However, Saudi Arabia was willing to absorb both cheating and error by reducing its production to try to balance the market.

This lasted until September 1985 when Saudi Arabia was neither willing nor able to continue this protective role, and it announced a major change in policy; in future they would produce at their quota level and use market related prices. At the December 1985 meeting of OPEC it was announced that in future its strategy was to go for a 'fair market' share. The market saw this as the start of a price war, and spot prices fell to below $10 per barrel. Thus oil was now being traded like any other commodity, and like other commodities the price became victim to the whims of perceptions and expectations in the market.

The officially declared purpose of the new policy was to reimpose discipline within OPEC and to force the non-OPEC producers to cut back their output thereby sharing the burden of restoring and protecting the price structure at its former levels. An alternative view is that Saudi Arabia now wished to pursue a long-term low-price strategy in order to reverse the world's move away from oil. Whatever the motives, the effect was that oil pricing had now returned to the situation that had existed prior to the 1928 ACHNACARRY AGREEMENT.

crude oil quality. *See* QUALITY, CRUDE OIL.

crude oil washing. *See* TANKER TYPES.

crude runs. The throughput of crude oil in a PRIMARY DISTILLATION UNIT. *See also* REFINING, PRIMARY PROCESSING.

crude short company. An oil company whose crude oil-refining capacity exceeds its crude oil-producing capacity. A company whose producing capacity exceeds its refining capacity is known as a crude long company. It is unusual for a vertically integrated (*see* VERTICAL INTEGRATION) oil company to have its producing capacity match its refining capacity. The interaction between the two types of company was very important in the 1950s and 1960s. The crude long companies could theoretically supply the crude short companies. To do so would have enabled the crude short companies to compete in product markets against the crude long companies. However, not to have done so would have, and did, encourage the crude short companies to go out and find more oil, thereby aggravating the excess supply situation.

crude slate (slate). The crude input into a refinery. Frequently it may be a blend of crudes to which the REFINERY CONFIGURATION has been designed. In some cases the slate may be a SPIKED CRUDE. Within the constraints set by the refinery configuration the refiner will seek the lowest cost slate to produce a given PRODUCT BARREL.

cryogenic plant. An industrial plant used to cool a gas in order to liquefy it. The most extreme example is a liquefied natural gas (*see* LNG) plant in which the gas — largely METHANE — must be cooled to -161°C to achieve a liquid state. The volume of the liquid is 600 times less that of an equal weight of gas, which makes container transport feasible. Because of the extremely low temperatures required the plant must operate to very narrow engineering tolerances, and many LNG plants have suffered from technical problems resulting in significant below-capacity operation.

cSt. Centistoke (*see* VISCOSITY).

CT. *See* CORPORATION TAX.

current account. The part of the BALANCE OF PAYMENTS which includes VISIBLE and INVISIBLE trade.

current assets. The part of a company's balance sheet which records cash, bank deposits and other items easily converted to cash. *See also* ASSETS.

current ratio. The ratio of the current liabilities to CURRENT ASSETS of a business. Current assets normally exceed current liabilities, the difference between the two being working capital which is normally financed from long-term sources.

customs duties. *See* TARIFF.

cutbacks. BITUMEN that has been made liquid by the addition of a dilutant.

cut-off rates. A company's target RATE OF RETURN on specific projects. They are usually set higher than the company's COST OF CAPITAL and provide the criterion by which the decision to carry out a project can be made. This is done either by using the cut-off rate as the DISCOUNT RATE or by comparing it with the INTERNAL RATE OF RETURN. *See also* INVESTMENT APPRAISAL.

cut points. The limits of the temperature ranges (*see* BOILING RANGE) at which different groups of compounds (i.e. CUTS) are obtained in the crude distillation process (*see* REFINING, PRIMARY PROCESSING).

cuts. A product or a group of products derived from a barrel of crude oil. *See also* REFINING, PRIMARY PROCESSING.

cyclic. A chemical term describing an organic chemical compound in which the carbon atoms are arranged in a ring, as distinct from an open chain (i.e. ACYCLIC) structure. *See also* CHEMISTRY.

cyclo-. A prefix denoting a compound in which the carbon atoms form a ring (i.e. cyclic) structure. *See also* CHEMISTRY.

cycloalkanes. *See* CYCLOPARAFFINS.

cycloparaffins (cycloalkanes, naphthenes). A subgroup of the paraffin series (*see* ALKANES) in which the carbon atoms are arranged in a ring (i.e. cyclic) structure. The other two main subgroups are straight-chain (i.e. normal or n-paraffins) and branched-chain (i.e. iso- or i-paraffins). *See also* CHEMISTRY.

cylinder stock. A heavy lubricating oil base stock of high viscosity derived by DEASPHALTING petroleum residue. It is used as the basis of STEAM CYLINDER OIL.

D

daily peak. The maximum volume of gas produced in a day.

daisy chain. In the FORWARD MARKET for crude oil, options on a PAPER BARREL are bought and sold. This may occur many times for a specific barrel, with (on occasions) the same firm buying/selling the same barrel. The sequence of buyers and sellers is known as a daisy chain. In late 1984, *PIW* traced one such chain where 24 trading entities bought and sold the paper rights to a cargo of crude 36 times.

darcy. The unit of measurement for permeability of the reservoir rock. It measures the ability of a fluid to move through the rock. *See also* GEOLOGY.

DCF. *See* DISCOUNTED CASH FLOW.

dead freight. A claim arising out of the non-fulfilment of a charter. It is met by a charge either on unused cargo space or deficiency in the weight of the cargo.

deadweight tonnage (dwt). The normal way of expressing the capacity of a tanker or bulk carrier. As well as the cargo it includes BUNKERS, crew, passengers, food, stores, fresh water, tanks and pipework. Traditionally, dwt is measured in long tons (i.e. 2240 pounds), but increasingly the metric tonne (i.e. 2204 pounds) is being used. *See also* DEEP-SEA VESSEL CAPACITY.

deasphalting. The removal of asphaltic constituents from residual stock in order to provide a base for manufacturing LUBRICATING OIL. It is a solvent refining process (*see* RAFFINATE) in which the asphalt is precipitated using liquid propane.

debentures. *See* BONDS.

debottlenecking. The removal of constraints on the DESIGN CAPACITY of equipment to allow greater use.

debt. Money or other property owed by one person or institution to another.

debt financing. The cost of debt. It includes INTEREST payments and repayments of the CAPITAL sum.

debtor turnover. The number of times the balance sheet figure for debtors can be divided into the company's TURNOVER.

debt ratio. *See* GEARING RATIO.

debt servicing INTEREST payments on a DEBT. *See also* OPPORTUNITY COST OF CAPITAL.

decant oil. A heavy AROMATIC oil resulting from CRACKING operations. When still contaminated with the CATALYST it is known as slurry oil. Decant oil is used as a FEEDSTOCK for coke production.

decline rate. The rate at which the flow of oil or gas from a field falls as PRODUCTION proceeds. It is a function of the geology of the field coupled with the level of investment in recovery methods. It is expressed in the production decline curve.

declining balance method. *See* DEPRECIATION.

decreasing returns to scale. A LONG-TERM phenomenon whereby if all factor inputs are increased by the same proportion output increases by less than that proportion. Economies of scale refer to falling long-term average costs which may be, in part, due to increasing returns to scale. In the SHORT-

TERM, diminishing returns (or diminishing productivity) is concerned with the situation in which if one factor is, for example, doubled output is less than doubled. The commonest explanation for decreasing returns to scale is that as the plant or company grows management becomes less efficient, thereby leading to a reduction in the efficient use of the factor inputs.

deep-sea vessel capacity. The carrying capacity of a vessel including the cargo, fuel, water and provisions. It is equivalent to the DEADWEIGHT TONNAGE.

default. The breaking of a contract to repay a DEBT and/or the interest on the debt. It is a matter of considerable concern in a situation of falling oil prices given the high levels of international debt owed by some of the major oil-producing countries.

deflators. Price indices used to convert money values into REAL VALUES.

degree day. A measure of how far the mean average daily temperature falls below an assumed base temperature. It can be used in two ways to determine the energy demand for space heating: (a) when evaluating trends in past demand it is used to disaggregate the other factors influencing changes in demand such as price and income; (b) it can be used to forecast demand, although this requires some forecast of actual temperature. Detailed degree day records for the major consuming countries are published in PLATT'S *Oilgram*.

delayed coking. A process in which coke is produced from a low-sulphur crude oil residue.

deliverability. The amount of NATURAL GAS that can be delivered in a specified time period. This may apply to a well, a field or a whole system.

delivered pricing. *See* BASING POINT PRICING.

Delphi technique. A survey method of FORECASTING in which those surveyed are selected because they are assumed to be expert in a particular field. A questionnaire is circulated to the experts. From their answers a summary is compiled and circulated to them so that they can express further views; the summary/feedback process may continue for further rounds. It is in effect an attempt to discover an expert consensus and has been used, for example, to gain a view of remaining world oil reserves.

demand. The quantity of a good or service that people wish to consume at a given price. An individual's demand for a particular good in a given period of time is a function of the price of that good, the prices of other goods, income and the consumer's tastes. A demand curve plots the quantity demanded against the price of the good, holding all the other variables constant, and is normally downward-sloping (*see* Fig. D.1). A change in the price of a good implies a movement along the demand curve. However, if any of the other variables changes, the demand curve will shift. For example, if the consumer's income rises, the demand curve will normally shift to the right, except in the cases of inferior goods. If the price of a substitute falls, then the demand curve will shift to the left. If the price of oil falls, normally the demand curve for coal would be expected to shift to the left (less demanded at each price). An Engel curve plots the quantity demanded against income, holding the other determining variables constant.

The responsiveness of consumer demand is of particular interest and is expressed by economists in terms of ELASTICITY.

price elasticity of demand. This is a measure of the responsiveness of the quantity demanded to small changes in the good's own price. Price elasticity of demand is usually negative because most demand curves are negatively sloped, by custom the minus sign is usually omitted.

cross-price elasticity of demand. This is a measure of the responsiveness of the quantity demanded of good A to small changes in the price of good B. If good A is a SUBSTITUTE for good B, then the quantity of A demanded will rise if the price of B rises. Consequently the cross-price elasticity of A with respect to the price of B will be positive. If the opposite were to happen and the cross-price elasticity were negative, A and B would be COMPLEMENTS.

income elasticity of demand. This is a measure of the responsiveness of demand to small changes in the consumer's income. It is positive for a normal or superior good and (some would say) greater than one for a luxury good. It is negative for an inferior good, where quantity falls as income rises. For example, some fuels become inferior as income rises in developing countries, and people switch from using wood as a fuel to using kerosine and electricity.

Considerable research has gone into attempts to measure the various elasticities for energy in general and oil in particular. This is because if such elasticities can be found, then as is apparent from the earlier discussion this enables forecasters to predict future oil demand, although this presupposes a forecast of income and prices. However, in practice, actual measurement is extremely difficult as suggested from the enormous variations in actual estimates of elasticity. The reason for this lies in a general problem associated with PARTIAL EQUILIBRIUM ANALYSIS. If the quantity demanded is a function of price, then other determinants must be held constant if the relationship between price and quantity demanded — the own price elasticity — is to be established. This is why only small changes can be considered. The implication is that if the price experiences a large change other variables will also change, making it virtually impossible to disentangle the effects of the different variables. An example will serve to illustrate. After both of the oil shocks of the 1970s (*see* FIRST OIL SHOCK, SECOND OIL SHOCK) oil demand fell. However, the sharp rise in oil price occurred almost simultaneously (probably with causal connections) with recession (i.e. lower income), changes in the price of other energy sources, changes in tastes (e.g., with respect to views of energy self-sufficiency), etc. In such a context measurement of elasticities is almost meaningless in terms of any practical value because it is impossible to assign cause-and-effect relationships between the determining variables and demand. Therefore, statements in the literature about price and income elasticities for oil and gas should be treated with the greatest circumspection.

It has been found useful to assist conceptual thinking to divide the effect of a change in the price of a good into two components: (a) substitution effect and (b) income effect. A fall in the price of A effectively means that the consumer can buy a bigger bundle of goods with his or her original income because there has been an increase in real purchasing power. If enough income is taken away so that the consumer could just purchase the original bundle of goods, the effect of the change in relative prices of A against other prices can be examined. This is the substitution effect, and is usually positive for a fall in the price of A. The income effect looks at the change in the demand for A with relative prices held constant. It is positive for a price fall in a good with positive income elasticity, but negative for a price fall in an inferior good. It is a theoretical curiosity that if a good is sufficiently inferior, the negative income effect can outweigh the positive substitution effect. This is the case for a Giffen good, and the demand curve will have a section with a positive slope. Economists sometimes distinguish between an ordinary demand curve and a compensated demand curve. The compensated demand curve shows only the substitution effect of price changes, whereas the ordinary (sometimes called marshallian) demand curve shows both income and substitution effects. In Fig. D.1 the horizontal distance between the ordinary and the compensated demand curve is accounted for by the income effect of a change from price *P*. The area under a demand curve (*see* Fig. D.2) can be used as a measure of WILLINGNESS TO PAY, and in COST–BENEFIT ANALYSIS the area under a compensated demand curve is thought to be more appropriate.

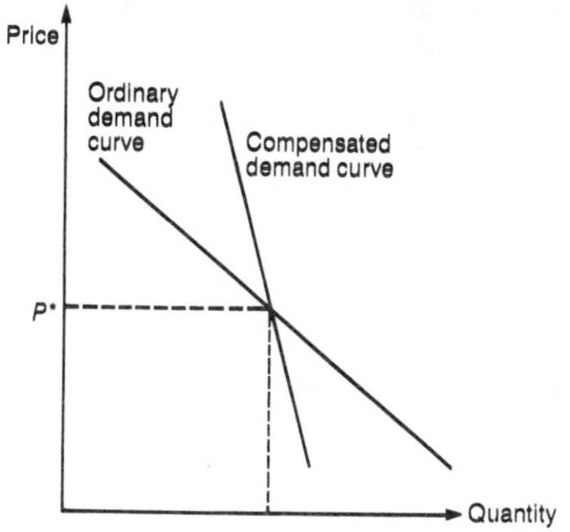

Fig. D.1. Ordinary and compensated demand curves

demand curve. A graphical representation of the DEMAND SCHEDULE with the price of the good on the vertical axis and the quantity demanded on the horizontal axis (*see* Fig.

D.1). A change in quantity demanded means a movement along the curve with other non-price demand determinants (e.g., income, tastes, price of other goods, number of consumers, etc.) held constant. A change in demand means the whole demand curve shifts in response to a change in a non-price variable such as income or taste. *See also* DEMAND.

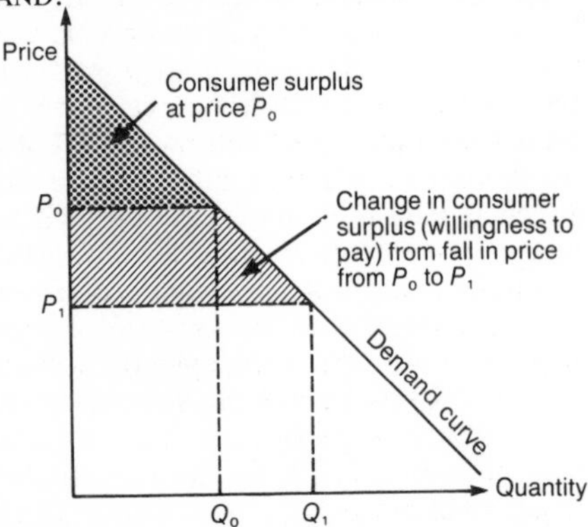

Fig. D.2. Willingness to pay and areas under the demand curve

demand function. A mathematical equation representing DEMAND, with demand on the left-hand side and demand determinants (e.g., price, income, etc.) on the right-hand side. ECONOMETRIC estimation techniques enable the replacement of the mathematical symbols with estimated numbers. If in log form the coefficient attached to each variable gives the ELASTICITY of that variable. Such functions are a key means of forecasting demand by simply inserting forecasts of the determining variable into the estimated equation and doing the sums.

demand schedule. Information (price and quantity demanded) in tabular form from which a DEMAND CURVE can be constructed.

demise charter. *See* BAREBOAT CHARTER.

demonstrated reserves. *See* RESERVES.

demurrage. A sum agreed by a charterer (*see* CHARTER) to be paid as damage for any delay beyond a stipulated time for loading or discharging.

depletable resource. *See* EXHAUSTIBLE RESOURCE.

depletion. Oil and gas reserves fall into the category of a natural non-renewable or EXHAUSTIBLE RESOURCES. Therefore, although very small quantities of HYDROCARBON deposits are being formed all the time, the stock of oil and gas reserves under the ground is effectively fixed. A barrel of oil or a cubic metre of natural gas produced 'now' cannot therefore be produced again in the future; if this loss of the marginal unit imposes a cost on future generations, there is said to be a USER COST of production in addition to the cost of extraction.

A company or government that produces from an oil or gas deposit depletes reserves over a period of time. For a large oil- or gas-field, the depletion period (i.e. from first to last production) would normally be of the order of 20–30 years. There is usually an initial 'build-up' period during which output rises as production and transportation equipment is installed, followed by a 'plateau' period when output remains approximately constant (usually longer for a gas-field than for an oil-field); finally, production declines during a 'drawdown' phase.

Deciding on the rate at which depletion will proceed means assessing both engineering and economic factors. Economists' theories about non-renewable resource depletion can be traced back to 19th century writers such as Ricardo and Jevons. Modern theory, however, begins from an article by Harold Hotelling in the *Journal of Political Economy* in April 1931; an updating of Hotelling theory appears in a paper by R.L. Gordon in the *Journal of Political Economy* in June 1967. As concern about natural resource shortages grew in the 1970s, there was a considerable increase in the literature on the subject both from economists and from physical scientists. Scientists have tended to be more concerned than economists about the prospect of absolute shortages of raw materials: some of their writings (e.g., D.L. Meadows *et al.*, *The Limits of Growth* Earth Island, London, 1972) (*see* CLUB OF ROME) anticipated severe resource scarcity and also rapid deterioration of the natural environment unless drastic remedial action was taken.

Most economists, following Hotelling-type theory, stressed the adjustment capacity of natural resource markets and their probable ability to cope with incipient shortages or surpluses. The essence of the economic theory of non-renewable resource depletion is that depletion decisions, within limits given by the state of technology, are determined primarily by price expectations and the rate at which resource owners discount the future. The resource owner has, in effect, to make an investment decision when deciding at what rate to deplete his finite stock over time. Leaving a marginal barrel of oil in the ground is, for instance, equivalent to investing in resources, and it has an OPPORTUNITY COST in terms of the return the owner could have obtained by producing the barrel immediately and investing the proceeds at the market rate of interest. Thus the opportunity cost of resource investment can approximately be measured by the owner's discount rate, assuming it is related closely to the market rate. Producing the marginal barrel — disinvestment in resources — also has an opportunity cost if the price of oil net of the cost of extraction is rising over time; the resource owner foregoes an increase in his margin over cost by producing at the present time. Only when the resource owner has so adjusted his depletion programme that his expected margin over cost is rising at a rate exactly equal to his rate of discount will he have an optimal programme in the sense that the expected net present value (*see* DISCOUNTED CASH FLOW) of his reserves is maximized. It is then no longer possible to increase the expected net present value of reserves by transferring output from one period to another.

Resource markets thus appear to have equilibrium tendencies. If future scarcity of some resource is perceived, owners of that resource will expect its price to increase relative to extraction costs. Therefore, assuming no change in the discount rate, they will tend to reduce current production so that the rate of depletion is lower; marginal units of the resource will be held in the ground so as to increase the net present value of reserves, in effect conserving some of those reserves for future generations. Demand side-effects, because of expectations of price increases, are also likely to have a conserving effect on consumption of the resource so that the effective life of reserves will increase. An expectation of scarcity therefore tends to set in motion forces that avoid or mitigate the anticipated shortage. If surplus is perceived, there will be an opposite effect, also because of attempts to increase the net present value of reserves. The depletion rate will tend to speed up so that more of the resource is consumed in the near term (when prices net of costs are expected to be relatively high) and less in the more distant future.

This kind of theory undoubtedly reveals some of the important driving forces in resource markets. However, even when uncertainty is introduced explicitly, it is difficult to simulate the amount of uncertainty to which a resource owner is subject. Apart from being very unsure how resource prices and costs (including taxes) will move and being uncertain of the appropriate discount rate, he will not know accurately the size of his reserves, nor will he be sure how extraction technology and other relevant factors may change during the depletion period. In practice, therefore, although depletion decisions are significantly influenced by discount rates and price expectations, in making such decisions there is bound to be a large element of managerial judgement about many different technical and economic variables.

Another complication is that real-world resource markets contain many imperfections and failures which prevent their operating to bring about a socially optimal rate of depletion. That is one reason why, in many countries, governments control rates of depletion of natural resources. Whether such regulated depletion rates are superior to the rates that would otherwise arise from imperfect resource markets is a controversial subject. Some economists maintain that governments control depletion rates mainly to pursue their own interests, which are not necessarily the same as those of society as a whole (*see* ECONOMIC THEORY OF POLITICS). *See also* ENERGY POLICY.

depletion allowance. A US tax allowance whereby a oil-producing company may offset a fixed percentage of the sales value of the oil or one-half the net profit of the oil indefinitely against taxable income.

depreciation. The tendency of ASSETS to become less valuable as time passes because of wear and tear, and advances in technology. Accounting conventions spread the cost of a physical asset over its assumed life. Two common conventions are the straight-line method (i.e. depreciation at a constant amount per period until the total cost of the asset, net of any residual value, is recovered) and the declining balance method (i.e. depreciation in each period at a constant proportion of the value of the asset so that depreciation in each period is a smaller amount than in the preceding period). Depreciation can normally be offset against profits for tax purposes, although depreciation rules laid down by the tax authorities may be different from those used by a company's accountants.

Calculations of depreciation can be based on HISTORIC COST — the original cost of the asset — or on REPLACEMENT COST. Historic cost depreciation does not provide sufficient funds to replace assets in times of general inflation, unless offset by cost-reducing technical advances for the asset in question. With replacement cost methods there is a regular revaluation of assets by a price index and a consequent recalculation of depreciation provisions.

Physical assets held by individuals, such as consumer durables (e.g., cars, domestic fuel-using appliances and other long-life household goods) tend to depreciate. Psychological obsolescence is a factor in the depreciation of such assets: if a more advanced version of a durable good appears on the market, owners of the less advanced versions may regard their assets as having depreciated. Depreciation is also used to denote the decline in value of a currency over time or at a given time against other currencies. *Compare* APPRECIATION. *See also* DEVALUATION.

depression. *See* RECESSION.

derived demand. The DEMAND for a factor input such as energy that is required not for its own sake, but as an input into a final product. Products facing a derived demand (e.g., oil and gas) tend to experience greater fluctuations in demand due to changes in the demand for the final product.

derived fuel. A fuel or energy that is produced from a basic fuel such as oil or gas. Examples include electricity, coke and TOWN GAS.

derrick. The mast-like structure positioned over a well. It houses the hoisting gear for raising and lowering the pipe. *See also* DRILLING.

derrick floor (drilling floor, rig floor). A raised base at the bottom of the DERRICK that carries the DRILLING TABLE.

derrick man (derrick monkey). A member of the drilling crew who works up the DERRICK.

derrick monkey. *See* DERRICK MAN.

DERV. An alternative name for DIESEL OIL that is used in the UK. It is an acronym for diesel engine road vehicle. *See also* FINISHED PRODUCTS, DIESEL OIL.

desalting. A process that precedes the main refining operation and removes any salt which may be contaminating the crude oil. *See also* REFINING, PRIMARY PROCESSING.

desander. *See* HYDROCYCLONE.

design capacity. The official CAPACITY of a piece of plant or equipment as specified by the engineers. The design capacity can often be exceeded, but in this case there is always a danger of overstressing the equipment, resulting in a breakdown. It is not uncommon for pipelines to be forced to operate above their design capacity.

desilter. *See* HYDROCYCLONE.

desulphurization. Generic name given to those processes used to remove, or reduce, the sulphur that is present in varying degrees in petroleum products. The type of process varies with the product being treated, with HYDROFINING, used for treating middle distillate products (e.g., gas oil, diesel oil, kerosine), being perhaps the most commonly used process, and RESIDUE (fuel oil) desulphurization (*see* HDS) being relatively uncommon. *See also* REFINING, SECONDARY PROCESSING.

detergents. *See* CALCIUM PETROLEUM SULPHONATES.

devaluation. The reduction of the official rate at which one currency is exchanged for another. In the context of oil its significance lies in the fact that internationally oil is traded for dollars, thus a government receives its revenue (sales or tax) in dollars. A devaluation increases the amount of revenue received in local currency. For example, an oil revenue of $10 billion to the UK at $1.40 to the pound sterling provides the UK with revenue of £7.143 billion. However, an exchange rate devaluation with a reduction in the exchange rate to $1.10 to the pound sterling increases the UK revenue to £9.091 billion. Devaluation results in exports becoming cheaper and imports becoming dearer, which carries implications for the balance of payments and domestic inflation. *Compare* REVALUATION.

development. The preparation of an oil or gas discovery for PRODUCTION. Development involves drilling DEVELOPMENT WELLS, which act as producing wells, together with the assembly of all the necessary aboveground equipment. It is this investment decision that creates PROVEN RESERVES. As one economist remarked 'oil in place is discovered, oil reserves are developed'. Therefore it is the investment act of development which creates producing capacity. The costs of developing a field can vary enormously depending upon the geology of the reservoir and its location (i.e. onshore/offshore, accessible/inaccessible, etc.).

development well (step-out well). A well within, or close, to the limits of a producing or producible POOL which on completion acts as a producing well. A development well is also used to establish the vertical and horizontal limits of the field and the precise nomenclature can be confusing. Drilling can lead to extension additions to the PROVEN RESERVES and enhances the knowledge of the geology of the field.

deviated drilling. *See* DEVIATED WELL; DIRECTIONAL DRILLING.

deviated well. A well that progressively deviates from the vertical. This may occur for one of three reasons: (a) to avoid JUNKED drilling equipment left in the original hole; (b) to produce from a large area using only one PRODUCTION PLATFORM; (c) to tap illegally into a neighbouring field. *See also* DIRECTIONAL DRILLING.

dewaxing. The removal of contaminating waxes from lubricating oil stocks. It is achieved generally by the filtration of a mixture of the oil and a solvent at a low temperature.

DHT. *See* DISTILLATE HYDROTREATING PROCESS.

diatomite. A soft rock that contains a considerable proportion of petroleum.

diesel engine road vehicle. *See* DERV.

diesel fuel. *See* DIESEL OIL.

diesel index. A mathematical ratio based on certain properties of DIESEL OIL. It is used to predict its combustion properties. *See also* QUALITY, DIESEL OIL.

diesel oil. A transportation fuel used mainly by commercial road vehicles and marine engines. *See also* FINISHED PRODUCTS, DIESEL OIL.

differential price. The price for a DIFFERENTIATED PRODUCT. It is particularly relevant in crude oil pricing (*see* CRUDE OIL PRICES). Once the price of the MARKER CRUDE is set other crudes can be priced on a differential basis. Thus the marker price is used as the starting point from which adjustments can be made for each crude to account for differences in quality and location. A differential price means there is no single price, but rather a structure of prices. *See* QUALITY DIFFERENTIAL.

differentiated products. Products that although essentially similar exhibit differences which may be real or imagined. Much of the work of advertisers is to try and create the illusion of difference. This is as opposed to a homogeneous product which is identical. For example, crude oil is a genuine differentiated product, whereas oil products of a given specification are homogeneous products, although advertisers of motor spirit try to make it a differentiated product. The difference lies in the implication for pricing. A differentiated product may be able to build up consumer loyalty which may allow the raising of its price above that of a similar product.

diminishing productivity. *See* DIMINISHING RETURNS.

diminishing returns (diminishing productivity). Sometimes known in economics as 'the law of diminishing returns'; an empirical generalization. In the short run, with at least one factor input into the production process fixed, as increasing quantities of the variable factor input are used with the fixed input, first the MARGINAL PRODUCT and then the AVERAGE PRODUCT of the variable input eventually diminish.

dip. A geological term that measures the inclination of rock strata with respect to the horizontal. *See also* GEOLOGY.

dipping. A procedure that is used to determine manually the depth of oil in a container or tank by means of a measuring rod. Dipping is a useful technique if the tank has accumulated BOTTOMS since this sludge cannot be detected using electrical measuring devices.

directional drilling (deviated drilling). A DRILLING technique that causes the well to deviate from the vertical. This is achieved by placing a WHIPSTOCK at the base of the hole which forces the bit and drilling pipe to deviate from their original direction. The result is a DEVIATED WELL.

direct marketing. Sales by a refinery company (integrated or otherwise) directly to the public without the involvement of a JOBBER or independent dealer.

dirty trade. The trade in crude oil and unrefined products (e.g., fuel oil) and in lubricating oils.

discounted cash flow (DCF). An approach to INVESTMENT APPRAISAL that includes several methods of evaluating the worth of prospective investments. A characteristic of all investment projects is that inflows and outflows of cash occur over a period of time, and all methods of appraisal should allow not only for the QUANTITIES of cash received and paid, but also for their timing.

The underlying principle of the DCF approach is that money has time value (*see* TIME PREFERENCE) so that money available at present is worth more, because it can be invested or used in some other way, than money available in the future. Thus the worth of $1 in the future should be discounted by some relevant rate of interest to bring it to a 'present value'. If that rate of interest were 10 percent, $1 next year has a present value of $1/1.1 (i.e. approximately $0.9091). More generally, the PRESENT VALUE (PV) of a stream of positive or negative future cash flows (x) is shown in Fig. D.3, where 1, 2, ..., n are time periods and r is the rate of interest (expressed as a decimal).

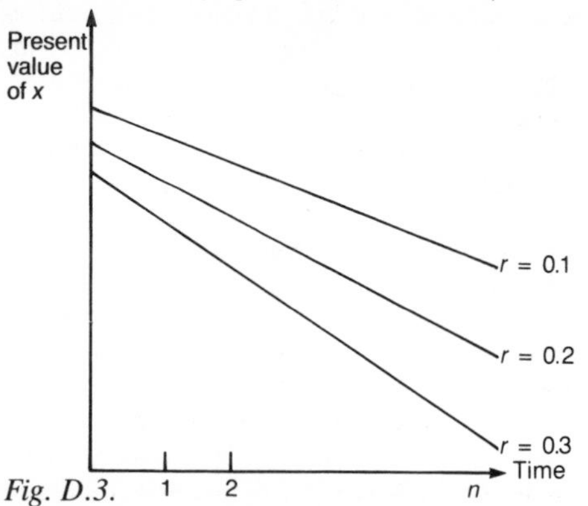

Fig. D.3.

Where the cash flows are net (i.e. revenues minus costs), their discounted sum is referred to as the net present value of the project. In all DCF analyses, revenues and costs are expressed in terms of cash flow (i.e. they relate to the period when cash flows into or out of the organization). Capital outlays, for example, are included when the money is spent in one go, rather than being spread over a period which is usually the case in accounting. Similarly, taxes are included as a cash outflow when they are paid, which is normally some time after the profits to which they relate have been earned.

net present value (NPV). This is the most commonly used of the DCF methods. In effect, it is a statistical standardization procedure that places a series of cash flows occurring at different time periods onto a common time base (the present). By discounting cash flows are given less and less weight the further in the future they lie. It is particularly important to take into account the timing of cash flows in the long-life FRONT END-LOADED capital projects that are typical of the oil and gas industries.

internal rate of return (IRR). This is another DCF method and is also known as investor's yield. The IRR of a project is the rate of discount (r in Fig. D.3 given above) that reduces the NPV to zero. In Fig. D.4, IRR is the intercept on the horizontal axis.

IRR is found by trial and error, using the principle that NPV will normally be negatively related to the discount rate (as in Fig. D.3). In cases where there are negative cash flows late in the project's life, however, there may be two or more discount rates which reduce the NPV to zero. In such cases of multiple roots where there are several IRRs, the method cannot be used except after modifications.

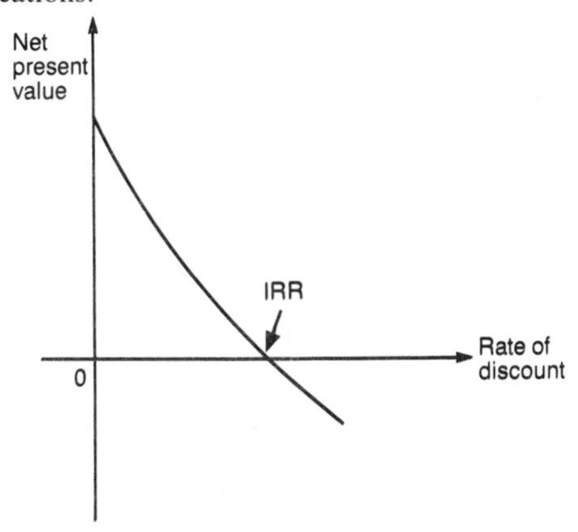

Fig. D.4.

annual capital charge (ACC). This is a third variant of DCF. It judges the acceptability of a project in terms of whether or not the net cash flows are sufficient to provide an adequate rate of return on the initial capital and to allow recovery of the full capital cost of the project at the end of its life (using a SINKING FUND).

A number of practical difficulties may arise in applying DCF methods.

(a) The costs and revenues of the project are in the future, therefore they must be forecast. Thus there is likely to be considerable uncertainty about the outcome of the project. All investment appraisal methods face this problem and so SCENARIOS or SENSITIVITY ANALYSIS are used to determine the likely variability of the outcome, assessing how robust the project is with respect to future uncertainty.

(b) There will also be uncertainty about the appropriate DISCOUNT RATE, although in principle the OPPORTUNITY COST of capital should be used.

(c) When competing projects are ranked (e.g., when there is CAPITAL RATIONING) NPV and IRR may give different rankings if there are significant differences in initial capital outlays, lives or cash flow timing as between two or more projects.

discounting. (1) The carrying out of a DISCOUNTED CASH FLOW exercise. (2) The allowing for the future effects of some anticipated event on security prices, commodity prices or exchange rates. For example, anticipation of a fall in the price of oil that might lead to a reduction in the value of sterling in relation to other currencies such that when the oil price falls the sterling exchange rate does not respond. In this case the foreign exchange market has already 'discounted' the fall in oil prices. (3) In US usage, the pledging of accounts receivable (i.e. sums owed by debtors) as COLLATERAL against a loan.

discount rate. The percentage rate used to compute the PRESENT VALUE of a CASH FLOW expected in the future (*see* DISCOUNTED CASH FLOW). The discount rate should reflect the OPPORTUNITY COST OF CAPITAL. For example, the $110 expected in one-year's time at a 10 percent discount rate has a present value of $100. This is because it is assumed that if $100 were invested today then the money could earn 10 percent thus giving the investor $110 in one year.

Discount rates are specific to projects because they must reflect the risks inherent in the project, including economic, technical and political risks. Clearly such risks vary between projects. The higher the risk, the higher the discount rate should be. Another way of looking at discount rates is that they reflect people's view of the future. In effect, a high discount rate means future income is viewed as less valuable than income now. Unfortunately, discount rates used by companies are usually closely guarded commercial secrets, thus making it difficult to discuss levels of discount rates as used in the oil and gas industries. However, it seems likely that in the development phase of the British North Sea real discount rates of between 15–20 per cent and even higher were used.

discovered resources. The amount of oil and gas in the earth, the presence of which has been confirmed by drilling. It is equivalent to OIL-IN-PLACE.

discovery well. A wildcat well (*see* WILDCAT DRILLING) that has discovered a COMMERCIAL FIELD.

discretionary system of licensing. A system of licence allocation whereby licences are awarded by the licensing authority based on various conditions and criteria determined by the government. Most parts of the world, including the UK, employ this system. In the UK, criteria such as the involvement of the applicant in UK companies for research and development work and the applicant's willingness to explore frontier areas are considered. UK EXPLORATION LICENCES are non-exclusive and last for three years; the licence applies to all areas of the UK Continental Shelf (UKCS), except where a PRODUCTION LICENCE has been awarded. Licences on the UKCS have generally been offered in licensing rounds which, at times, have included auctions (*see* AUCTION SYSTEM OF LICENSING) as well as experiments involving the nomination of blocks by companies. Licences may cover all or part of a single block (approximately 250 square kilometres) or several blocks. Conditions of licences include relinquishment (or surrender) terms (*see* RELINQUISHMENTS) which, with the exception of licences awarded in the fifth and sixth rounds, require the licensee to return 50 percent of the licence area to the government after six years while keeping the remaining area for up to 40 years. A rental (or AREA RENTAL) payment on the licence is due. This is a fixed sum per square kilometre and increases yearly to a maximum payment, which in the Eighth Round was set at $5400. Further licence conditions are for a ROYALTY payment (applying to fields granted development consent before April 1982) and control over the disposal of oil whereby UKCS oil must be landed in the UK before being exported unless a waiver is issued by the government. In addition, the licensee was obliged to sell 51 percent of the oil produced (*see* PARTICIPATION OIL) to BNOC at the official price prior to BNOC's demise.

To ensure GOOD OIL-FIELD PRACTICE, work programmes must be approved by the government. Before development may take place, approval must be granted by the government. The ANNEX A (Field Development and Production Programme) is an overall description of the oil-field project and is generally required by the Secretary of State for Energy after the draft ANNEX B has been accepted. The Annex B (Background Information) is much more detailed than the Annex A and comprises a comprehensive checklist including, for instance, manpower resources and anticipated well performance. Annex A is used in conjunction with the STAGED APPROVAL PROCEDURE, which is a three-phase process covering the life of the oil-field. The first stage roughly covers the development period and initial production, up to plateau production, the second stage covers plateau production, and the final stage is concerned with the field's remaining life including abandonment.

discriminating monopolist. A monopolist (*see* MONOPOLY) who charges different prices in different markets. To be able to do this the markets must have different elasticities of demand (*see* ELASTICITY). It must be possible to keep the markets separate in order to prevent ARBITRAGE whereby the buyer in the lower-priced market resells in the higher-priced market at a price that undercuts the monopolist. It has been argued that the development of VERTICAL INTEGRATION can assist in the development of discriminating monopoly in the following ways: (a) by providing the seller with information about the existence of differential elasticities; (b) VERTICAL INTEGRATION into the lower-priced market can be used to prevent arbitrage. *See also* DISCRIMINATORY PRICING.

discriminatory pricing. The charging of different prices FOB for the same product. This has been a very common occurrence in the international oil industry. *See also* DISCRIMINATING MONOPOLIST.

diseconomies of scale. *See* ECONOMIES OF SCALE.

disequilibrium. *See* EQUILIBRIUM.

dispersant. A chemical that is used to reduce the surface tension between two liquids. Dispersants are commonly used to break up oil slicks.

distillate fuels. A collective term for liquid fuels from the MIDDLE DISTILLATES range, mainly GAS OIL and KEROSINE, which are used for domestic and industrial power and heating. *Compare* DISTILLATES. *See also* FINISHED PRODUCTS; FINISHED PRODUCTS, AVIATION FUELS; FINISHED PRODUCTS, FUEL OIL; FINISHED PRODUCTS, GAS OIL; FINISHED PRODUCTS, KEROSINE.

distillate hydrotreating process (DHT). A process that is used to remove sulphur and nitrogen compounds from DISTILLATES including KEROSINE. The process involves subjecting the distillates to a catalytic HYDROGENATION process. *See also* REFINING, SECONDARY PROCESSING.

distillates. The products obtained by condensing vapour during the FRACTIONAL DISTILLATION process. *Compare* DISTILLATE FUELS. *See also* REFINING, PRIMARY PROCESSING.

distillation. A chemical process that involves the boiling of a liquid with the collection by condensation of the resulting vapours. The distillation of crude oil is the basic refining process which precedes all others. *See also* ATMOSPHERIC DISTILLATION; FRACTIONAL DISTILLATION; REFINING, PRIMARY PROCESSING.

distillation column (distillation tower). The refinery equipment in which DISTILLATION is carried out. *See also* REFINING, PRIMARY PROCESSING.

distillation curve. The volume of different products obtained from a crude oil by a simple DISTILLATION process expressed in graphical form.

distillation cuts. The different products obtained from crude oil by a simple distillation process. *See also* FINISHED PRODUCTS; REFINING.

distillation range. *See* BOILING RANGE.

distillation yield. *See* FINISHED PRODUCTS.

distressed cargo. A cargo of crude oil or products that the seller has been forced to sell in the SPOT MARKET at whatever price he can get. This may arise because the owner has storage problems or, more likely, because he is faced by financial embarrassment. Distressed cargoes usually sell well below the going rate. If there is little volume being traded and the sale is reported, it can have an effect on price expectations out of all proportion to reality.

district heating (collective heating). The provision of domestic heating within a specific locality. District heating is normally associated with the use of heat that is a by-product of, for example, electricity generation (i.e. the cooling water). Before the introduction of district heating this heat was frequently wasted and caused THERMAL POLLUTION.

diurnal swing. Fluctuations in the demand for gas within a single day in which the base demand load is exceeded by the peak demand load. Local storage of gas may be needed to meet such daily changes in demand.

diverter. An inflatable torus (i.e. ring) that can be used to seal the ANNULUS, causing the DRILLING FLUID to be diverted to the MUD PITS on a drilling rig. *See also* DRILLING.

divestiture. The breaking up of a vertically integrated (*see* VERTICAL INTEGRATION) company into separate companies each involved with a specific stage of the industry. Calls for divestiture frequently recur in the USA where it is seen as a possible mechanism to reduce the power of the oil companies.

dividends. Profits that are not retained in the business, but are distributed to ordinary shareholders. A company's dividend may be expressed as an absolute amount of money per share or as the percentage which the total sum for distribution bears to the nominal value of the company's ORDINARY SHARE.

doctor test. A test carried out on finished MOTOR SPIRIT to establish whether the sulphur content has been reduced to acceptable levels in the refining process. *See also* QUALITY, MOTOR SPIRIT.

DOE. An abbreviation used in the USA and in the UK for the Department of Energy.

dog house. A small shelter on the rig floor (*see* DERRICK FLOOR) for use by the drilling crew.

dog-leg. *See* SIDE-TRACK HOLE.

domestic heating oil. MIDDLE DISTILLATES used for home heating. After the mid-1950s there was a very rapid expansion in the consumption of domestic heating oil because of the falling real price of oil (*see* FUEL SWITCHING). *See also* FINISHED PRODUCTS, GAS OIL.

domestic income. *See* DOMESTIC PRODUCT.

domestic product (domestic income). The total value of output produced actually within a country. This is as opposed to national product, which is the output produced by the nationals of a country whether working in that country or not. The distinction between domestic product and national product can be extremely important in the context of large oil- and gas-producing countries in which foreign companies play an important role. Hence domestic product is very much higher than national product. *See also* GROSS DOMESTIC PRODUCT.

dominant producer model. *See* SWING PRODUCER.

doodlebugger. A member of a SEISMIC EXPLORATION crew.

dope. A lubricant for the threads on a DRILL PIPE.

double. Two sections of DRILL PIPE joined together.

double taxation. A situation in which the same tax base is taxed twice. Frequently this is avoided by double taxation agreements. This concept was particularly important when, in the late 1940s, producing countries led by Venezuela introduced a profits tax on the oil companies' profits attributable to crude oil. US companies were able to offset the taxes paid to a producer government against their US tax liability, thereby avoiding double taxation. Initially no such arrangement existed for UK-domiciled firms, although this was very rapidly changed. *See also* CRUDE OIL PRICES.

downcomer. A duct in a DISTILLATION COLUMN that carries the liquid flowing down the column from one tray to the tray below.

down dip. A situation in which one well is below another well due to the slope of the strata that contains the RESERVOIR. *See also* GEOLOGY.

downhole. *See* WELL BORE.

downhole safety valve. An emergency valve fitted to the production tube (*see* PRODUCTION STRING) of a well below ground or seabed level. The valve can be used to stop the flow of oil or gas.

downside risk. The likely effects of a bad outcome when there exists a range of possible outcomes to a situation, ranging from bad to good. Thus the expression 'the downside risk is low' means in effect if a bad outcome were to occur, the negative effects would be small. Clearly projects in which the downside risk is low are usually quite attractive. *See also* RISK.

downstream. (1) Describing crude oil or gas processing (e.g., refining), petrochemicals, gas liquefaction, etc. (2) In the context of the economics of VERTICAL INTEGRATION, it means moving into activities closer to the FINISHED PRODUCT and away from the raw material. *Compare* UPSTREAM.

downtime. The period during which plant or equipment is not being used because of repair, routine maintenance or any other reason. In effect, downtime is the difference between DESIGN CAPACITY and the actual CAPACITY. In the oil industry it is often described as the difference between calendar days and STREAM DAYS. Some elements of downtime can be foreseen (e.g., scheduled maintenance), whereas others are random in nature and cannot be predetermined. *Compare* LOST TIME.

draw-down. (1) The difference between the flowing and static bottom hole pressures in a well. *See also* BOTTOM HOLE DIFFERENTIAL PRESSURE. (2) The total amount borrowed under a specific loan agreement.

drawworks. The hoisting gear used in DRILLING operations for the raising and lowering of the DRILL PIPE.

drill collar. The sections of DRILL PIPE immediately above the bit. These sections are thicker and heavier than the standard sections. *See* DRILLING.

drilling. The drilling of a hole, or well, is necessary at a number of stages.
(a) At the exploration stage, drilling is required to establish if oil or gas is present in a particular geological formation. Exploratory drilling of this type is known as WILDCAT

DRILLING, with the resulting wells being known as wildcat wells.
(b) Further drilling is carried out as a follow-up to successful wildcat drilling, known as appraisal drilling. Analysis of APPRAISAL WELLS provides further information about the size and characteristics of the accumulation. Such data are necessary before a decision as to the commercial viability of the discovery can be taken.
(c) Before the oil can be produced, development drilling is required. A number of DEVELOPMENT WELLS are drilled, from which the oil will actually be produced, since even if it is possible to produce oil from the wildcat and appraisal wells, there will normally be insufficient for production purposes.

Operationally, wildcat drilling is the most complex stage, since it proceeds on the basis of theoretical calculations and projections rather than being based on empirical knowledge of the formation (*see* GEOLOGY). Nevertheless, the basic essentials of the drilling process are the same regardless of the particular phase of drilling.

Primitive wells, which were used to produce water or brine, were dug by hand, with a sharp, chisel-like tool, or bit. The rock debris, or cuttings, which tended to accumulate, were removed periodically by placing them in a container that could be hauled out of the hole. As the hole got deeper, the walls had to be supported with wood, or brick, to prevent them from caving in. The three essential features — a sharp cutting tool, removal of cuttings and strengthening of the walls of the hole — still underlie the technical complexities of modern drilling techniques.

The essential drilling tool is the bit, which is a sharp tool capable of biting into the ground and making a progressively deeper hole. There are basically two types of drilling: (a) ercussion drilling, in which the bit is raised and lowered to pound out a hole; (b) rotary drilling, in which a bit that is capable of crushing the rock is rotated at the end of a pipe.

In the early days of oil exploration, in the mid-19th century, a type of percussion drilling known as cable drilling (or cable tool drilling) was used. This consisted of a heavy bit suspended on a wire rope (a cable) which was pounded up and down by power from a steam engine. Cuttings were removed periodically by hoisting out the bit and inserting a bailer, a hollow tube with a valve attachment which could scoop out the debris. The walls of the hole were consolidated by inserting steel pipes known as CASING. Although the cable tool method was reasonably effective for shallow wells, it was slow, and in the early 20th century, ROTARY DRILLING methods were developed and gradually superseded cable drilling.

In rotary drilling, the bit is attached to the end of a hollow piece of pipe (a drill pipe) which is made in uniform lengths of approximately 30 feet. They are designed to screw together so that progressively longer lengths of pipe (called a drilling string) can be achieved as the hole deepens. The sections at the top and bottom of the drilling string are of different design from the standard drill pipe. The topmost section, called the kelly, is square or hexagonal in cross-section, being designed to fit into an aperture of similar section in the drilling table, which provides the rotary power for the drilling operation. The engine powers the drilling table, which rotates and, via the kelly, transmits torque to the drilling string and hence to the bit. To counteract the stresses at the bottom of the string, the sections of pipe immediately above the bit, called drill collars, are heavier and have much thicker walls than the ordinary drill pipe sections. The heaviest of all these sections, at the very bottom of the string, is called the shoe.

The hoisting equipment necessary to raise and lower the drilling string consists basically of a block and tackle system, which is supported by a derrick, or mast. The derrick is the main visible feature of a drilling operation. Hoisting gear (called the drawworks) consists of a wire rope wound round a drum and connected to a crown block (the fixed part of the system) in the top of the derrick, and to a travelling block with a hook to which the drill pipe is attached. The system is normally driven by diesel–electric power. Progress in drilling is made by lowering the hook to apply weight to the drill string.

As drilling progresses, cuttings and debris have to be removed from the hole. This is effected by flushing drilling fluid down through the hollow drill pipe and back up to the surface through the space — annulus — between the drill string and the wall of the bore hole. Originally, drilling fluid consisted of muddy water, and it is still referred to as mud, although today there is a wide range of complex fluids specifically designed to cope with different drilling conditions. As well as its primary purpose of bringing drilling cuttings to the surface, the drilling fluid also cools and lubricates the bit, controls subsurface pressure, consolidates the side of the hole, etc. The 'mud' must be properly formulated for all of these functions.

The sequence of operations involved in drilling a well begins with 'spudding in', the traditional name given to the start of drilling. Drilling begins with a large-diameter bit (normally 24 inches) and proceeds with new sections being added to the drill string every 30 feet (initially the drill collar and later ordinary drill pipe).

As the hole gets deeper, it is necessary to consolidate it, to prevent it from caving in. This is done by inserting lengths of steel pipe, known as casing, down the hole at regular intervals. Because each successive batch of casing has to be inserted down through the previous one, the hole progressively becomes smaller in diameter, and smaller bit sizes are used as it gets deeper (*see* Table D.1).

Table. D.1.

	Casing diameter (in.)	Approx. depth (ft)
Surface string	18.125	500
Anchor string	13.375	3000–5000
Intermediate string	9.625	8000–10 000
Oil string	7.0	12 000–15 000

When the casing has been lowered, it is secured by cementing it into place. For this, the space between the casing and the wall of the hole (i.e. the annulus) is filled with cement, or slurry, which is pumped down the casing, followed by mud, which forces the liquid cement up the annulus, where it is allowed to set. This procedure is carried out each time a change is to be made to a smaller diameter drill pipe and casing.

As drilling proceeds, the bit becomes worn and needs to be replaced at regular intervals (normally about every 24 hours). This involves removing the entire drilling string from the hole, by unscrewing a section at a time, replacing the bit and then performing the reverse operation to insert the string back into the hole. The complete operation is known as a round trip.

When casing has been cemented to a depth of 5000 feet, the permanent wellhead is installed, consisting of a casing head and a CASING HANGER. The service wellhead, consisting of high-pressure BLOW-OUT PREVENTERS, is also installed at this stage. This is an essential means of controlling pressure during the drilling phase, but is replaced by a production wellhead (i.e. CHRISTMAS TREE) if the well subsequently achieves production.

If the well is to produce, then the narrowest diameter casing (7 inches), called the oil string, is cemented into place, and the well is ready for completion by the replacement of the service wellhead by the production wellhead.

In the course of exploratory drilling, a comprehensive record of the production potential is kept for each well. In addition the technical details of the drilling operations are recorded. These records are known as the well log, a general term embracing a variety of appraisal techniques and processes described below.

coring. A cylindrical sample, or core, of the various rock formations being drilled through is obtained by substituting a core barrel for the normal drilling bit. The core barrel is hollow, with a ring-shaped diamond head. As it penetrates the rock, it forces the core sample up the barrel, which is some 60 feet long. The samples can then be removed and examined.

stem testing. This is carried out to obtain a sample of any fluid contained in a formation. Evaluation of the sample, which is frequently water rather than oil, provides information about pressure and flow rates which is useful in evaluating the reservoir.

electric logging. This method is used for evaluating subsurface formations. It involves inserting a system of electrodes into the hole. The electrodes simultaneously measure certain properties relevant to production potential and record them on a chart at the surface. The general terms logging, or logging the well (or hole), normally refer to this process.

flow testing. This involves the controlled production of oil from a well for a period in order to estimate the production potential and likely performance of a reservoir. It is carried out by perforating the casing at a selected depth, using a perforating gun, which fires charges through the casing, allowing the oil to flow to the surface.

The results of appraisal techniques are used to determine the optimum location of appraisal or development wells.

drilling bit. *See* BIT; DRILLING.

drilling delays. *See* DRILLING HAZARDS.

drilling fluid (drilling mud). A fluid used in DRILLING operations for flushing out cuttings from the hole.

drilling hazards. In the course of DRILLING operations, various interruptions can occur as a result of either mechanical failure or unanticipated subsurface physical conditions.
(a) Mechanical failure can result in a part of the DRILL PIPE or DRILL COLLAR fracturing and being left in the hole when the string is raised to the surface. The lost part is known as a fish and attempts to recover it (i.e. fishing) can be made by attaching to the string one of a range of fishing tools, such as a magnet (or junk basket), or an overshot, which can grip the fish and pull it free. Since delays in drilling are expensive, the decision may be taken to abandon fishing attempts and drill a SIDETRACK HOLE (dog-leg) which bypasses the abandoned fish.
(b) Lost circulation occurs when more mud (DRILLING FLUID) is being circulated down the hole than is returning at the top. This occurs if a porous or fissured formation is penetrated, and mud escapes into it. Since such formations are the potentially prospective ones, tests will be made to evaluate it. The loss of further quantities of mud can be prevented by the addition of various thickening materials to the circulating fluid.
(c) The inadvertent sucking-in of oil or gas into the well can occur if, for example, the bit is raised too rapidly from the hole. If gas rises to the surface as a result of SWABBING, it may be necessary to seal off the well temporarily, in order to prevent a BLOW-OUT, until corrective action can be taken.

drilling mud. *See* DRILLING FLUID.

drilling string. The length of pipe used in DRILLING which is made up of a number of sections of DRILL PIPE joined together.

drilling table. The part of the surface DRILLING equipment that is rotated by the engine and hence rotates the DRILLING STRING or DRILL PIPE.

drill pipe. A hollow pipe to which the drilling BIT is attached and which is extended by the addition of further sections as the hole deepens. *See also* DRILLING.

drill stem testing. The obtaining of a sample of any fluid encountered in a formation during DRILLING.

dry gas. Petroleum gas from which the more easily liquefied components have been removed (i.e. stripped) either naturally or by processing. It consists mainly of METHANE. *See also* NATURAL GAS.

dry hole. A well drilled without finding oil or gas in commercial quantities.

dry tree. A sub-sea wellhead where the equipment is encased in a water-tight chamber.

dual basing point system. The pricing system which followed the US GULF PLUS system (*see* CRUDE OIL PRICES). It came about during the World War II when UK government objected to paying for fuel oil for the Royal Navy at a price as though the fuel oil had been shipped from the Gulf of Mexico. The result was that a second basing point was introduced based on the GULF. This, in effect, gave Middle East oil a natural market and enabled a rapid expansion of this market after 1945. *See also* BASING POINT PRICING.

dumping (predatory pricing). The export of a good at a price that is below marginal cost (*see* INCREMENTAL COST). It is regarded as unfair competition by the importing country and is viewed with alarm since it can lead to local producers being forced out of business, leaving the exporter in a MONOPOLY position. In the context of oil and gas, the dumping issue has arisen with respect to some oil producers' downstream output of oil products and petrochemicals, and has been used by lobbies within the importing countries as an argument for restricting imports of these products.

duopoly. A market structure in which there are only two sellers of the product. It is therefore a very simplified version of OLIGO-

POLY in which uncertainty about the rival's behaviour dominates market behaviour. *Compare* MONOPOLY.

Dutch disease. A process whereby the discovery and exploitation of a hydrocarbon such as oil or gas leads to the overvaluation of a country's exchange rate. This implies that the country's exports become more expensive and the imports cheaper. This in turn leads to a contraction of the non-hydrocarbon traded goods sectors in the economy as exports fall and imports replace domestic production. For an industrial country it implies a decline in the manufacturing sector whereas, for a developing country, agriculture may well suffer. The term was coined as a result of the experience of The Netherlands following large-scale gas finds. The phenomenon is increasingly a subject of research in economics.

dwt. *See* DEADWEIGHT TONNAGE.

dynamic positioning. A computer-controlled system of THRUSTER PROPELLERS which are used to keep a ship on station. It is an alternative to the use of anchors.

E

earnest money. A deposit placed by bidders for a lot as an indication of their seriousness.

East Texas. One of the oil industry's largest fields developed during the 1920s. Because of the LAW OF CAPTURE in the USA, which allows drillers to 'capture' oil located in others' acreage (because of its fluidity, *see* CRUDE OIL CHARACTERIZATIONS), it is often cited as an example of excessive drilling on a field. It was also used to argue that the oil industry was one of INHERENT SURPLUS. Provided producers covered their variable costs (which would be low) they had a vested interest in pumping the oil as rapidly as possible before their neighbours did likewise, thus leading to overproduction. Such situations gave rise to the need for regulatory bodies such as the TEXAS RAILROAD COMMISSION to control production partly to prevent damage to the field, but also to limit price competition.

econometrics. The branch of economics concerned with the measurement of economic relationships. It encompasses aspects of economics, mathematics and statistics that provide the framework and techniques required for the empirical investigation of these relationships.The methodology used in econometrics consists of three important stages: specification, estimation and verification.

(a) Specification is the first stage, in which the econometrician relies on economic theory to provide an initial understanding of the relationship he wishes to investigate. As an example, consider the demand for natural gas by households. Economic theory tells us that the demand for a product is determined by the income of the consumer, the price of the product and prices of other products that may be substitutes or complements for the product (e.g., the demand for natural gas in the household sector will depend on the income of the household, the price the household has to pay for the gas and, perhaps, the prices of electricity and oil). This is somewhat simplistic; other factors such as the number of gas appliances owned by the household, and possibly the price of these appliances, may be equally important in determining the demand for natural gas. Having decided what the variable factors in the relationship are, and distinguishing between the dependent or ENDOGENOUS variable — in the example this would be natural gas — and the independent or EXOGENOUS variables, income and prices in the example, the econometrician then turns to mathematics to express the relationship in a functional form. This functional form may be a simple linear relationship or a more complicated non-linear formulation. Having specified what can now be termed an econometric MODEL the researcher needs to gather the empirical evidence which he is going to use to measure the relationship expressed by the model specification. Once the data have been collected, and if necessary transformed to meet the requirements of the model, the next stage is to estimate the PARAMETERS of the model.

(b) The estimation stage is based on a choice of statistical techniques appropriate to the structure of the model. This may be a simple linear single equation technique such as ordinary least squares or a more complicated procedure that incorporates the interactions between two or more equations. Prior to estimation there are several assumptions that the econometrician has to make regarding the specification of the model. One of these states that the exogenous or explanatory variables must be totally independent of each other. If they are not, the variables are said to exhibit multicollinearity. This can pose a problem for the econometrician when he comes to interpret the results of the estimation stage, the seriousness of which depends on the degree of multicollinearity between the variables.

(c) Once the model has been estimated, the final stage is the verification of the model. This is an important stage in econometric model building since it can provide further information about the accuracy of the specification of the model. The estimated parameters are checked for consistency with economic theory. Economic theory provides the econometrician with *a priori* expectations regarding the sign and magnitude of the parameter estimates. In the example given above the parameter associated with the price of natural gas would be expected to be negative. This is because according to economic theory the demand for a product falls as its price rises. There are also certain statistical criteria to be satisfied at this stage. Against each parameter estimate there is a maintained hypothesis that has to be tested. This hypothesis states that the estimated parameter is significantly different from zero. If a parameter is not significantly different from zero then the econometrician will have some doubt as to the importance of that variable in the relationship. A t-ratio test is normally used for this purpose, with the t-ratios being calculated at the same time as the parameter estimates. As a rule of thumb, a t-ratio of magnitude greater than 2 implies that the maintained hypothesis holds, and the parameter estimate is significantly different from 0. However a t-ratio of less than 2 does not always imply that the variable is not significant. The existence of multicollinearity may reduce the t-ratio and give the appearance that the variable is not significant, whereas in fact the effect of this variable may have been captured by another variable. Although all the parameter estimates may be significant and consistent with economic theory, there may be a danger that when the model was specified another important variable was omitted from the relationship. The coefficient of determination (R^2) measures the extent to which the model actually describes or fits the performance of the endogenous variable. The R^2 may range from 0 to 100, when $R^2 = 100$ would suggest a perfect fit, therefore another rule of thumb states the closer the value of R^2 is to 100 the better the model is performing. A very low value for R^2 implies that an important explanatory variable has been omitted from the specification. The verification of the model enables the econometrician to identify any problems that may be inherent in the model specification or in the data itself. At this stage the econometrician may decide to go back to stage one and re-specify and re-estimate the model.

Once the estimated model has been accepted it will generally be applied to one or more of the following purposes: structural analysis; forecasting; policy evaluation. Structural analysis is concerned with the quantitative measurement of economic relationships, such as measuring price and income ELASTICITIES. Forecasting uses the estimated model to predict the value of the dependent variable at some time outside of the data sample period, normally the future. For example, a macroeconomic model of the UK economy may be used to predict the level of economic activity or GROSS DOMESTIC PRODUCT (GDP) under different SCENARIOS regarding, for example, the price of oil. Policy evaluation allows alternative policy options to be tested using the estimated model, and the outcome of the different policies to be compared. For example an econometric model of the demand for petrol may be used to assess the effect of different indirect taxation levels on demand.

economic development. *See* ECONOMIC GROWTH.

economic growth. The growth of output measured by GROSS DOMESTIC PRODUCT (GDP) or gross national product (GNP). During the 1950s and 1960s it was used as the criterion by which to measure the progress of an economy, and it tended to be used as a synonym for economic development. However, during the 1960s growing disillusionment caused doubt to be placed on the concept. This disillusion stemmed from two sources. (a) Because of technical problems associated with measurement of economic growth increasing doubt began to be cast on its accuracy, especially in the context of less-developed countries. (b) Many began to question whether it reflected welfare in a society either because it ignored distributional issues or because it tried to measure only material aspects of well being.

In the oil-producing countries economic growth, as measured by GDP or GNP, is particularly misleading. This is because growth of output stems largely from oil, and it is doubtful if oil (or any depletable resource) can be counted as 'output' since its production simply implies switching assets below the ground (i.e. oil) into assets above the ground (i.e. dollars).

economic model. *See* MODEL.

economic rent. The sum over and above the amount necessary to induce economic activity. For instance, where the PRESENT VALUE of revenue exceeds the present value of all economic COSTS, including NORMAL PROFIT, that amount is economic rent. This is equivalent to riccardian differential rent, but differs from a transfer earning, which is that sum needed to keep a factor employed in its existing function.

In practice economic rent has always played a crucial role in the oil industry. The economic rent associated with oil production has generally been large especially during and after the 1970s. This raises the problem of how the rent is to be divided between the owner of the oil (normally the government) and the oil company that produces it. In theory, capture of all of the economic rent by the government should not inhibit the oil company's level of activity, since, as defined above, economic rent is extra above the amount necessary to induce economic activity. In practice, however, it is far from clear where the normal profit of the company ends and economic rent begins. The companies have always argued that a large part of economic rent is in fact required as a reward for the risks involved in oil exploration (i.e. it is part of normal profit).

A further complication over the division of the rent is that it is set initially (i.e. before oil is discovered) in ignorance of its eventual size. The setting of the division is the result of a bargaining process. However, over time the relative bargaining power of the two parties will change, leading to the creation of pressures to renegotiate the division. This to a large extent explains why oil concession agreements are inherently unstable. *Compare* RENTAL.

economic theory of politics. Much traditional economic theory starts from the assumption that governments (i.e. politicians and bureaucrats) when intervening in economic activity do so from altruistic and selfless motives with the objective of maximizing the long-term welfare of the society. The economic theory of politics, which had its origins in the 1950s and began to gain popularity in the 1970s, challenges this assumption. It starts from the basic assumption that politicians and bureaucrats are selfish maximizers (like anyone else in the market place). The subject therefore concerns itself with addressing the question what is maximized and what effect does this maximization have upon government policies? In effect it introduces the notion of 'government failure' as an antidote to the idea that MARKET FAILURE requires government intervention.

In the oil and gas industry, the economic theory of politics has played an influential role in the debate as to whether the rate of production should be determined by the market (i.e. private oil companies) or by the government. For example, the argument that governments have a longer time perspective than private companies, and would therefore be more likely to conserve reserves, is challenged because the politician with a view to maximizing votes is looking only to the next election.

economies of scale. Economies of scale are said to occur where higher levels of output are associated with lower COSTS per unit of output in the LONG TERM. Diseconomies of scale occur where larger outputs lead to higher average costs. For any given company, economies of scale can be technical (or real) and pecuniary (or financial).

(a) Technical economies are associated with reductions in the physical quantities of inputs required per unit of output, and it is these reductions that lower costs. They are a property of the firm's PRODUCTION FUNCTION, economies resulting from increasing RETURNS TO SCALE and diseconomies resulting from decreasing returns to scale in the production function. Examples of technical economies include the specialization of labour and capital equipment, and can occur in areas such as electricity generation and the transport of oil in tankers and pipelines.

(b) Pecuniary economies result from paying lower prices per unit of inputs as output expands. Price discounts for bulk buying of inputs are a good example.

The economies and diseconomies of scale referred to above are internal to the separate companies in an industry. However, they can also be external to the separate companies, but internal to the industry as the industry's output expands. In this case each company's costs become a function of the industry's output, as well as of its own output.

An external technical economy occurs when the output that a company can produce with given inputs rises as the industry's output expands. This effect on the company's production function lowers its cost curves; an external technical diseconomy would raise them. An example is a situation in which the expansion of a transport industry's output has resulted in the congestion of roads or shipping lanes or airports, thus increasing journey times and raising each company's cost curves. There can also be external pecuniary economies (diseconomies), where as the industry's output expands the prices of scarce inputs decrease (increase). In the oil and gas industries, because both oil and gas move in three-dimensional space and can flow continuously, technical economies of scale have always been very large. A classic example is the POINT SIX RULE associated with refineries. This has carried two important implications for the way in which the industry behaves. (a) To obtain technical economies of scale the equipment must run at or close to full capacity which implies pressures to oversupply if excess capacity is in place. (b) It implies large capital outlays on (relatively) large pieces of capital equipment (e.g., tanks, pipelines, etc.). This in turn implies both a low marginal cost of supply aggravating the tendency to oversupply and that access to capital may act as a barrier to entry, preventing competitors entering the market place. For these reasons, economies of scale are one of the factors which explain why the oil industry is characterized by an oligopolistic structure (i.e. relatively few companies, or governments, dominate the industry).

ECU (European Currency Unit). A basket currency composed of fixed amounts of European currencies, officially created by the resolution of the Council of Ministers of the Economic Community (EC) No 3180/78, 18 December 1978.

Persuant to the EC Council Resolution No 2626/84, 15 September 1984, the ECU is defined as the sum of the following amounts of the EC currencies:

ECU = 0.719 Deutsche mark
+1.31 French franc
+0.0878 Pound sterling
+140.00 Italian lire
+0.256 Dutch guilder
+3.71 Belgian franc
+0.14 Luxembourg franc
+0.219 Danish krone
+1.15 Greek drachma
+0.00871 Irish punt

Except for the pound sterling all the above currencies participate in the European Monetary System (EMS).

When the basket is defined, the quantity of each component currency within the ECU is fixed to reflect weights, which are agreed by the Council of EC Ministers on the basis of country's economic indicators, mainly GROSS DOMESTIC PRODUCT and inter-Community trade.

After definition, the constituent currency quantities remain constant, whereas the weight of each currency varies daily according to the fluctuations in the value of individual currencies in relation to each other. The weight of a currency within the ECU is defined as equal to the fixed quantity of the currency divided (or multiplied in the case of the pound and Irish punt) by its exchange rate against the ECU.

The official role of the ECU within the EMS can be summarized as follows:

(a) The ECU is the NUMERAIRE for the exchange rate mechanism of the EMS.
(b) The ECU is the basis for the indicator of divergence between currencies in the EMS.
(c) The ECU is the denomination for intervention and the credit mechanism.
(d) The ECU is a means of settlement among monetary authorities.

The composition of the ECU basket is not invariable since, as long as a flexible exchange rate system exists, the weights of currencies within the ECU can vary over the years to such an extent that they may no longer reflect the economic significance of the countries in the EC. Therefore, the Council's Resolution of 1978 stipulates that the ECU definition must be re-examined every five years or on request if the weight of any of the currencies has varied by at least 25 percent.

edge water. Water that underlies the oil/gas RESERVOIR. *See also* GEOLOGY.

elasticity. A measure of the responsiveness of an economic variable y to a change in some other variable x. Defined as the ratio of the percentage change in y to the percentage change in x, the formula for 'arc' elasticity is:

$$\frac{\text{small change in } y}{(y_1+y_2)/2} \div \frac{\text{small change in } x}{(x_1+x_2)/2}$$

where δ indicates a change in the variable. When δx becomes very small, the measure is called a 'point' elasticity and becomes:

$$\frac{\text{proportionate change in } y}{\text{proportionate change in } x} \div \frac{\delta y/y}{\delta x/x} = \frac{\delta y}{\delta x} \cdot \frac{x}{y}$$

The elasticity measure is used instead of the slope $\delta y/\delta x$ because the slope is not independent of the units in which y and x are recorded.

The price elasticity of demand is the percentage change in the quantity of good A demanded divided by the percentage change in A's own price. Because the DEMAND CURVE normally has a negative slope, this elasticity is normally negative. However, for convenience, the negative sign is often omitted.

The cross-price elasticity of demand is the percentage change in the quantity of good A demanded divided by the percentage change in the price of some other good B. It is positive for SUBSTITUTES and negative for complements (*see* COMPLEMENTARITY).

elastomer. An elastic POLYMER (e.g., STYRENE–BUTADIENE RUBBER).

electric logging. A method of evaluating subsurface rock formations. *See also* DRILLING.

elevators. Clamps on the TRAVELLING BLOCK attached to the DRILL PIPE to enable it to be raised or lowered. *See also* DRILLING.

ELFO. The German name for light GAS OIL used for domestic and smaller commercial central heating. *See also* FINISHED PRODUCTS, GAS OIL.

ELSBM. *See* EXPOSED-LOCATION SINGLE-BUOY MOORING.

EMS. (European Monetary System). A monetary system that went into operation in March 1979 with the main objective of creating a 'zone of monetary stability in Europe'. At the heart of the EMS is a system of fixed, but adjustable exchange rates between eight European currencies: Belgian franc, Danish krone, Deutsche mark, Dutch guilder, French franc, Greek drachma, Irish punt and Italian lire. Each currency has a central rate expressed in terms of ECU. These central rates determine a grid of bilateral central rates, around which fluctuation margins of ± 2.25 percent (6 percent for the Italian lira) have been established. At these margins, intervention by the participating central banks is obligatory.

During the oil market uncertainty of 1982-85 many argued that the UK should join the EMS to minimize fluctuations in the sterling exchange rate which arose because of uncertainty over oil prices.

emulsion. An intimate mixture of two liquids that do not normally mix such as oil and water.

endogenous. Describing a variable in a MODEL or system that is determined by or within the system. For example, in a model of the demand for fuel oil, the endogenous variable would be fuel oil demand which is dependent on other explanatory variables such as the price of fuel oil relative to the price of other fuels and the stock of fuel oil using appliances. An endogenous variable may also be referred to as a dependent variable, although this may be somewhat ambiguous since a dependent variable may appear as an independent or explanatory variable elsewhere in the system. *Compare* EXOGENOUS.

energy accounting. *See* ENERGY ANALYSIS.

energy analysis (energy accounting). At its simplest level, the various methods by which the energy flows in an economy can be measured. For example, an energy balance matrix can show at various levels of sectoral disaggregation the flows of primary energy into (and out of) an economy, their conversion to useable energy and their final use (both direct and indirect). Many variants on the simple matrix have been developed for use as tools by forecasters and policy makers such as the Brookhaven Energy Reference System and LEAP. Such energy 'budgets' have played a role in the natural sciences for many years. However, following the FIRST OIL SHOCK and the growing interest in energy in general and oil in particular, some began to argue such methodologies could be used to provide an energy theory of value on which resource allocation could be based rather than the monetary flows of economic theory. This has proved an extremely contentious area.

energy balance matrix. *See* ENERGY ANALYSIS.

energy coefficient. The ratio of the energy input and output; this may be in absolute or incremental terms. The coefficient can be measured at a regional level (e.g., in the INTERNATIONAL ENERGY AGENCY countries), at a national level or within a country at a more disaggregated level by sector. The energy inputs, which may be total energy or due to a specific fuel, are normally measured by physical units or heat value. These inputs may be in terms of primary energy or useful energy (*see* ENERGY CONVERSION). When the output is a regional or national aggregate GROSS DOMESTIC PRODUCT or gross national product is used, although the former is technically more accurate since the national product reflects output produced outside the regional or national boundaries. If the coefficient refers to disaggregated sectoral information, it is also possible, depending on the homogeneity of the output, to measure the output in physical terms (e.g., tons of steel, tons of cement, etc.). The coefficient is essentially a measure of energy or fuel efficiency, and a lowering of the coefficient reflects conservation or possibly in the case of a fuel-specific coefficient fuel switching. The coefficients vary between countries and over time in the same country.

The following generalizations can be made. Based on past experience, over time the general ENERGY INTENSITY of an economy initially increases (i.e. the coefficient rises) reflecting a change in economic structure towards industrialization coupled with increasing consumption. However, at some point energy intensity begins to decline. This reflects the congenital tendency of engineers, irrespective of any economic consideration, to seek more technically efficient ways of using inputs to produce a given output. Thus some of the decline in energy intensity observed after the FIRST OIL SHOCK and the SECOND OIL SHOCK was simply a continuation of the trend of the 1950s and 1960s. However, the first part of the generalization requires an important qualification. In the early stages of economic development TRADITIONAL ENERGY (e.g., fuelwood) tends to dominate. This is true in both the last century and the present century. The use of such fuels tends to be very inefficient when it comes to conversion from useable to useful energy (*see* ENERGY CONVERSION). Thus, for example, if traditional energy is included in the energy consumed in THIRD WORLD countries, many such countries exhibit greater energy intensity of output than the industrialized countries. Until recently the tendency has been to omit traditional energy from energy consumed because of the associated measurement problems. This has led to gross understatements of the energy coefficients of the Third World. *See also* ENERGY ELASTICITY.

energy conversion. A two-stage process: (a) the conversion of primary energy supplies into useable energy (e.g., the refining of crude oil into products, the production of electricity from hydrocarbons); (b) the conversion of useable energy into useful energy (e.g., heat, light, etc.) for use in an appliance, although the appliance can range from a fireplace comprising three stones to a jet engine. Both stages involve a loss of energy. The extent of the loss reflects the energy efficiency of the conversion unit from primary to useable and from useable to useful.

energy efficiency. *See* ENERGY CONVERSION.

energy elasticity. Another name for ENERGY COEFFICIENT when the coefficient is measured in incremental form.

energy intensity. A measure of the amount of energy or fuel used to produce a specific output. *See also* ENERGY COEFFICIENT.

energy policy. Fashions for explicit energy policies vary. At times, such as during the 1970s, when energy problems appeared particularly intractable, governments tended to formulate energy policies since there appeared to be a public demand for action to be taken. In the mid-1980s when, in the industrialized countries, energy issues no longer seemed so pressing many of the energy programmes (e.g., promoting home energy production or stimulating energy conservation) on which governments had embarked ten years previously had been greatly scaled down or had disappeared.

Although explicit energy programmes designed to promote policies regarded as desirable are less in vogue than they were, virtually all countries have some form of energy policy, if only to give some limited protection — by taxes, subsidies or controls — to indigenous energy producers at the expense of imports. Many countries also have taxes designed to collect substantial revenue from the energy industries. For example, there are generally heavy taxes on motor spirit for which the market demand is inelastic, and there are often special taxes on oil and gas production intended to appropriate the ECONOMIC RENT from exploitation of those resources.

In economic terms, an explicit policy towards energy may be justified on traditional welfare economics grounds that the relevant markets are imperfect (e.g., because of monopolistic forces) or fail to produce goods in the 'right' quantities. Some cases of imperfections and failures in energy markets which can be held to justify policy measures of a market-replacing (direct controls) or market-improving (stimulating competition and internalizing external costs) nature are as follows:

(a) Oil, gas and other energy markets are imperfect in the economic sense since there are often very large organizations on the supply side such as OPEC and the major oil companies. In terms of traditional economic theory, the presence of MONOPOLIES justifies action by national governments or international organizations to prevent exploitation of the consumer because monopolists tend to restrict output and raise prices, due to the lack of competition. Action to promote energy conservation by governments may also be justified on the grounds that there are no organizations comparable in size to those on the supply side of the market which have an incentive to persuade consumers to use energy sparingly.

(b) One case of energy market failure is that markets when unaided will fail to deal adequately with the environmental problems associated with the production, transportation and consumption of energy, because some or all of the costs fall not on those causing the pollution, but on others. Thus there will be excessive pollution unless the government intervenes by, for example, 'internalizing' the external costs or setting environmental standards so that they are taken into account in organizational decision-making.

(c) Another example of market failure concerns the lack of attention to security of supply. For example, a company installing a stand-by electric generator will not be able to appropriate all the benefits of its action. Its investment in the generator will take some of the strain off the electricity supply system as a whole. In such circumstances, there will be a tendency to underinvest in security of supply which governments could remedy by such measures as holding excess stocks of fuels and keeping excess capacity in the electricity supply system.

(d) Markets are likely to fail to provide adequate information for decision-making. For example, they may not provide consumers with information on how to use fuel efficiently, on the relative efficiencies of fuel-using appliances and on what price movements to expect in the future. Since the information required for decision-making is always about the future, it is not possible to provide perfect information, but, in principle, in some cases governments may be able to improve on the information provided by markets.

(e) It is sometimes suggested that energy markets are short-sighted, failing to take a long-term view and, in particular, failing to safeguard the interests of distant generations. In depletion decisions, for instance, private companies may discount the future at a rate higher than society would wish, thus concentrating on a short-term view of resource production, and may make insufficient allowance for tendencies towards long-term scarcities of certain non-renewable materials. Government action to change discount rates, to alter price expectations or to control depletion rates directly may thus be justified.

There may also be macroeconomic reasons for energy policy measures. Development of large oil or gas discoveries might be held back to avoid too sharp an impact, via the exchange rate, on traditional industries or development might be accelerated to bring quick benefits to the balance of payments.

The above arguments are contentious partly because some of them depend on simplified versions of the relevant economic theory. The anti-monopoly argument, for instance, is not conclusive because monopolies may in some circumstances promote technological advances that would not otherwise occur (or would not occur so quickly). More generally, the case for government action in energy markets, as elsewhere, depends on the view that governments are capable of identifying the social interest in particular cases and are willing, and able, to pursue that interest even to the extent of subordinating their own interests (such as the personal promotion prospects of individuals or the interest of the administration as a whole in being returned at the next election). If governments are not far-sighted and benevolent towards society as a whole (i.e. if they are imperfect), the results of their actions will not necessarily be better for society than the outcome of imperfect markets. *Compare* ECONOMIC THEORY OF POLITICS

Engel curve. A curve that relates the level of demand for a good or service to different levels of income for the consumer, assuming other factors influencing consumption such as price are held constant. *See also* DEMAND.

engine sludge. *See* SLUDGE.

Engler degree. A measure of viscosity. *See also* QUALITY, DISTILLATE AND RESIDUAL FUELS.

enhanced oil recovery (EOR). Less conventional means of artificially extending the productive life of an oil field by, for example, using chemicals, solvents or the application of heat; formerly called tertiary recovery. *Compare* SECONDARY RECOVERY. *See also* PRODUCTION, ENHANCED OIL RECOVERY.

ENI (Ente Nazionale Idrocarburi). The Italian state oil company. Created in 1953 by Enrico Mattei, it was based upon Agip, which remains its major operating company.

Ente Nazionale Idrocarburi. *See* ENI.

Enterprise des Recherches et d'Activités Petrolieres, L'. *See* ERAP.

environmental pollution. Simply, something in the wrong place, in the wrong quantity, at the wrong time; the bringing by humans into the environment of substances or energy likely to cause hazards to human health, harm to living resources and ecological systems, damage to the structures or amenities, or interference with legitimate uses of the environment (after M.W. Holdgate). *See also* AIR POLLUTION; THERMAL POLLUTION.

EOR. *See* ENHANCED OIL RECOVERY.

equilibrium. A state of economic balance for companies or individuals or markets. An economic model (of a market, for example) in static equilibrium consists of 'a constellation of selected interrelated variables so adjusted to one another that no inherent tendency to change prevails in the model which they constitute' (F. Machlup). All the internal (or ENDOGENOUS) variables chosen for the model are in a simultaneously compatible state of rest, whereas the external (or EXOGENOUS) variables and parameters are treated as constants. Economists often carry out 'comparative static analysis' by changing the parameters or exogenous variables of a model, thus creating disequilibrium. They then compare the new equilibrium configuration with the original one.

In a single market, equilibrium occurs when the price is such that the quantity suppliers wish to supply at that price is equal to the quantity demanded. The market 'clears', and there is no excess demand or supply. There is no inherent tendency for price or quantity to change (i.e. there is an equilibrium price and quantity). However, if, for example, one of the determinants of DEMAND such as income were to change then the market would move out of balance into disequilibrium until a new equilibrium price and quantity were obtained as shown in Fig. E.1.

If there were several interrelated markets, then two sorts of analysis are possible. PARTIAL EQUILIBRIUM ANALYSIS would analyze the behaviour of one market in isolation, holding other variables and interactions constant. General equilibrium analysis, by contrast, would seek simultaneously consistent equilibria in all markets, such that there would be no excess demand in any market; all the equations in the model would be simultaneously satisfied.

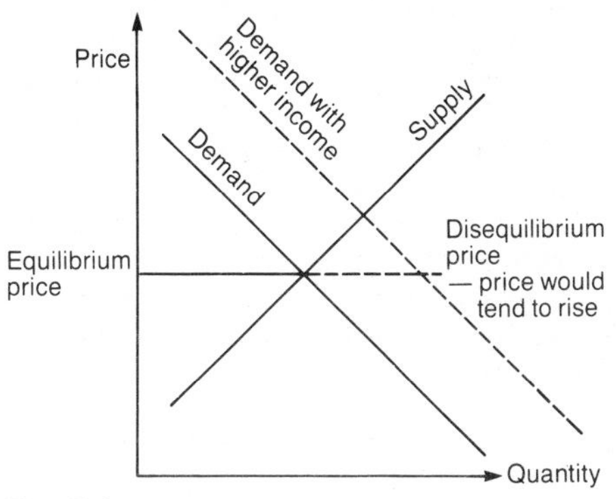

Fig. E.1.

The equilibriating properties of a model are important. In particular, it is possible to specify the conditions under which an equilibrium will exist, will be unique (there could be several equilibria) and will be stable. For example, in the sense that a perturbation from an equilibrium price initiates forces that move the market back to equilibrium, rather than away from it.

The concept of equilibrium and its derivatives is important for economists. It is only a guide to thinking about the analytical framework used to address economic problems. There is a danger that non-economists will use the terms of equilibrium analysis without understanding their meanings. This is particularly noticeable when the media talks about an equilibrium price of oil as though there is some price, which once attained, will be sustained. There is, however, no such price because the determining variables of the price are continually changing. Most obviously and importantly expectations about future price movements change which influence current behaviour and hence current price.

equities. Ordinary shares in a company which give the holder voting rights and the entitlement to receive DIVIDENDS. Holders of debentures and preference shares (*see* BONDS) have prior claims on the profits of a company, and so these forms of investment are generally regarded as less risky than equities.

equity. The value of a company's ASSETS after all outside liabilities (except the shareholders) have been subtracted.

equity capital. Finance raised by a company through the issue of shares to investors. *See also* EQUITIES.

equity crude. A crude oil that a company LIFTS by virtue of its having an equity share in an oil-producing company. *See also* EQUITY MARGIN.

equity margin. The difference between the TAX-PAID COST of EQUITY CRUDE and its value on the open market. It is in effect the reward which accrues to the company for being involved in the producing operation.

ERAP (L'Enterprise des Recherches et d'Activités Petrolieres). France's national oil company wholly owned by the French government. It was created by merging two other companies and came into being in 1966.

errors and omissions. *See* BALANCING ITEM.

escalator clause. A clause in a contract to allow for changes in the value of money as a result of INFLATION. It creates an automatic linkage between a monetary obligation and the price level.

escrow. Normally a written document, although it could be money or goods, providing evidence of obligations between two or more parties. It is given to a stranger (i.e. a person who is not party to it) on condition that he delivers it to the other party only if certain conditions have occurred (e.g., on the death of a person or on the payment of a price). Once the delivery of an escrow has taken place, it takes effect as a deed. However, if the conditions are not performed, it never becomes a deed.

ethane. A gas present in some NATURAL GASES and in REFINERY GASES. It is an ALIPHATIC compound and is the prime FEEDSTOCK for the production of ETHENE. *See also* CHEMISTRY.

ethanol (ethyl alcohol). An ALCOHOL that can be used as an intermediate in the the production of other chemicals or as a solvent. A further use is as a motor fuel, either alone or mixed with motor spirit (gasoline) to produce GASOHOL. It is normally produced by synthetic means.

ethene (ethylene). An ALKENE that is used as a basic building block in the petrochemical industry. It is produced in a refinery by the THERMAL CRACKING of paraffin hydrocarbons and steam. *See also* CHEMISTRY; FEEDSTOCKS; PETROCHEMICALS.

ethyl alcohol. *See* ETHANOL.

ethylene. *See* ETHENE.

eurocurrency market. *See* EURODOLLAR MARKET.

eurodollar market. An international market for inter-bank and non-bank dollar deposits and short- and medium-term bank dollar credits. These activities are conducted by banks that are not resident in the USA with both resident and non-resident customers. This market is mostly a wholesale money market, where only large amounts are transacted between borrowers and lenders, mainly commercial and central banks, industrial and commercial enterprises and institutional investors. The interest rates on eurodollar deposits and loans can be different from the equivalent dollar interest rates in the domestic market, since the eurorates are set in the eurocurrency market which is a free market. In the absence of disruptions, although US banks ARBITRAGE between the euro- and the domestic money markets and the two appear to be highly integrated. Changes in domestic interest rates and reserve requirements influence eurodollar rates, since, by its very nature, the eurocurrency market is more open to external influences than domestic markets. The misleading term eurodollar, which is used to describe activities that are international in nature, derives from the original European-based development of the market at the end of the 1950s, when two important changes affected international financial markets: (a) restrictions were placed on the use of sterling in external transactions in 1957; (b) the disappearance of the large US foreign trade surplus was compensated by capital inflows into the USA and increasing foreign holdings of dollars, which became increasingly available in Europe and were used both as a reserve currency and as an international vehicle currency in financing and invoicing international transactions. European banks — in particular UK banks which were restricted in their use of sterling — became more active in providing dollar facilities for their customers. The oil price rises in 1973–74 made a significant impact on the development of the eurodollar market. A considerable proportion of the large current-account surpluses of the oil-exporting countries was kept in the form of foreign exchange reserves, mainly denominated in dollars. Capital flows from the surplus oil-exporting countries were channelled to the deficit oil-importing countries by the international banking system, which quickly adapted to these new pressures and positively answered the increased demand for loans, especially through the market for syndicated medium-term euro-credits.

Europa point. A fixed point in the Straits of Gibraltar that is used to measure distances to northern European ports.

European Currency Unit. *See* ECU.

European Monetary System. *See* EMS.

exchange rates. The relative price of one currency in terms of another. Exchange rates are usually expressed as the number of domestic currency units per unit of foreign currency, an exception being the sterling exchange rate which is expressed in the number of foreign currency units per pound sterling. Exchange rates may be fixed, when the monetary authorities in either or both of the countries concerned intervene in the currency markets to keep the exchange rate at a specific level, or they may be floating, when their level is determined by market forces. Other possibilites are: managed exchange rates, when intervention is not continuous, but periodic; or the existence of a fixed range of allowed fluctuations.

To an extent the BALANCE OF PAYMENT surpluses and deficits affect exchange rates. The exploitation of North Sea Oil has had a noticeable impact on sterling, which has been allowed to float since 1972. Furthermore, the discovery of North Sea oil has boosted incomes expectations in the UK, increasing the real demand for money and thus exerting upward pressure on sterling. The sudden oil price increase in 1979-80 (*see* SECOND OIL SHOCK) enhanced this process as the UK became a net exporter of oil. *See also* DUTCH DISEASE.

excise tax. A tax generally imposed on a specific, domestically produced good, commonly levied on units sold or produced rather than as an AD VALOREM TAX. A ROYALTY is an example of an excise tax.

exhaustible resources (depletable resources, non-renewable resources). Natural resources (e.g., oil and natural gas) that do not reproduce themselves. Thus once an exhaustible resource has been consumed it is lost forever. Several points are relevant in the context of oil and gas.

Firstly, in theory the supply will eventually run out. During the 1970s this led to a widespread belief that one day the last barrel of oil (or cubic foot of gas) would be produced (*see* CLUB OF ROME), and considerable time and effort was wasted in determining when this would occur. In reality, as the resource increases in scarcity the price will rise thereby curtailing consumption.

Secondly, the fact that oil and gas are exhaustible considerably complicates the decision of when to produce (i.e. now or later). Normally in considering whether to produce a good or not the decision-maker looks at marginal costs (*see* INCREMENTAL COSTS) and MARGINAL revenues. If marginal revenue exceeds marginal costs the decision-maker will produce because by doing so he adds more to total revenue than to total cost. If he does not produce, time dictates the opportunity is lost forever. An exhaustible resource produced today, however, cannot be produced tomorrow, and if not produced today it is conserved for production tomorrow. Hence the oil or gas production decision must relate the present gain of producing today (i.e. marginal revenue less marginal cost) with the expected future gain if the production decision is delayed. This is further complicated by the time value of money (*see* TIME PREFERENCE) which means DISCOUNTED CASH FLOW methodology must be used to compute the present value of the future gain for comparison with the current gain. This means that the production decision depends on the decision-maker's perception of future prices, costs and discount rate. For example, many would argue that the FIRST OIL SHOCK was the result of a supply reduction following expectations of higher future prices coupled with the takeover by producer governments, leading to a change in discount rates. The problem this causes is that perceptions are subject to whim and fancy and can accordingly change very rapidly. This makes any forecasting of supply, and hence price, extremely uncertain. *Compare* DEPLETION. *See* CRUDE OIL PRICES.

exogenous. Describing a variable that is used to explain the behaviour of another variable — an ENDOGENOUS variable — in a MODEL or system, but is not itself determined within that system. Such a variable may also be referred to as an explanatory or independent variable, but this may be misleading because an independent variable may appear as a dependent variable elsewhere in the system. (In this case it would in fact be an endogenous variable.)

exotics. A term that is used, generally in a disparaging way, to describe various alternatives to hydrocarbons which are RENEWABLE ENERGY sources. Examples include solar, wind, tidal and geothermal energy. Although such sources are hardly new, interest in their development expanded rapidly following the oil price rises of the 1970s.

expectations. Views pertaining to the future or future events. In economics, the role of expectations may be of paramount importance in determining economic behaviour because decisions are often made according to some expectations regarding the future. Unfortunately expectations are of a very subjective nature; they are not readily observable and consequently cannot be measured. However, many attempts have been made to incorporate formally expectations into econometric models (*see* ECONOMETRICS). The adaptive expectations model attempts to overcome the problem of non-observability by assuming that current expectations are determined partly by past expectations and partly by modifying past expectations in the light of current achievements, since expectations are seldom fully realized. The rational expectations approach covers a variety of interpretations, but is essentially based on the assumption that expectations are in fact the same as predictions made according to some relevant economic theory.

In the oil and gas industry, expectations play a crucial role in determining events. Because oil and gas are EXHAUSTIBLE RESOURCES, expectations of future prices and costs play a key role in the exploration, development and production decisions. Equally, on the demand side, expectations play a role in determining the choice of fuel-using appliances and the decision by refiners on whether to restock or destock. The role of expectations in decision-making in the industry is complicated by the long TIME LAGS and LEAD TIMES which characterize many of the projects in oil and gas both on the supply and the demand side. Hence expectations must relate to periods in excess of five to ten years. To predict events, it is necessary to know both what current expectations are and how they may change.

expensing. In the oil and gas industry, this is usually linked to the way in which ROYALTIES are treated for tax purposes. In the 1950s, in the Middle East, royalties paid by oil companies were effectively included in the 50 percent profits tax. However, under expensing arrangements introduced in the mid-1960s royalty is treated as an operating cost deductable from gross income to determine taxation. The effect is to increase the GOVERNMENT TAKE.

explicit costs. COSTS that appear directly in the financial accounts of a company. Explicit costs are used to make a distinction from implicit costs which, although an OPPORTUNITY COST for the company, would not appear in the accounts. For example, a company that uses its own capital to finance a project would not incur the cost of borrowing, but the use of the capital would 'cost' the company what it may have received had it lent the money outside the firm.

exploration. The search for oil and/or gas. Three stages are involved.

(a) Property acquisition requires the company to obtain the right to explore for oil on a specific acreage by purchase, lease or some other method. Normally this right (embodied in the exploration licence) also gives the company some rights with respect to any oil found such as the right to develop and produce any commercial discoveries, although the nature and extent of these rights can vary considerably in different countries.

(b) The second stage is to survey the area using geological and/or geophysical survey techniques (*see* SEISMIC EXPLORATION) to establish where the oil and gas might be. Frequently some survey work is undertaken prior to the property acquisition stage in order to establish the likely prospects of the acreage since this would obviously play an important role in determining the terms of the exploration licence. The owner of acreage with good prospects could command more favourable terms than those for poor prospect acreage. If the company itself carries out such survey work this is done under an exploration permit. Frequently a number of such permits would be given to different companies for the same acreage. Alternatively the owner of the acreage (normally the government) may commission an independent survey of the acreage, the results of which would be made available to companies with a potential interest in acquiring an exploration licence.

(c) Because the oil and gas is hidden underground (in the absence of surface seepage, which is rare) the final stage requires the drilling of a well as the ultimate test of the presence of oil or gas. Such exploratory wells are called wildcat wells.

The treatment of costs associated with exploration — property acquisition costs, survey and wildcatting costs — varies between companies. They may be fully expensed (i.e. they are simply entered on the cost side of the company's balance sheet to be offset against tax). Alternatively, they may be capitalized (i.e. they are added to the capital of the company). Some combination of these two processes may also be adopted.

Exploration is expensive, although the cost can vary enormously depending on the accessibility of the acreage. Thus exploration OFFSHORE or in remote areas is clearly more costly than exploration onshore in accessible areas. Furthermore, for most countries, it involves the commitment of a high proportion of the costs in foreign exchange. Exploration is also highly risky in the sense that oil or gas may not be found, although risk can be reduced by more comprehensive survey activity prior to drilling. The result of these two features is that many governments have avoided sole involvement in exploration, preferring instead to get the companies to explore under JOINT VENTURE or SERVICE CONTRACTS whereby the company supplies the foreign exchange and takes the risk.

The nature of exploration can help in explaining why exploration licences (be they CONCESSIONS, joint ventures or service contracts) have an inherent tendency to be unstable and tend to lead to conflicts with the owner of the acreage. The terms of the exploration licence are set in ignorance of the profitability (or otherwise) of the investment. These terms are the outcome of a bargaining process between the company and the acreage owner. In the exploration phase the company has the stronger relative bargaining power (subject to the degree of competition for the acreage), but at the end of the exploration phase, assuming commercial discovery, relative bargaining power swings sharply in favour of the owner if the owner is a government with the power to invoke FORCE MAJEURE. This frequently leads to demands for renegotiation and explains why oil companies are often unwilling to explore in certain countries.

The output of exploration is essentially knowledge as to where oil and gas are. This tends to complicate the economic analysis of exploration activities mainly because of the problems of valuing knowledge. Thus there is considerable discussion in the economic literature as to what factors to consider when determining optimum exploration levels. *See also* DRILLING.

exploration licence. *See* EXPLORATION.

exploration permit. *See* EXPLORATION.

exponentially weighted moving averages. *See* ADAPTIVE FORECASTING.

export refinery. A REFINERY located away from the market for refined products whose purpose is to export products rather than crude oil. During the 1970s many of the THIRD WORLD oil producers developed grandiose plans to build such refineries. Many of the plans, however, were poorly thought out and lacked coordination and relevance to the market situation with the result that many were quietly dropped. *See also* REFINERY LOCATION.

exposed-location single-buoy mooring (ELSBM). A floating chamber attached to a PRODUCTION PLATFORM to serve as a flexible connection to a tanker for loading. Such a system has no storage capacity.

extender oil. An oil used to change the physical characteristics of synthetic rubber.

external capital rationing. A situation that occurs where there is a capital market restraint which prevents an organization from undertaking all those investment projects with rates of return which exceed that rate at which it can borrow. The organization thus chooses between a number of mutally exclusive investment projects, all of which have prospective returns in excess of its borrowing rate. The use of internal rate of return (*see* DISCOUNTED CASH FLOW) for ranking projects may in such circumstances give misleading results. The NET PRESENT VALUE approach is generally preferred, although a modified internal rate of return method (based on differences between projects) can be used. External capital rationing can be imposed, as it is by the UK goverment on the nationalized industries, since it limits the funds available to those industries. *Compare* INTERNAL CAPITAL RATIONING.

externalities. Costs or benefits associated with a project, but which a private company would not include in their accounts because the cost imposed or benefits obtained are external to the company although they are tangible to the society. An example is legally allowed air pollution. If legislation were imposed, forcing a company to stop emitting the pollutant(s) then the externality would have become internalized since the cost of preventing the pollution would have to be met by the company and would enter its financial accounts. The existence of externalities means the resources allocation of an economy may not be efficient because costs and benefits (i.e. revenues) may not reflect their real levels in the economy. *See also* COST–BENEFIT ANALYSIS.

extrapolation. A process of extending actual measures of a variable, usually over time, into the future. The extension can be done by 'eye', or it can be derived statistically from use of the linear regression equation which 'fits' the most representative line through the data points.

In the past simple extrapolation was a common forecasting technique, but it suffers from serious problems and is now little used by forecasters. The two main problems are that is is often difficult to fit a line which really represents the trend. For example, the two jumps in oil price in the 1970s (FIRST OIL SHOCK; SECOND OIL SHOCK) would make any trend line of oil prices meaningless, although many of the oil price forecasts of the 1970s were obtained by extrapolating such a meaningless trend line. Also the forecast assumes the future will simply be more of the same. Not only is this usually incorrect it also leads to the ignoring of key determining influences behind any trend. Thus the extrapolation approach in its simple form using a time trend is both meaningless and counter-productive since it tends to obscure the dynamics of change. Such techniques were used by the CLUB OF ROME in its *Limits to Growth* study in the early 1970s.

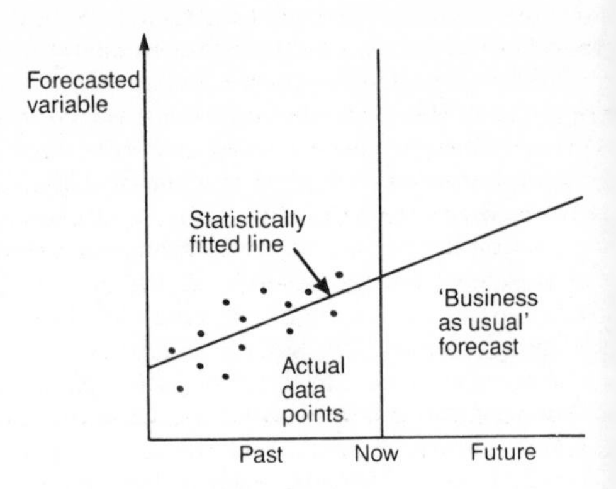

Fig. E.2.

Exxon. The current name (as of 1972) of what used to be the Standard Oil Company of New Jersey, which had traditionally been viewed as the world's largest oil company. It started life as the parent company of STANDARD OIL, which was broken up in 1911 by a US Supreme Court decision (*see* ANTI-TRUST LEGISLATION.

F

factors of production. *See* PRODUCTION FUNCTION.

farm-in. A concept originating in the USA, now common in the UK. A licensee (strictly, a 'participator' as defined under PETROLEUM REVENUE TAX regulations), may farm-out part of a tract whereby a company farming-in gains an interest in acreage and in return incurs the due proportion of past costs up to the date of the farm-in and of future costs. In return the company farming-in receives a proportion of the oil subsequently produced. There are three broad types of farm-in — exploration, development and production farm-ins — and they have sometimes been undertaken in the UK for reasons of tax efficiency.

farm-out. *See* FARM-IN.

FAS (free alongside ship). Freight rates in bulk trade are often quoted FAS at the loading port.

fast line. The end of the drilling line fixed to the reel on the DRAWWORKS. *See also* DRILLING.

fault trap. A type of oil-bearing rock formation. *See also* GEOLOGY.

FCC. *See* FLUID CATALYTIC CRACKING.

feedstock. (1) The input of crude oil, liquefied natural gas or natural gas to a refinery or petrochemical plant. (2) Already processed products that are being used as an intermediate input to a further process.

field. A geographic area under which there are oil and gas POOL(S), but precise definition of a field can be problematical. A key characteristic is that a pool contains a single natural pressure system. Although a pool may contain more than one PAY ZONE with different characteristics which affect drilling and/or operating costs, production from one zone affects RESERVOIR PRESSURE in the whole pool. A field consists of a pool or pools on the same geological structure, but the latter term is difficult to define precisely. A series of fields in a region are called a trend or basin. Uncertainty over the definition of a field can create problems of ownership and also adds to uncertainty when trying to identify reserve volumes.

Field Development and Production Programme. *See* ANNEX A.

field mapping. A process whereby a geologist attempts to outline the underground rock structure based upon what is visible above the ground rather than the geophysical techniques which attempt to measure what is under the ground unseen.

FIFO. An accounting practice in which stocks are valued on a first-in first-out basis. *Compare* LIFO.

fifty–fifty taxes. An even split of profits between the companies and the oil/gas-owning government. It is usually associated with the introduction of POSTED PRICES in the international industry which began in the late 1940s. At first sight it has the comforting appearance of 'fair shares', but this is not the case because earnings should reflect inputs.

final boiling point. The maximum vapour temperature recorded during the distillation of a crude or petroleum fraction (*see* FRACTIONAL DISTILLATION) under test conditions (i.e. a liquid turned to vapour for condensation and collection). Different crudes have different final boiling points depending on their quality. *See also* REFINING, PRIMARY PROCESSING.

final good. *See* INTERMEDIATE GOOD.

final product. A product from a refinery or petrochemical process that undergoes no further processing, but is used by the consumer. The lack of further processing is determined by market conditions and not by technical constraints. For example, heavy fuel oil can be a final product or it can be a FEEDSTOCK for a CRACKING process. *See also* FINISHED PRODUCTS.

financial asset. *See* ASSETS.

finding costs. The cost of finding COMMERCIAL OIL or gas. It includes all expenses involved in acquiring acreage, survey work and the cost of drilling. Together with the cost of development which involves converting OIL-IN-PLACE into RECOVERABLE RESERVES it represents the cost of 'replacing' a barrel produced. In terms of economic theory at the margin it determines the LONG-TERM competitive price. A problem with the concept is that finding costs can only be known after the event. Prior to successful exploration the finding cost per barrel is determined by EXPECTATIONS.

fingering. The situation in which pockets of oil are isolated in a reservoir by channels of water and so cannot be recovered. *See also* PRODUCTION.

finished products. Petroleum products are used mainly as sources of energy, to provide power for transport or for space and process heating and lighting, either directly or by transformation into a secondary energy source (e.g., electricity). The bulk of refinery output consists of a fairly small range of products, called the main products, which fulfil these energy end uses. Examples of main products include motor spirit, aviation fuels, kerosine, gas oil, diesel oil and fuel oil. NAPHTHA (light distillate feedstock), which is used for the manufacture of petrochemicals, is not classed as a finished product and is discussed elsewhere (*see* PETROCHEMICALS).

A wide range of petroleum products is also used in non-energy applications, and although most of these products are comparatively low-volume in industry terms they are highly specialized in their manufacture and application, and constitute a much more heterogeneous range than the energy use products. They are collectively known as special products and include bitumen, wax, solvents, lubricating oils, etc.

It is conventional in the oil industry and in many statistical sources to group products according to their specific gravity (*see* QUALITY). This corresponds to the distillation yield, with light products such as motor spirit being produced at low boiling points, whereas heavy fuel oils require much higher temperatures (*see* REFINING, PRIMARY PROCESSING). For this reason, it is useful to know where individual products lie within this barrel split (*see* Table F.1).

Table F.1.

Boiling range	Barrel split	Individual products
0	Gases	LPG
0–180	Light distillates	Motor spirit, aviation spirit
180–350	Middle distillates	Kerosine, aviation turbine kerosine, gas oil, diesel oil
350	Residue	Fuel oil, bitumen

The categorization summarized in Table F1 is useful for making a rough assessment of how product demand compares with any given crude distillation yield. It also corresponds roughly to a sliding scale of value, since, as a broad generalization, the lighter the product, the higher the value. This is because there are less light products in each barrel of crude and they undergo more refining processes. More importantly, it also reflects the fact that the main product names do not refer to precisely defined substances, but in most cases to a blend of several components. The discussion of crude distillation (*see* REFINING, PRIMARY PROCESSING) shows that the distillation cuts from which each product is made can be varied in response to different marketing and refining considerations. To keep the boiling point spectrum as an implicit frame of reference makes it easier to visualize the areas of overlap between adjacent products assumed in such terms as wide-cut gasoline (which includes kerosine) or light fuel oil (which includes gas oil). In addition to differences of CUT POINT, each finished product may be required to be manufactured to meet different quality specifications according to differing market requirements. The end result is that the fairly small range of main products is manufactured in a very large number of grades. This means that, although for general purposes it is quite meaningful to talk about the generic main product (e.g, motor spirit), there are contexts in which characteristics need to be specified more precisely, an example being OCTANE NUMBER and lead content in the case of motor spirit. In some cases complete technical precision is required, in which case it is necessary to define the detailed product quality specification. *See also* FINISHED PRODUCTS, AVIATIONS FUELS; FINISHED PRODUCTS, BITUMEN; FINISHED PRODUCTS, DIESEL OIL; FINISHED PRODUCTS, FUEL OIL; FINISHED PRODUCTS, GAS OIL, FINISHED PRODUCTS, KEROSINE; FINISHED PRODUCTS, MOTOR SPIRIT; FINISHED PRODUCTS, SOLVENTS; FINISHED PRODUCTS, WAXES.

finished products, asphalt. *See* FINISHED PRODUCTS, BITUMEN.

finished products, aviation fuels. These have a confusing array of product names, but they fall mainly into two types: (a) those required for piston-engined (propeller-driven) aircraft; (b) those required for gas turbine-engined aircraft (jets).

Early aircraft were piston-engined and consumed a gasoline-based fuel similar to that of the car engines of which they were adaptations. At the present time, apart from those in the lighter range, most aircraft have gas turbine engines and consume mainly, although not exclusively, kerosine-based fuels. With this distinction in mind, the individual product names become easier to comprehend.

aviation carrier turbine fuel (avcat, JP5). A kerosine-type jet fuel with a high flash point (*see* QUALITY, AVIATION FUELS). It is used mainly in aircraft employed by navies.

aviation spirit (aviation gasoline, avgas). A high-octane product that is consumed by piston-engined aircraft. This now represents a very small market, being confined to older aircraft or light aircraft of less than 500 horsepower.

aviation turbine gasoline (avtag, JP4, wide-cut gasoline). A wide-cut gasoline which overlaps the light and middle distillate boiling point ranges. It was developed mainly for military use, but a civil version (Jet B) is also available.

aviation turbine kerosine (ATK, avtur, Jet A). The world's principal aviation fuel, being consumed by almost all the jet fleets of the major commercial airlines.

finished products, bitumen (asphalt). The very heavy residue which remains after vacuum distillation (*see* REFINING, PRIMARY PROCESSING). It is usually solid at normal temperatures. The main use of bitumen is in road surfacing, but there are other, more specialist applications (e.g., impregnation of heavy-duty fabrics and felts such as roofing felts and for waterproofing).

finished products, detergents. Calcium petroleum sulphonates.

finished products, diesel oil (DERV (UK)). A transportation fuel for two main types of engine: (a) the automotive high-speed diesel engine, which is the main type of engine used in commercial road transport (e.g., buses and lorries) as well as in some private cars; (b) marine diesel engines.

For automotive use, ignition characteristics for diesel oil are less critical than for motor spirit (*see* QUALITY, DIESEL OIL; QUALITY, MOTOR SPIRIT), which makes high-cost upgrading processing, apart from sulphur removal, largely unnecessary (*see* MOTOR SPIRIT UPGRADING). This means that diesel oil has a lower added value than motor spirit. In the case of marine diesel, fuel quality is even less restrictive because the engines in which it is used require lower rotational speeds than do automotive engines. In many cases, it is possible to include in marine diesel a proportion of residual fuel oil (*see* FINISHED PRODUCTS, FUEL OIL), which has the lowest added value of all the main-line products, thereby further reducing its value.

As with motor spirit (*see* FINISHED PRODUCTS, MOTOR SPIRIT), no effective substitutes exist for the transport applications of diesel oil, apart from other oil products. Some cost advantage can be obtained, however, by substituting down the barrel (i.e. by replacing a lighter fuel with a heavier, less valuable one). For example, motor spirit be replaced by diesel, or marine diesel by residual fuel oil. *See also* FINISHED PRODUCTS, GAS OIL.

finished products, fuel oil. Residual or heavy fuel oil, which is literally the residue remaining after light and middle distillates have been separated out. It is perhaps the least specific of all the main product names, since it covers such a wide range of product grades. Fuel oils are manufactured by blending components together and not by undergoing secondary or upgrading processing as do most light and middle distillate products. Fuel oils have a variety of applications which differ according to fuel oil quality.

Although product quality is discussed in detail elsewhere (*see* QUALITY, DISTILLATE AND RESIDUAL FUELS) it is necessary here to introduce the concept of fuel oil viscosity, since this underlies many of the differences between the diverse grades of fuel oil. Viscosity is a measure of the rate of flow of the fuel oil. A high-viscosity fuel oil is difficult to pump at ambient temperatures, largely as a result of the concentration of waxy and asphaltic compounds in the heavy residue remaining after distillation. The blending of gas oil or kerosine into the heavy residue improves the rate of flow and produces a lower-viscosity fuel oil. Many fuel oils are therefore blends of distillate and residual oils, and the term fuel oil applies to a spectrum of fuels, ranging from almost pure distillate to heavy, undiluted residue.

The main end uses of fuel oils are in industry and as ships' bunker (*see* BUNKERING). In industry, oil-fired steam-raising plant provides process heat and large-scale factory/office central heating. It can also be used in power generation, either on-site or centrally. The fuel oil used in such boiler plants does not normally have a very restrictive quality specification (other than sulphur content), and these types of application represent a major outlet for some of the heaviest fuel oils. Oil-fired industrial furnaces are widely used in a variety of manufacturing industries including metals, glass, ceramics, bricks, cement, chemicals, etc. for a range of melting, heating and drying processes; some of the more specialist of these applications may require fuel oil of a specific quality. For most industrial turbines, gas and distillate fuels (*see* FINISHED PRODUCTS, DIESEL OIL; FINISHED PRODUCTS, GAS OIL) are the preferred fuels because residual fuels can lead to corrosion problems.

There are a few industrial processes (e.g., in certain high-temperature processes in glass manufacture) where fuel oil is considered technically superior to other fuels by virtue of its heat radiation characteristics. In most industrial applications, however, it can be effectively replaced by one or more of the alternative energy forms and must therefore compete on the basis of relative cost and ease of handling and storage. Demand for fuel oil is therefore relatively elastic with respect to price, although only in the medium to longer term, since energy-consuming appliances tend to be fuel-specific. Conversion to an alternative fuel can involve several years LEAD TIME for investment appraisal and the installation of new facilities. Dual-fired facilities, currently more common in power generation, are found increasingly in industry, and this trend will result in much greater short-term ELASTICITY of fuel oil demand with respect to price.

The type of fuel oil used by ships depends on the size and engine power. Smaller ships are powered by marine diesel engines (*see* FINISHED PRODUCTS, DIESEL OIL) and the largest by heavy fuel oil-powered steam turbines. The latter group represents by far the greater volume of bunkers and is a major outlet for the heaviest, high-sulphur grades. As with other transport applications, demand is relatively inelastic with respect to price in the absence of any effective substitute fuel.

finished products, distillate fuels. *See* FINISHED PRODUCTS, DIESEL OIL; FINISHED PRODUCTS, GAS OIL.

finished products, gas oil. The main middle distillate together with diesel oil (*see* FINISHED PRODUCTS, DIESEL OIL). Gas oil and diesel oil are broadly similar products, in the boiling range 200–300°C. As a general rule, the terms diesel oil and gas oil are used to distinguish the automotive and non-automotive applications, respectively. The lighter gas oils, which are close to and overlapping the kerosine fraction, are used mainly for domestic and smaller commercial central heating. In this application they are termed heating oil (UK), ELFO (Germany), fod (France) and Number 2 fuel (USA). Heavier gas oils are used as industrial fuels in cases where their greater cost in relation to residual (heavy) fuel oils (*see* FINISHED PRODUCTS, FUEL OIL) can be justified, either because of their technical superiority for a particular application (e.g., in industrial gas turbines) or because their cleanliness and ease of handling and storage are a major benefit. Middle distillates used as liquid fuels are referred to as distillate fuels, to distinguish them from residual fuel oil.

Although the correct designation of middle distillates used as fuels is distillate fuels, they are sometimes referred to as fuel oils, which can cause confusion with heavier residual fuel oils, for which the latter term is normally reserved.

finished products, kerosine (paraffin (UK), burning oil). The fuel that is used in the industrialized world mainly in central heating and domestic heating stoves. Smaller-scale applications include fuel for camping stoves and blow-lamps. In the developing world, it is use widely as a fuel for cooking stoves and as lamp fuel. Kerosine and aviation turbine kerosine (*see* FINISHED PRODUCTS, AVIATION FUELS) are virtually the same product.

finished products, liquefied petroleum gas. *See* LPG.

finished products, motor spirit (motor gasoline, mogas, petrol (UK)). The fuel consumed by the majority of private cars. The fuel's chemical energy is converted into useful power in the internal combustion engine. In order to ensure smooth combustion and efficient engine performance, the fuel must conform to stringent quality specifications (*see* QUALITY, MOTOR SPIRIT), which normally require additional processing to be carried out (*see* MOTOR SPIRIT UPGRADING). Most upgrading facilities have high capital and operating costs, which result in the finished product having the highest added value of all the main-line products.

Since there is currently no effective substitute for the petrol-fuelled internal combustion engine (other than the diesel engine), any increase in the price of motor spirit tends to be countered in the short term by reduced consumption and in the longer term by more efficient engine design. Demand for motor spirit is therefore relatively inelastic with respect to price in the short to medium term.

finished products, light distillates. *See* FINISHED PRODUCTS, SOLVENTS.

finished products, solvents. Light distillates that are often manufactured from within a very narrow boiling range (e.g., 100–120°C) which accounts for their alternative name of special boiling point spirits (SBPs). It is a low-volume, but highly specialized range of products in terms of refining and marketing. The best known solvent product is white spirit (turps substitute), but others have a variety of applications in dry cleaning and the manufacture of printing inks, polishes, insecticides, edible oils, etc.

finished products, special boiling point spirits. *See* FINISHED PRODUCTS, SOLVENTS.

finished products, waxes. A fraction extracted from the heavy gas oil (waxy distillate) obtained by VACUUM DISTILLATION. As with many other special products, there is a diverse range of applications, of which the main one nowadays is as a sealant for treating paper and cardboard for packaging. Apart from its use in making candles, other applications include polishes, greases, lubricants and petrochemical feedstocks (*see* PETROCHEMICALS).

fireflood extraction. *See* *IN SITU* COMBUSTION.

first-in first-out. *See* FIFO.

First Oil Shock. The term used generally to describe the sequence of events in the last quarter of 1973. During 1973, the oil market showed signs of continued strength. In this context, the producing countries began to feel that the oil price agreements of 1971 (i.e. the TEHERAN AGREEMENT and the TRIPOLI AGREEMENT) were being overtaken by events. On 8 October 1973 the six Gulf members of OPEC met in Vienna with representatives of the oil company to open negotiations on prices. This was two days after the start of the October War between Israel and Egypt. At this point the POSTED PRICE of Arabian light was $3.011 per barrel. The negotiations stalled, and the oil companies requested an adjournment. On 16 October the same six ministers announced a unilateral price increase to $5.119 per barrel. This was announced in Kuwait following, but not part of a routine OAPEC ministerial meeting, and before the announcement of the ARAB OIL EMBARGO. The other OPEC members rapidly followed the Gulf decision.

The main significance of the 16 October price increase was that it was the first time that the producing countries unilaterally determined prices. The effect was electric given the size of the increase coupled with the panic engendered by the embargo. The subsequent market was characterized by panic. In mid-December an auction of Iranian crude produced a price of $17.04 per barrel, and it was reported that a bid of $23 per barrel was rejected. In this context of half panic/half euphoria OPEC met again in Teheran on 22 December. At that meeting the Shah of Iran was pushing for very high price increases supported by the majority, with only Saudi Arabia resisting. In the event, the decision was taken to raise the price as of 1 January 1974 to $11.651 per barrel. This was the First Oil Shock.

It was a shock for three reasons. (a) Because of the size of the increase (287 percent) in the price of the major commodity in world trade, it provided a shock to the international economic system. This took the form of a massive redistribution of income and resources from the oil importers to the oil exporters. (b) The speed of the change (less than 10 weeks) caught everybody (including the producers themselves) by surprise. (c) It was a shock because it exposed the industrial countries to their vulnerability. The previous 20 years had seen the governments of industrial countries content to leave their oil supplies in the hands of the oil companies. Suddenly, with no warning and very limited understanding, the governments were forced to face the possibility of oil shortages and, at best, huge import bills. *See also* CRUDE OIL PRICES.

fiscal policy. The taxation and spending policies of a government.

Fischer–Tropsch process. A chemical process that is used to liquefy coal. The process, which was developed in 1925, involves the gasification of coal followed by condensation under high pressure using a catalyst. Its use is only economic in the presence of cheap and abundant coal supplies. Even then it is generally only used when there are political considerations which either make oil unavailable (as in the case of Germany during World War II) or require self-sufficiency (as in the SASOL plant in South Africa). *See also* COAL LIQUEFACTION.

fish. Any piece of drilling equipment accidentally left in the hole. *See also* DRILLING.

fishing. An attempt to recover a FISH. *See also* DRILLING.

Five, The. *See* CLUB OF ZURICH.

five-sevenths rule. The system used by the Iraq Petroleum Company (IPC) in the 1950s to determine the LIFTING arrangements between the parties. According to this rule, which was one option open to the partners, the total production was determined by adding up everyone's demands and then reducing the total if the highest single demand (nomination) exceeded five-sevenths of the sum of the two lowest nominations. The revised total was then divided by equity share with the underlifters (*see* OFFTAKE NOMINATIONS) in a position to supply overlifters. In the same way as the APQ system in Iran, the five-sevenths rule was designed primarily to restrict supply.

fixed assets. Property owned by a company (or government) in the form of land, buildings, plant, machinery, vehicles and furniture. In the balance sheet they are normally valued at COST less DEPRECIATION. In the context of the oil and gas industry the term covers the above ground assets less stocks. *See also* ASSETS.

fixed-bed catalytic cracking. A form of CRACKING. As the name implies, the CATALYST remains stationary in the reactor during the process. *See also* REFINING, SECONDARY PROCESSING.

fixed bonuses. Terms of an oil or gas production licence whereby at certain points of the operation (e.g., specified production levels) the operator must pay to the owner a fixed sum. *Compare* AREA RENTAL.

fixed costs. *See* SUNK COSTS.

fixed exchange rate. *See* EXCHANGE RATES.

fixed rate loan. A loan in which the rate of interest payable is fixed for the period of a loan. *Compare* FLOATING RATE LOAN.

flag of convenience. The flag of a country where vessels are registered in order to take advantage of low registration fees, taxes and other charges. The most well-known flag of convenience is that of Liberia, but others are Greece, Cyprus and the Bahamas. Generally, because of lower charges and less stringent operating rules associated with flags of convenience, more tankers have been switched to flags of convenience.

flame arrester. A device fixed to a vent pipe from equipment containing flammable material. Its function is to prevent any ignition of the vented contents from flashing back into the equipment causing the ignition of the contents.

flammable limits. The range of percentage volume of a gas/air, or gas/oxygen, mixture within which an ignition would cause burning or an explosion.

flange-up. The act of making the final connection on a piping system.

flared gas. *See* FLARING.

flare stack. *See* FLARING.

flaring. The burning-off of unwanted gas. Flaring is done at the end of a flare stack whose purpose is to maintain a safe distance between the flaring gas and other potentially flammable material. Such flaring in the industry is common in two situations. The first is in refineries when small amounts of gas are produced as a by-product and are uneconomic to process and/or market. The second is the flaring of ASSOCIATED GAS in oil-fields. Prior to the 1970s very high proportions of associated gas were flared. This flaring provided a classic example of the difference between private and social costs. To the companies the gas had no value. It was uneconomic to export it as an energy source, given the high handling charges and low gas price, and the cost of reinjecting the gas to maintain reservoir pressure was not justified by the additional barrels produced in the future when the companies would no longer be the operators. However, from the point of view of the country the gas represented a valuable natural resource either as a potential energy or FEEDSTOCK source or as a means of extending the life of the oil-fields. In the 1960s, gas flaring was the subject of considerable argument between the companies and producing governments. During the 1970s, two factors changed the situation. (a) The general increase in energy prices following the oil price shocks (*see* FIRST OIL SHOCK, SECOND OIL SHOCK) and the perceived energy shortages appeared to improve significantly the economic prospects for using the gas. (b) Many producer governments took over the oil operations from the companies, thus the decision of whether to flare or not devolved to the governments. The result was the introduction of large-scale projects to utilize the previously flared gas, with the result that in many countries the percentage of associated gas flared dropped dramatically in the following years.

flash point. The lowest temperature at which a liquid produces sufficient vapour to produce a flash when ignited. More specifically it refers to a product quality specification for aviation turbine kerosine that defines a safe storage and handling temperature for the fuel (*see* QUALITY, AVIATION FUELS).

floating exchange rates. A regime of EXCHANGE RATES in which the monetary authorities of the countries concerned do not intervene directly in the currency markets. The result is that market forces determine the level of exchange rates.

floating rate loan. A loan in which the rate of interest is allowed to change. The change is achieved by indexing the interest rate to some agreed benchmark such as another rate of interest or the rate of inflation. *Compare* FIXED RATE LOAN.

floating storage. The practice of using oil tankers for storing crude oil rather than for transporting the oil. As freight rates fall, tankers become attractive for storage compared with sites on land, provided that there is adequate anchorage. The amount of tanker capacity used as storage has declined from 25.5 million dwt in 1982 to 16.6 million dwt in June 1985. During the mid-1980s, Saudi Arabia, faced with acute shortages of ASSOCIATED GAS began to build up large volumes of such storage to ensure sufficient crude production to meet immediate gas needs. The sale of these stocks was under the control of NORBEC.

floor man. A member of the drilling crew whose work station is on the floor of the DERRICK.

floor price. The lowest price below which an actual price is not sustainable. If the actual price for any reason falls below the floor price market forces would automatically come into operation to push up the actual price. In economic theory the floor price is the competitive equilibrium price represented by the marginal cost (*see* INCREMENTAL COST). In the context of the oil and gas industries the concept is simple to define, but more complex to use. Thus the floor price is simply the price below which the market perceives the actual market price cannot go. Once that price is reached, users of oil and gas, arguing that the price can only rise, would then proceed to buy for stock purposes, allowing for the cost of storage. Other things being equal, this increased demand would cause prices to rise, thereby confirming that the perception of the floor price was accurate. The problem is that in the absence of understanding and accurate information and in the presence of uncertainty the perception of a floor price can and will change.

When the INTERNATIONAL ENERGY AGENCY (IEA) was created one of its provisions was the adoption of a $7 per barrel floor price — described as a minimum safeguard price. How such a price was to be maintained was not made clear, and the reasons behind the level chosen were complex and concerned politics rather than economics. Its purpose was essentially to provide some economic protection for projects designed to replace OPEC oil. This was based on the assumption that the collapse of OPEC could lead to a fall in oil prices which might embarass economically indigenous energy investments within the IEA.

flotation collar. A raft used to float a platform JACKET into position.

flotel. *See* ACCOMMODATION PLATFORM.

flow testing. The controlled production of oil or gas from a newly discovered source that is used to estimate production levels and the likely performance of a reservoir. *See also* DRILLING.

fluid catalytic cracking. One of the most common forms of CRACKING; a process that converts a heavy, low-value feedstock (e.g., FUEL OIL) into lighter, higher-value products (e.g., MOTOR SPIRIT). *See also* REFINING, SECONDARY PROCESSING.

fluid coking. A process that converts the residue from certain crude oils (*see* RESIDUAL FUEL OIL) into lighter, more valuable products (e.g., MOTOR SPIRIT and GAS OIL). Some coke is also formed. *See also* REFINING, SECONDARY PROCESSING.

flush production. PRODUCTION from an oil well in the initial stages (*see* PRODUCTION, NATURAL DRIVE MECHANISMS).

flux oil. An oil (usually a RESIDUAL FUEL OIL) that is blended with BITUMEN to increase its fluidity or produce a softer consistency.

fob (free on board). The valuation of goods up to the point of embarkation. It is the convention to measure exports of merchandise fob.

fod (fuel oil domestique). A GAS OIL used for domestic central heating in France. It is equivalent to ELFO. *See also* FINISHED PRODUCTS, GAS OIL.

force majeure. A clause in a contract that allows either party to suspend the operation of the contract as the result of unexpected events (e.g., political upheavals, strikes, etc.). Most oil and gas purchase contracts contain such clauses. During the SECOND OIL SHOCK, such clauses were used extensively by governments to take over the marketing of crude oil from the companies. The result was that generally long-term contracts were discredited.

forecasting. Forecasts are sometimes made purely by 'judgement' without the forecasters specifying how he has reached his conclusions. There is an element of subjectivity in virtually all forecasts, but the more explicit methods can be classified as follows:

(a) Surveys that collect opinions about the future. For example, consumers can be asked how much of a product they expect to buy (this is particularly applicable to durables since people often have buying plans for such items); salesmen can be asked how much they expect to sell; 'experts' in some field can be surveyed. The last method could be employed to ask people knowledgeable in the energy field when they expect various new energy technologies to reach commercial exploitability. One such opinion survey method is the DELPHI TECHNIQUE.
(b) Observations of behaviour. For example, a company wishing to assess the price sensitivity of demand for its product could reduce or increase price in a small region in order to observe the response of consumers and to draw general conclusions. Alternatively, the company could use a test market in a region to try out an advertising campaign or a new product, thus testing the reactions of the consumers in a limited area.
(c) Analysis of historical data to draw conclusions relevant to the future. There are two broad categories within this approach. The first is to construct mathematical models that do not explicitly try to isolate causal variables. For example, a trend, such as a simple linear relationship, an exponential, a modified exponential or an EXPONENTIALLY WEIGHTED MOVING AVERAGE, can be fitted to past data and then projected into the future. Such an approach is sometimes distinguished from true forecasting and described as projecting. The second category is 'causal' model building in which the forecaster attempts to specify the 'independent variables' which determine the 'dependent variable' he is trying to forecast (*see* ECONOMETRICS). In energy demand modelling, in consumer markets for example, the independent variables might be taken as real personal disposable income, energy prices relative to the general price level and the energy appliance stock. From an econometric model or by less formal means the coefficients (ELASTICITIES) of energy demand with respect to each of the independent variables would be estimated. Then forecast values of the independent variables would be used to derive a forecast of energy demand.

Because of the uncertainty of the future, forecasts are usually accompanied by statements of the margin of uncertainty which may surround them. *See also* SCENARIOS; SENSITIVITY ANALYSIS.

forward contract. *See* FORWARD MARKET.

forward market (futures market). A market in which traders exchange forward contracts. These are individually tailored contracts stipulating a future date and price — forward price — at which a stated quantity of a commodity will be exchanged for money. Therefore, through forward commodity markets, traders can look into future commodity prices. Security and currencies can be exchanged in forward financial markets, in which traders can look into future interest rates or currency values. The purpose of such markets is to reduce the uncertainty for the final users of the commodity in a context where the daily price of such a commodity may fluctuate significantly. However, such markets also attract speculators who normally have no interest in using the commodity, but whose sole purpose is to buy cheap and sell dear. These can be full-time professionals or the so-called 'doctors and dentists' who play the market on a casual basis. Final users may also trade speculatively. In theory these speculators should help in smoothing out price fluctuations by acting counter to the market trend. For example, if demand (and price) falls, speculators should increase demand in anticipation of a later increase in demand (and price), hence slowing the fall in price. In practice the reverse appears to be the case due to the lemming-like response of the speculators, especially by the 'doctors and dentists'. Thus price speculation is accentuated rather than dampened.

In the oil industry prior to the SECOND OIL SHOCK no such futures market existed. However, subsequently in a market where increasing volumes were being sold on a spot basis or on short-term contracts futures markets for crude oil and some products began to emerge. This emergence of future markets tended to provide a quantifiable view of expectations of the market which, if anything, added to the volatility of the markets. In effect it was further proof that oil by late 1985 was for the first time being traded like any other primary commodity. The likely future growth of such markets remains uncertain.

forward price. *See* FORWARD MARKET.

fraction. A group of compounds collected

by FRACTIONAL DISTILLATION that condenses within the same temperature band. (In the oil-refining context, a fraction is referred to as a cut or distillation cut. *See also* REFINING, PRIMARY PROCESSING.

fractional distillation (fractionation). The process of separating (by heating) various groups of compounds contained in a substance (e.g., crude oil) and collecting those which condense at broadly similar temperatures. Distillation in the oil-refining context is normally referred to as CRUDE OIL DISTILLATION or ATMOSPHERIC DISTILLATION. *See also* REFINING, PRIMARY PROCESSING.

fractionating column. The column used in the process of FRACTIONAL DISTILLATION, with intermediate points along its length at which groups of compounds can be collected as they condense. *See* REFINING, PRIMARY PROCESSING.

fractionation. *See* FRACTIONAL DISTILLATION.

fracture pressure. The amount of pressure required to split a rock to allow fluid to enter.

free alongside ship. *See* FAS.

free on board. *See* FOB.

freezing point. A product quality specification for aviation fuels designed to safeguard high-altitude performance of the fuel, by defining the point at which it starts to solidify. *See also* QUALITY, AVIATION SPIRIT.

freight costs (freight rates). Tanker freight rates are expressed as a percentage of a flat rate applicable to the particular voyage for which the vessel is contracted. The flat rates are published as WORLDSCALE and are designed to equate the returns to a standardized vessel on all routes. Rates vary depending on whether they are spot (i.e. short-period prompt-start contracts), medium-term or long-term. Medium- and long-term charter rates are related to spot rates through the effects of spot rates on expectations. Long-term rates tend to be lower than spot rates and reflect costs more closely since eventually new building may be substituted for very-long-term charter.

freight rate measurements. *See* TANKER FREIGHT RATES.

freight rates. *See* FREIGHT COSTS.

front end-loaded. Refers to projects in which the bulk of the costs occur in the early stages; normally associated with large capital-intensive projects. *See also* LEAD TIMES.

fuel oil. A term normally used to describe HEAVY FUEL OIL or RESIDUAL FUEL OIL, which is the residue remaining after CRUDE OIL DISTILLATION, and which is used mainly to provide heat and power in industry and for ships' BUNKERS. Because many fuel oils are blends of heavy residue oil and distillates, such as gas oil, the term is frequently used less specifically to refer to any liquid fuel, including very light fuel oils which may contain little, if any, residue. Fuel oil is therefore a rather imprecise description and normally needs to be qualified. *See also* FINISHED PRODUCTS, FUEL OIL.

fuel oil domestique. *See* FOD.

fuel switching. The substitution of one fuel for another. Since the early 1970s fuel switching has been responsible for an important part of the decline in the demand for oil. For example, a key area has been the switching from oil to coal as the basis for thermal generation of electricity. The extent of fuel switching is a function of three factors: (a) the suitability of alternative fuels to do the job; (b) the present and expected levels of relative fuel prices; (c) the technical flexibility to change the fuel-using appliance (i.e. the cost of converting existing equipment or the costs of acquiring completely new fuel-using appliances).

furfural. An organic solvent used in the production of LUBRICATING OIL base stocks to remove AROMATIC, naphthenic (*see* CYCLOPARAFFINS), olefinic (*see* ALKENES) and unstable hydrocarbons. The effect is to improve the VISCOSITY INDEX and stability characteristics of the lubricating oil.

futures market. *See* FORWARD MARKET.

G

game theory. *See* THEORY OF GAMES.

gas. *See* NATURAL GAS.

gas cap. A volume of gas present above the oil in many oil-bearing formations. *See also* GEOLOGY; PRODUCTION.

gas cap drive. A type of natural recovery mechanism (*see* NATURAL DRIVE) enabling oil to be produced without artificial assistance. *See also* PRODUCTION, NATURAL DRIVE MECHANISMS.

gas coning. The tendency of the gas/oil contact to move down more rapidly near to a well.

gaseous fuels. The light and highly volatile distillates obtained from refining crude oil. They are used to produce motor spirit, AVIATION SPIRIT and LPG. *See also* FINISHED PRODUCTS; REFINING, PRIMARY PROCESSING.

gas-field. *See* FIELD.

gas gathering. A process whereby the ASSOCIATED GAS from a number of wells is collected in one place for transportation or further processing.

gas grid. A system of gas pipes (similar to an electricity grid) designed to match supply from different areas with demand from different areas. Its main function is to increase the base load demand for gas and decrease PEAK DEMAND thereby allowing greater capacity utilization.

gasification. *See* MANUFACTURED GAS.

gas injection (gas lift). A type of secondary recovery mechanism which maintains RESERVOIR PRESSURE in order to permit further recovery of oil. *See also* PRODUCTION, SECONDARY RECOVERY MECHANISMS.

gas lift. *See* GAS INJECTION.

gas liquids. Refers to BUTANE, PROPANE and pentane (*see* NATURAL GASOLINE). *See also* NATURAL GAS.

gasohol. A term created in the USA for a mixture of 10 percent ETHANOL and 90 percent gasoline (*see* MOTOR SPIRIT). However, it has increasingly been used to denote any combination of ethanol or METHANOL mixed with motor spirit. Between 10–15 percent of the additive can be used with only minor modifications to the internal combustion engine. Above this level more significant changes are required to the engine. There have been two major gasohol programmes. One in Brazil, which was begun in 1975, was based on the large-scale fermentation of sugar-cane and cassava, The other in the USA was based on a much smaller scale. Gasohol has never been an economic project and invariably requires heavy subsidization. *See also* ALCOHOLS.

gas oil. A MIDDLE DISTILLATE product used as a liquid fuel. Lighter gas oils are used mainly for domestic and small-scale commercial central heating. Heavier gas oils are used as industrial fuels. *See also* FINISHED PRODUCTS, GAS OIL.

gas/oil contact. The boundary surface between an accumulation of oil and an overlying accumulation of gas. *See also* GEOLOGY.

gas/oil separator plant. *See* GOSP.

gasoline. The term used for motor spirit in the USA. It is frequently abbreviated to 'gas'. *See also* FINISHED PRODUCTS, MOTOR SPIRIT.

gas saturation pressure (bubble point). The pressure at which the ASSOCIATED GAS begins to come out of solution of crude oil at a given temperature.

gas-to-oil ratio (GOR). The volume of ASSOCIATED GAS at atmospheric pressure produced in a given volume of crude oil. Conventionally it is measured as cubic feet of gas per barrel of oil.

gas treatment plant. A plant whose purpose is to remove 'impurities' (i.e. water and sulphur) and/or to split the gas into different gas streams. An example is the stripping of the liquids from WET GAS to produce GAS LIQUIDS and DRY GAS. *See also* NATURAL GAS; SCRUBBING.

gathering lines. The pipelines that run from the wellhead to a central storage or treatment centre.

GDP. *See* GROSS DOSMESTIC PRODUCT.

gearing ratio (debt ratio, leverage ratio (USA)). The ratio of fixed-interest DEBT to shareholders' interest plus the debt. At the risk of over-simplification, the higher the gearing ratio, the more vulnerable is the equity shareholder, although if the company can earn more on the borrowed capital than it costs this could increase his dividends.

General Agreement on Participation. This was the agreement signed in October 1972 between the oil companies and the Arab Gulf states which gave each government an equity share in its OIL OPERATING COMPANY. *See also* PARTICIPATION.

general equilibrium analysis. An economic tool of analysis that examines simultaneously all the markets in an economy and their interaction. Thus a change in one market will 'shunt' itself through to other markets. For example, a change in the oil market affects other markets which in turn will affect the oil market. As can be imagined the analysis is extremely complex, but it attempts to provide a more realistic description of the way an economy works than PARTIAL EQUILIBRIUM ANALYSIS which tends to assume away linkages between markets. *See also* EQUILIBRIUM.

general-purpose tanker (handy-sized tanker). A term used by AFRA to define a relatively small oil tanker in the 16500–24999 dwt range. Such tankers are mainly used for the carriage of oil products as opposed to crude oil.

Geneva I. The first of two agreements concerned with exchange rates between the oil-producing governments and oil companies. It was reached in January 1972. The agreement was to adjust the oil price terms of the TEHERAN AGREEMENT and TRIPOLI AGREEMENT to take account of the fall in the value of the US dollar in which oil was priced. Geneva I raised POSTED PRICES by 8.49 percent which would raise the GOVERNMENT TAKE by 8.57 percent. This matched the rise in gold price in relation to the dollar. It also allowed for future changes by linking the oil price to an index based on the exchange rates of nine industrialized countries vis-à-vis the dollar. *Compare* GENEVA II.

Geneva II. The second of the two exchange rate agreements between oil-producing governments and oil companies (*compare* GENEVA I). It was reached in June 1973. Geneva II provided an increase of 11.9 percent in POSTED PRICE following the official US devaluation of the dollar in February 1973. It also revised the index formula. Geneva I and Geneva II were examples of the growing ability of the producer governments to involve themselves in administering the price of oil.

geochemistry. A branch of GEOLOGY.

geology. The science of the earth. With certain exceptions (i.e. occasionally when oil seeps from cracks in the earth's surface) oil is found deep underground, and the final proof of its existence is to drill a well to bring it to the surface (*see* DRILLING). Petroleum geology is concerned with locating the areas in which oil is most likely to be found (prospective areas) and where drilling is most likely to prove successful.

The science of geology includes within it many specialisms: stratigraphy, the study of the distribution of different rock types; sedimentology, the study of the rocks themselves; palaeontology, the study of fossils;

geochemistry, the chemistry of rocks and the fluids contained in them. Geophysics, a related, although separate science, studies the physical properties of rocks, from which information about the likely distribution of different rock types can be inferred.

The fact that so many related disciplines are involved in the study of petroleum geology illustrates the complexity of the processes which led to the accumulation of petroleum (oil and gas) deposits. Petroleum is thought to have originated from organic matter — both plant and animal — which, after decomposing millions of years ago, sank to the bottom of the sea to become embedded in layers of mud or sediment. Areas in which sediment accumulated more thickly, either because of the existence of natural depressions, or as a result of subsidence, are known as sedimentary basins. Over a long period of time, and with the action of surf, currents and earthquakes, the sediment was gradually compacted into rock — the source rock — in which petroleum was originally formed.

Rocks differ in the extent to which they contain pores or cavities. The types of rock that made good potential source rock for the formation of petroleum were of a relatively porous type, capable of retaining fluid in their pores and cavities. In the process of compaction, the source rock underwent a reduction in its amount of pore space and, as a result, fluid was expelled. A process of migration then followed, in which petroleum (or possibly petroleum mixed with water) progressed through various rock strata or was prevented from doing so, according to the rock type encountered. To permit the passage of fluid through it, rock must be not only porous (*see above*), but also permeable, which means it must have cavities and spaces which are sufficiently interconnected to allow fluid movement as well as retention. Impermeable rock layers are those which do not permit the passage of fluid, thus when these are encountered the movement of petroleum is halted.

In the process of migration, therefore, petroleum is expelled from the source rock in which it formed and moves through various porous, permeable strata (e.g., sandstone or limestone) until it encounters impermeable rock (e.g., shale, clay or salt). When progress of the petroleum is halted it forms an accumulation or oil reservoir. The porous rock in which it is retained is known as reservoir rock. As a very broad generalization, therefore, three features which might be expected to be associated with the presence of oil are: (a) sedimentary basins, which have been relatively undisturbed by subsequent geological movement; (b) porous rock structures in which oil could have accumulated; (c) impermeable rock strata overlay the porous rock structure and act as a trap preventing further migration of the petroleum.

The two main forms of petroleum accumulation are stratigraphic traps and structural traps; the latter category, which accounts for over 50 percent of known petroleum revenues, is subdivided into anticlinal traps and fault traps.

The commonest form of petroleum accumulation, which is typical of many Middle East structures and is also the simplest structurally, is the anticlinal trap. In this formation, the rock layers lie in folds resembling an arched, dome-like structure in which the oil has accumulated. From above, the different types of rock strata form closed, eliptical or circular structures, thus retaining the oil within the formation. When water, oil and gas all occur in a formation, their relative densities cause the oil to lie with the water below and the gas above it. The interfaces between them are known as the oil/water contact and the gas/oil contact, respectively.

Fault traps are rock formations in which the oil has been formed by a shift in the layers of rock relative to each other, so that the oil-bearing layer has been left adjacent to an impermeable layer. It frequently happens that anticlinal traps and fault traps occur together, either side by side, or in combination in the same formation. It is then referred to collectively as a structural trap.

Stratigraphic traps are those in which the oil accumulation has been trapped as a result of the characteristics of the rock strata themselves (i.e. from stratigraphic causes) rather than for structural reasons, such as folding or faulting within the strata. Thus the porous and permeable reservoir rock has been sealed within impermeable rock as a result of the sedimentary process in which oil was originally generated. Examples of stratigraphic traps are reef traps, when a former coral reef was sealed off by impermeable rock, or salt domes.

The most commonly used geological investigations designed to establish prospective areas are geological FIELD MAPPING and

GEOPHYSICAL SURVEYS (e.g., GRAVIMETRIC SURVEY).

geophysical survey. Since the 1920s geologists have been increasingly able to use GEOPHYSICS to try to map the unseen geology below the surface in order to decide on the most likely prospects for drilling exploratory wells. Three main methods are used: (a) magnetic surveying measures variations in the magnetic field of the subsurface structures; (b) gravity surveying measures the effect of gravity on rock density; (c) seismic surveying measures the speed of travel of seismic waves through the rocks. Of the three methods, the seismic approach is the most widely employed. In general, gravity and magnetic surveys are used to locate possible acreage, whereas seismic surveys are used to locate the actual site of the drilling rig actually. From the data derived from such surveys it is possible to create an underground cross-section map showing the rock formation in terms of thickness and formation, which in turn can indicate where oil may be. As yet the only way to confirm the accuracy of the underground map and the existence of oil or gas is to drill a well.

geophysics. A science related to GEOLOGY that is concerned with the properties of the earth as represented by the gravitational, electrical, magnetic and mechanical characteristics. *See also* GEOPHYSICAL SURVEY.

geosyncline. A large sedimentary basin where very deep layers of sediment have accumulated. *See also* GEOLOGY.

geothermal gradient. The measure of the rise in rock temperature below the surface. The gradient is normally measured in degrees fahrenheit per 1000 feet.

Giffen good. *See* DEMAND.

gilt-edged. (1) Describing fixed-interest government SECURITIES traded on the stock exchange. Fluctuations in the market determine the effective rate of interest. (2) Describing any secure investments. *See also* BONDS..

Gini coefficient. A measure of concentration which can be used to assess such things as industrial structure or income distribution. It is derived from the LORENZ CURVE and is expressed as a ratio of the area between the 45° line and the Lorenz curve divided by the area above the 45° line. The coefficient therefore ranges between 0 and 1. The closer the coefficient is to 1 the greater the concentration. Thus a coefficient of 1 means that all the market is supplied by one company, whereas a coefficient of 0 means all supplying companies have identical market shares. A weakness of the measure is that it cannot alone say anything about the companies' control of the industry since it cannot take account either of cartelization (*see* CARTEL) of the firms or the existence of interlocking directorships whereby the same directors (e.g., nominated by banks) sit on the boards of different companies. *See also* CONCENTRATION RATIOS.

glycol. A constituent of antifreeze.

GNP. Gross national product (*see* GROSS DOMESTIC PRODUCT).

go-devil. A type of PIG with self-adjusting blades that is pumped through a pipeline in order to clean it.

going in hole. The lowering of the DRILL STRING into the well bore. *See also* DRILLING.

good oil-field practice. A legal term (with technical connotations) that is used in production licences. Oil- and gas-fields can be rate-sensitive, which means that the amount of oil and gas ultimately recovered depends upon the speed at which the reserves are depleted. If production is too rapid the fields can be 'damaged' (i.e. potential future production will be lost). There are other ways in which oil reservoirs can be damaged. Hence good oil-field practice means not damaging the fields. As with most legal concepts, the precise meaning is a matter of debate. *See also* CAPACITY.

GOPA (Government Oil Pipeline Agency). It was announced in March 1985 that BNOC would be abolished and replaced by the Government Oil Pipeline Agency. GOPA would be a small organization with a much reduced brief concerned largely with managing pipelines and handling royalty monies.

GOR. *See* GAS-TO-OIL RATIO.

GOSP (gas/oil separator plant). A plant that removes ASSOCIATED GAS from crude oil before the crude oil is transported. A GOSP forms part of the gathering system on an oilfield.

GOS price (government official sales price). The publicly announced price at which the government would sell when governments began to market their own crude oil and products. Thus during and after the 1970s when the OIL OPERATING COMPANIES were taken over it was the GOS price which formed the basis of the international price structure for crude oil. In practice, during periods of market weakness, governments would offer discounts on the GOS price (sometimes publicly and sometimes secretly), whereas when the markets were tight premiums were added to the GOS price. *See* CRUDE OIL PRICES.

government official sales price. *See* GOS PRICE.

Government Oil Pipeline Agency. *See* GOPA.

government take. The revenue of an oil-producing government from the oil operations within its frontiers. It normally refers specifically to the revenue which accrues from taxation of oil production (including ROYALTIES), but on occasions it also refers to the value of crude oil to which the government has access by virtue of PRODUCTION SHARING agreements. In effect it measures the government's share of the revenue from oil operations. During the 1970s much was made by the producing governments of the fact that their tax take on refined products (by virtue of their 'taxes' on crude) was less than the indirect taxes (*see* INDIRECT TAXATION) received by the consuming governments on the sale of the products. However, this was misleading. The producer government take was value-added and therefore an addition to GROSS DOMESTIC PRODUCT, whereas the taxes received by the consumer governments were simply transfer payments from the consumer to their governments.

GPW. *See* GROSS PRODUCT WORTH.

gravity survey. A method of geophysical survey in which measures of the variations in the surface gravitational field are used to deduce the nature of the underlying rock. *Compare* MAGNETIC SURVEY; SEISMIC EXPLORATION. *See also* GEOPHYSICS.

gravity, specific. A measure of the weight per unit volume of a substance, thus in terms of crude oil, the higher the specific gravity, the heavier the crude. The common unit of gravity for crude oil is that of API GRAVITY, which is inversely related to specific gravity, so that the heavier the crude, the lower the API gravity. *See also* QUALITY, CRUDE OIL; APPENDIX.

gravity differential (gravity premium). The value, and hence the price, of crude oils varies in relation to their gravity, since the lighter crudes yield a higher proportion of the lighter, more valuable products, after refining. A specific formula was applied to determine the gravity differential. Thus deviations in gravity (*see* API GRAVITY) from the marker crude (i.e. Saudi Arabian light crude 34–34.9°API) attracted a fixed amount for each degree of API difference from the marker API — added to the market price for each degree above and subtracted for each degree below. In theory the value per API degree differential should have reflected the difference in the GROSS PRODUCT WORTH accounted for by the difference in API number. In practice, it tended to reflect the general state of the market and was used as a means to fine tune prices to reflect market conditions. *See also* QUALITY, CRUDE OIL.

gravity platform. An offshore PLATFORM (usually of reinforced concrete) that is held in position on the sea bed by virtue of its own weight.

gravity premium. *See* GRAVITY DIFFERENTIAL.

greenhouse effect. The burning of HYDROCARBONS such as oil and natural gas releases carbon dioxide into the atmosphere. If the carbon dioxide is not removed (i.e. by photosynthesis in plants), the increased levels of carbon dioxide would allow solar radiation to be absorbed by the earth, but would reduce the reverse transmission to space of long-wavelength infrared radiation. This would result in the earth's temperature tending to rise. The greenhouse effect has given rise to growing concern about the environmental implications of the ever-increasing use of hydrocarbons as an energy source.

gross cash flow. *See* CASH FLOW.

gross domestic product (GDP). A measure of the total flow of goods and services produced in an economy in a specific period (normally one year). It is measured by aggregating the VALUE ADDED produced by each sector of the economy at market prices. This avoids the double counting which would occur if the market value of the output of each sector were to be aggregated. For example, if the output of the crude oil sector were simply added to the output of the refining sector the value of crude oil would be counted twice. If indirect taxes (*see* INDIRECT TAXATION) are removed and subsidies added then GDP at market price becomes GDP at factor cost. If the output of capital goods used to replace capital used is removed (i.e. DEPRECIATION) then the gross figure becomes net.

GDP can be measured in one of three ways: (a) by measuring the value added produced by each sector; (b) by measuring the income or the returns to the factors of production (i.e. land, labour, capital and entrepreneurship) which occur as a result of producing the output; (c) by measuring the expenditure of those in the economy which acts as an equivalent to income.

In practice some of the output in an economy is produced by foreigners, whereas some nationals produce output in other countries. If income (output) which accrues to nationals from abroad is added to GDP and income (output) remitted abroad is subtracted then the domestic product becomes the national product (gross national product, GNP, or net national product). In oil-producing countries in which foreign oil companies play an important role GDP is always significantly higher than GNP because of the oil company profits remitted abroad.

Traditionally measures of GNP per capita have been used as a measure to identify the standard of living of a country and its place in the international league tables of development. However, there has been growing unease with this use of GNP for a number of reasons. (a) As a mean average, GNP per capita takes no account of the actual distribution of income. (b) GNP excludes the measurement of items that influence people's standard of living (e.g., increased leisure) and includes items such as defence which it can be argued contribute little to peoples' welfare. (c) In many less developed countries there are seriously distorting problems with the statistical measurement of GNP. (d) International comparisons of GNP require conversion to a common currency unit which can pose serious difficulties over which exchange rates should be used (many countries have several 'official' exchange rates) and how far the exchange rate used reflects differences in purchasing power of income between countries.

In the oil- and gas-producing countries there exists a further problem. Given that oil and gas are both EXHAUSTIBLE RESOURCES their production and sale does not represent income, but rather the LIQUIDATION of a country's assets. There is a school of thought which argues that the output of oil and gas, as well as other exhaustible resources, should be excluded from GDP.

gross national product. *See* GROSS DOMESTIC PRODUCT.

gross product worth (GPW). A measure of the value of the refined products from a barrel of crude oil at the refinery gate. It is in effect a weighted average composed of the physical output of products (usually as a result of a simple distillation process) multiplied by their respective market values at the refinery gate (i.e. excluding indirect taxes). If the cost of transporting and refining the crude is subtracted from gross product worth, this produces the net product worth or netback. The netback provides a market evaluation of the crude oil. If the cost of crude oil received at the refinery is subtracted from the gross product worth this provides the refinery margin. If the cost of refining is also removed this gives a figure for the profitability (or otherwise) of the refinery.

gross profit. *See* PROFIT.

gross tonnage. The volume of the interior of a crude oil tanker expressed in tons per 100 cubic feet.

Gulf. A US-based MAJOR. It was formed in

1907 and was financed by the Mellon Banking Group. Generally thought of as similarly sized to BP, in 1984 the company was taken over by SOCAL.

Gulf, The. A term used by those who do not wish to enter the argument as to whether the sea between Iran and Arabia should be called the Arabian Gulf or the Persian Gulf.

Gulf plus. The pricing system used in the 1930s and 1940s (*see* CRUDE OIL PRICES) whereby products were priced throughout the world as though they had been imported from the USA via the Gulf of Mexico irrespective of their actual origin. *See* BASING POINT PRICING.

gum content test. A test for the stability of MOTOR SPIRIT, which detects the presence of by-products formed in any deterioration process that may have occurred during storage. *See also* QUALITY, PRODUCTS.

H

handy-sized tanker. *See* GENERAL-PURPOSE TANKER.

HD oil. *See* HEAVY-DUTY OIL.

HDS. *See* HYDRODESULPHURIZATION.

heating oil. Alternative name, employed in the UK, for light GAS OIL that is used for domestic and smaller-scale commercial central heating. *See also* FINISHED PRODUCTS, GAS OIL.

heave. The vertical movement of a semi-submersible drilling PLATFORM or drill ship.

heavy crude. A crude oil that has a high yield of heavier, less valuable products. For example, Kuwait crude, with an API GRAVITY of 31.1°API, is one of the heaviest crudes and yields 50 percent or more of FUEL OIL after refining. *See also* QUALITY, CRUDE OIL.

heavy-duty oil (HD oil). Oil that was originally developed for the lubrication of some high-speed diesel engines as well as spark ignition engines subject to high piston and crankcase temperatures.

heavy ends. *See* LIGHT ENDS.

heavy fractions. *See* LIGHT ENDS.

heavy fuel oil (HFO). A fuel oil consisting of the largely undiluted residue (*see* RESIDUAL FUEL OIL) which remains after crude distillation (*see* REFINING). Heavy fuel oils are used mainly for industrial applications (power and heat) and for ships' BUNKERS. *See also* FINISHED PRODUCTS, FUEL OIL.

hedge. Action taken by participants in a market to offset the effects of fluctuations in prices. The entry of a buyer or seller in a futures market is a form of hedging. Thus a seller who must meet a contract to supply oil in one month's time at a fixed price can protect himself against a rise in oil price by buying the oil now and selling a one-month future in oil for the same quantity or by buying himself a one-month future in oil.

Herfindhal index. A measure of industrial concentration (i.e. how much of the market is controlled by how many firms). It measures the size of a company (in terms of sales, employment, etc.) as a proportion of the industry total. The maximum value of the index is 1, which means that one company has all the sales, employment, etc. *See also* CONCENTRATION RATIOS.

HFO. *See* HEAVY FUEL OIL.

high absorbers. *See* ABSORPTIVE CAPACITY.

high-octane. Refers to fuel (usually MOTOR SPIRIT) with an OCTANE NUMBER in excess of 100. This implies the fuel has a lower tendency to ignite spontaneously (i.e. knock) in the combustion chamber. *See also* QUALITY, MOTOR SPIRIT.

historic cost. A measure of the actual costs at the time they have been incurred. An asset shown at historic cost in a balance sheet may therefore be under- or over-valued, depending upon what the asset would now sell for. *See also* DEPRECIATION.

hole. *See* WELL BORE.

homogeneous product. A product that is identical. For example a refined product of a specified quality is a homogeneous product. The term is used to distinguish it from a DIFFERENTIATED PRODUCT, which although performing the same function does have differences. An example of a differentiated product is crude oil. In economic theory, the significance of a homogeneous product is that no brand loyalty on the part of the buyer can exist, hence only price determines from which supplier the buyer will buy.

honeycombe coke. *See* PETROLEUM COKE.

horizontal integration. The situation in which companies involved in the same activity (e.g., refining) join together willingly by merger or unwillingly by takeover. Horizontal integration tends to reduce the number of firms in an industry which can be seen in the measurement of CONCENTRATION RATIOS. It therefore has important implications for levels of competition in the industry. In the international oil industry, a particular form of horizontal integration that was of enormous importance concerned the joint control of crude oil supplies by the MAJORS. This arose as a result of the joint ownership by these companies of the various OIL OPERATING COMPANIES (e.g., ARAMCO, KOC, IPC). The joint ownership was the result of intervention by the home governments of the majors at the time that the original concessions were granted together with attempts by the companies to allow new sources of crude oil to enter the market without causing excess supplies. Thus crude long companies were encouraged to take as partners those companies short of crude oil (i.e. the CRUDE SHORT COMPANIES). The result of this joint ownership was two-fold. (a) Each company knew precisely just how much the other companies were producing. Thus a major problem in an oligopolistic market — uncertainty — was effectively removed (*see* OLIGOPOLY). (b) Because of the way in which production levels were determined (*see* APQ; FIVE-SEVENTHS RULE) any tendency by partners to overproduce, thereby weakening prices, could be controlled. These two factors meant that the major companies could orchestrate supplies of crude oil in order to contain the potential excess producing capacity. This control also meant that new sources of crude oil could be brought into the market in the 1950s and 1960s in a relatively orderly fashion by cutting back production in other areas. *Compare* VERTICAL INTEGRATION.

horsepower (hp). A unit of mechanical power. One horsepower is equal to 746 watts. *See also* APPENDIX.

Horton sphere. A spherical pressure tank used for the storage of volatile liquids. Its function is to reduce the evaporation loss associated with conventional storage tanks.

hp. *See* HORSEPOWER.

hundred-year storm. A technical criterion for the building of offshore platforms, etc. It represents the combination of storm conditions (wave height and wind speed) which on average should occur only once every 100 years. Offshore structures should be designed to withstand such a storm.

hybrid platform. A GRAVITY PLATFORM in which the base and storage area are made of reinforced concrete, but whose upper sections are made of steel.

hydrocarbons. Chemical compounds in which carbon and hydrogen are the only elements. In the petroleum context, the term is used for any compound with carbon and hydrogen as the major, although not the exclusive elements. In this sense, it is often synonymous with oil and gas products. *See also* CHEMISTRY.

hydrocarbon solvents. *See* SOLVENTS.

hydrocracking. A refining process that converts heavier, less valuable products (e.g., FUEL OIL) into lighter, more valuable ones (e.g., MOTOR SPIRIT or GAS OIL). The use of hydrogen results in paraffinic products (*see* ALKANES) low in sulphur. Since it involves the addition of hydrogen, for which an additional plant is required, it tends to be more expensive, and hence less common, than CATALYTIC CRACKING. *See also* REFINING, SECONDARY PROCESSING.

hydrocyclone (desander, desilter). Equipment that is used to remove sand and/or silt from DRILLING FLUID by centrifugation.

hydrodesulphurization (HDS). A process that removes sulphur compounds from petroleum products by treating them with hydrogen in the presence of a CATALYST. *See also* REFINING, SECONDARY PROCESSING.

hydrofining (hydroheating). The most commonly used refining process for the removal of sulphur from products in the middle distillate range (e.g., GAS OIL, AVIATION TURBINE KEROSINE). *See also* REFINING, SECONDARY PROCESSING.

hydroforming. The first CATALYTIC REFORMING process to be developed. It was first used in the USA in 1939. *See also* MOTOR SPIRIT UPGRADING.

hydrogenation. A chemical reaction (*see* HYDROFINING) in which the hydrogen contents of fuels is increased. *See also* REFINING, SECONDARY PROCESSING.

hydrogen sulphide. *See* ACID GAS.

hydroheating. *See* HYDROFINING.

hydroliquefaction. The conversion of coal into liquid fuel. *See also* COAL LIQUEFACTION.

hypoid lubricants. Types of extreme-pressure lubricants (i.e. lubricants with very high film strength which increases their ability to maintain the surfaces of highly loaded gears). These lubricants are used in hypoid gears.

I

IBRD. International Bank for Reconstruction and Development; otherwise known as the World Bank.

ICOR. *See* INCREMENTAL CAPITAL–OUTPUT RATIO.

identified resources. Oil or gas resources whose location, quality and quantity have been specified. *See also* RESERVES.

IEA. *See* INTERNATIONAL ENERGY AGENCY.

ignition point. The temperature to which a fuel must be raised before it will burn.

IGS. Inert gas system (*see* TANKER TYPES).

illuminating oil. A fuel suitable for use in wick-fed or mantle-type lamps. The most commonly used form of illuminating oil is kerosine. *See also* FINISHED PRODUCTS, KEROSINE.

IMF. *See* INTERNATIONAL MONETARY FUND.

imperfect competition. A generic term for market structures that are not competitive. Most commonly these include MONOPOLY and OLIGOPOLY.

impermeable. A geological term describing rock that does not allow the passage of fluid, and hence oil. *See also* GEOLOGY.

implicit costs. *See* EXPLICIT COSTS.

incidence of tax. *See* TAX INCIDENCE.

inclinometer. An instrument that is used to measure the size and direction of any deviation of the WELL BORE from the vertical. *See also* DIRECTIONAL DRILLING.

income. The flow of money (or goods) that accrues to an individual, a company or an economy during a specified period of time. It can be expressed either in NOMINAL or REAL terms to allow for the effect of INFLATION.

income effect. *See* SUBSTITUTION EFFECT.

income elasticity of demand. A measure of the responsiveness of DEMAND to a change in INCOME, with other factors determining the amount bought held constant. In effect it is a measure of the slope of the ENGEL CURVE. At its simplest, it is measured by the percentage change in demand divided by percentage change in income. If the resulting coefficient is less than 1 then income elasticity is inelastic (i.e. demand is not responsive to income changes), if it is greater than one then the income elasticity is elastic. For oil and gas, CONSERVATION is a major determinant of the income elasticity of demand over a period of time. *See also* ELASTICITY.

incoterms. A set of international rules that are used to interpret many of the terms used in foreign trade contracts. It is produced by the International Chamber of Commerce.

incremental barrel (marginal barrel). The last barrel produced. Its significance is that in economic theory its cost of production in a competitive market determines the FLOOR PRICE of oil.

incremental capital–output ratio (ICOR). A measure of the additional investment required to produce an additional unit of output. Since an ICOR is usually a ratio of monetary values, its level is determined by technology (e.g., the engineering relationship between physical inputs and outputs) and the prices of inputs and outputs. In capital-intensive industries, such as those involving oil and gas, the ICOR plays an important role in planning.

incremental costs (marginal costs). The cost of producing the last unit of output.

Independent Petroleum Association of America. *See* IPAA.

independents. A term that began to be widely used in the 1950s to describe the non-MAJORS (i.e. the relatively small oil companies) commonly based in the USA.

index number. A single number which gives the value of a number (normally in a time series) expressed as a percentage of a number in the chosen base period. Index numbers can measure price, volume or value. Their prime function is to help in the identification of changes in a variable. The calculation and interpretation of index numbers can be both complex and controversial.

indicated reserves. Oil and gas RESERVES that are known to be capable of production, but lack the precision of proven reserves. *Compare* INFERRED RESERVES.

indirect taxation. A TAX not levied directly on a specific person or institution. Instead it is a tax levied on expenditure by being directed at specific goods. It then rests with the supplier to determine how much of the tax is paid by the consumer (i.e. where the TAX INCIDENCE falls) subject to the constraints of the market. If a government introduces indirect taxes to raise revenue then it will impose the tax on a good (e.g., motor spirit) with an inelastic demand (*see* ELASTICITY). In most countries, motor spirit attracts high levels of indirect taxation, although the revenue that accrues should never be equated with the revenue of the oil-producing government (*see* GOVERNMENT TAKE). Sometimes, however, a government may use indirect taxation to deter the consumption of a good, an example being a tax on heavy fuel oil in order to protect an indigenous coal industry. To be effective, in this case, the good must have an elastic demand.

indivisibilities. Indivisibilities arise when a factor input or consumer good is only available in a discrete unit of a minimum size larger than that at which a producer or consumer would choose most efficiently to use it, were continuously variable quantities available. As an example, oil tankers are available in minimum sizes and the AVERAGE COST per barrel of transporting a cargo in a half-full tanker is normally greater than the average cost of the larger cargo which would fill the tanker.

Industry Supply Advisory Group (ISAG). A group of oil companies that supplies information and advice to the INTERNATIONAL ENERGY AGENCY.

inert. Describing a substance that does not react chemically. For example, an inert gas is one that does not support combustion and does not burn.

inert gas system. *See* TANKER TYPES.

inferior good. *See* NORMAL GOOD.

inferred reserves. RESERVES with an even greater degree of uncertainty over their producibility than INDICATED RESERVES.

inflation. A rise in the general level of prices as measured by a representative 'basket' of goods. It is important not to confuse inflation with changes in relative prices. A rise in oil prices is not in itself inflation (although such a rise may aggravate an inflationary situation due to its knock-on effects) since oil is only one of the items in the basket of goods, and at the time when the price of oil rises the price of other items in the basket may fall.

Inflation is important because, among many other things, it draws a distinction between NOMINAL and REAL values. For example, if the price of oil rises in one year by 10 percent and inflation is running at 10 percent the real price of the oil (i.e. what it will actually buy in terms of other goods and services) is unchanged. If, however, the price of oil has risen by 10 percent and inflation is running at 5 percent then the real price of oil has risen by 5 percent (i.e. the oil will now buy 5 percent more goods and services than before its price rose).

There is little doubt that the oil price rises of the 1970s (*see* CRUDE OIL PRICES; FIRST OIL SHOCK; SECOND OIL SHOCK) helped to generate inflation by virtue of their effects on costs. However, studies suggest that their contribution to world inflation was less than many imagined.

inherent surplus. A concept, prevalent in the 1950s and 1960s, which implies that because it was assumed that the oil industry was an industry of decreasing costs — an incorrect assumption (*see* CRUDE OIL COSTS) — an automatic mechanism is created whereby output continues to expand irrespective of demand. This forces down price and forces out of business higher-cost producers, in theory leading to the creation of a NATURAL MONOPOLY. It is possible to develop a modern version which argues that while the price of oil remains above long-run marginal cost (*see* INCREMENTAL COST) there is an incentive to develop excess producing capacity which, because of low short-run marginal costs, will be produced thereby leading to excess supply in relation to demand. This forces the price of oil down.

inhibitor. A substance added to petroleum products in order to prevent or retard undesirable changes in the quality of the product or the condition of the storage equipment.

initial boiling point. The temperature at which a petroleum fraction begins to distill in a test situation. *See also* BOILING RANGE; REFINING, PRIMARY PROCESSING.

injection well. A well drilled in order to inject a substance (e.g., water or gas) into a reservoir for the purpose of producing oil by secondary recovery methods. *See also* PRODUCTION, SECONDARY RECOVERY MECHANISMS.

INOC (Iraqi National Oil Company). The Iraqi national oil company created in 1964.

inorganic trade. Sales (normally on a spot basis) that contain no guarantee of further sales to the same source. Because they are in effect 'one-off' transactions, they normally involve relatively large quantities since small single sales would not justify the TRANSACTIONS COSTS involved.

in-place reserves. The total amount of oil and gas in a reservoir whether recoverable or not. *See also* RESERVES.

input–output analysis (I–O analysis). A method of assessing the structure of production by taking account of the interrelationships between the producing sectors in an economy. It has been described by its originator and prime exponent, Nobel prize winner Wassily Leontief as '... an adaptation of the neoclassical theory of general equilibrium to the empirical study of the quantitative interdependence between interrelated economic activities'.

I–O analysis represents the structure of an economy in algebraic form via a set of simultaneous linear equations. These equations show how the output of each sector (or industry) is distributed between 'intermediate output' (i.e. sold to other sectors to be used up in their current production) and 'final output' (i.e. sold to final demanders for private or public consumption or for exports or to add to the capital stock). The coefficients of these equations represent the specific characteristics of the economic structure and are usually derived from an 'input–output table'. This depicts, in the form of a matrix, the flows of transactions between all the sectors over a given period of time, normally one year. The rows of the table show the distribution of each sector's output to the other sectors and to final demand. The columns show how each sector obtains from the other sectors the inputs that it requires for its own production. The sectoral disaggregation in the table can be considerable and is determined by data availability and the purposes for which the table is being prepared. Between 30 and 100 sectors are common. I–O tables are available for many countries. However, their preparation demands a large amount of data and they are expensive to produce, thus they are usually only prepared at intervals of three to five years.

The matrix of I–O coefficients is usually derived after making the simplification that the inputs to and outputs from each sector are linearly related in fixed proportions. For each sector, the inputs into that sector are divided by the sector's output to yield a column vector of coefficients relating inputs to output. The equation system can then be written in matrix notation as:

$$x = Ax + f$$

where x is the column vector of sectoral outputs, A is the matrix of I–O coefficients and f is a column vector of sectoral final demands. These equations can be rewritten as:

$$(I - A)x = f$$

where I is an identity matrix, and solved:

$$x = (I - A)^{-1}f$$

It is now possible to ask questions of the system, for example, by solving the set of simultaneous equations to find how any given set of final demands will require a consistent set of total outputs x from all sectors in the economy. Because of the simultaneity, the solution takes account not only of the direct requirements of each sector necessary to increase final output by a given amount, but also of the indirect requirements. For example, an increase in the demand for electricity requires not only inputs of oil, but also extra inputs (including more electricity and oil) in order to enable that oil to be produced, etc.

The great strength of I–O analysis lies in its ability to capture these complex interrelationships through a set of simultaneous linear equations which can be solved easily with the aid of a computer. To the extent that the underlying economic relationships are significantly non-linear, a price has to be paid in terms of the precision of the results.

I–O analysis has been widely used in national and regional planning where it has been used not only to work out projected levels of gross output, but also to determine impacts on employment and on incomes. It has been used to estimate the energy requirements of different development strategies in both developed and developing countries. In the case of energy, outputs and inputs have been measured in physically commensurable units, as well as in money values. I–O analysis can also be used to estimate the impact on the prices of the outputs of all sectors of the economy as the result of changing one or more energy prices. In addition it can be used to account for the output of pollutants by different sectors and for inputs required to reduce emissions of pollutants, thus enabling different pollution control strategies to be examined.

input–output tables. *See* INPUT–OUTPUT ANALYSIS.

***in situ* combustion** (fireflood extraction). A type of enhanced oil recovery involving the generation of controlled burning of some of the oil in the reservoir. *See also* PRODUCTION, ENHANCED OIL RECOVERY.

installment system. The option for smaller companies to pay bonuses in installments in a situation of cash bonus bidding for exploration licences. It is intended to make such auctions as competitive as possible. *See also* AUCTION SYSTEM OF LICENSING; BONUS BIDDING.

intangible assets. ASSETS whose valuation is highly subjective, such as the goodwill attracted by a company or the ownership of a patent.

Intascale (IS). A schedule of normal tanker freight rates first published in 1962 by the International Tanker Nominal Freight Scale Association Ltd, London. Reference rates (flat rates) designed to cover operating and capital costs of a 19 500 dwt vessel were for the first time calculated for voyages between all the principal world ports. Rates were quoted both in pounds sterling and in US dollars. Actual contract rates were negotiated in terms of percentages above or below the flat rate. The Intascale procedure was considered simpler than working in terms of pounds or dollars per ton of cargo. Confidence in Intascale ended with the devaluation of sterling in November 1967, and it was replaced by WORLDSCALE in 1969.

integration. A term used in the context of discussions about the structure of a particular industry. In the international oil industry aspects of integration have always been of central importance to the way in which the industry has behaved. Three types of integration can be identified.

vertical integration. This refers to the joining together by merger, acquisition or starting a new AFFILIATE at a different stage of the industry (e.g., an oil-producing company developing transportation and refining interests). Thus a fully vertically integrated oil company is one which produces crude oil, refines it and distributes products.

horizontal integration. This refers to the joining together or the creation of firms or affiliates in the same stage of the industry (e.g., when the OIL OPERATING COMPANY is owned by a number of oil companies or when two oil companies link activities for the distribution of their products).

conglomerate integration. This refers to the expansion of a company into areas outside its traditional spheres of activity. For example, during the 1970s many oil companies bought into different activities including hotels and supermarkets. This form of integration increases the economic power of the company in the economy in general rather than in a specific industry.

The significance of both vertical and horizontal integration (and each has its own specific implications) is that they represent a concentration of economic power within the industry. The greater the extent of the integration the greater the economic power which accrues to the integrated activities. There are benefits to the consumer that can arise from the existence of integration. In economics (both generally and specifically with respect to oil) there is considerable debate as to whether the gains to the consumer are offset by the reduced competition which the resulting concentration of economic power can lead to. *See also* CONGLOMERATE INTEGRATION; HORIZONTAL INTEGRATION; VERTICAL INTEGRATION.

interest. In economy theory the INCOME earned by CAPITAL. *See also* RATE OF INTEREST.

interest cover. A financial ratio that reflects the number of times the fixed-interest payments of a company to service its loan capital (*see* DEBENTURES) are exceeded by earnings. In effect it shows the decline in earnings that would have to occur before interest owed could not be paid out of current income. It is a useful guide to the solvency of a company.

interlocking directors. A situation in which directors from a single source sit on the boards of different companies. This generally arises when single sources (e.g., banks or pension funds) have the right to appoint directors to a large number of companies. This right of appointment may arise because of EQUITY shareholding or by virtue of the provision of loaned funds. A result of this is that companies that may appear to be separate entities are in fact controlled, or at least influenced, by a single company. Thus an industry that may be viewed as 'competitive' when conventional CONCENTRATION RATIOS are applied (i.e. showing a large number of independent companies with no one dominating) may in reality be much less competitive. This issue was of some importance in the USA during the late 1970s when there was debate concerning the competitiveness of the domestic oil industry. At that time, there was strong pressure to force the oil companies to divest themselves of affiliates (i.e. break up the VERTICAL INTEGRATION of the industry).

intermediate good. The output of one production process, which is subsequently used as an input into another production process. Such a commodity may be used further DOWNSTREAM by the same company, or it may be sold to another company within the industry or to another industry. Finished steel products are examples of intermediate goods since they are used by several other industries such as vehicle manufacture, shipbuilding and engineering. The term intermediate good also includes oil and gas which are used as inputs into the production processes of all industries.

Some commodities may be both an intermediate good and a final good (i.e. one that is consumed immediately without further processing). Gas is an example of such a commodity; it is used as an intermediate good in the generation of electricity and as a final good for residential heating.

intermediate string. *See* DRILLING.

internal capital rationing. The allocation of budgets to departments or AFFILIATES within a company. The budgets may not be exceeded even if the department or affiliate has profitable opportunities for investment. The result is that projects must be ranked according to some criteria of profitability and then executed until the budget constraint is reached. *Compare* EXTERNAL CAPITAL RATIONING.

internal combustion engine. An engine in which the heat produced by ignition of a fuel is converted into mechanical energy. Normally it refers to an engine powered by motor spirit or diesel oil.

internally generated funds. Sources of finance for investment that come from within a company as opposed to financing by external borrowing. Normally internally generated funds are drawn from the RETAINED EARNINGS of a company. They are not costless since there is the OPPORTUNITY COST of such finance. Thus the funds used for the project could have been invested elsewhere to provide an alternative profit or income. During the 1950s and 1960s the MAJORS tended to finance the bulk of their investment by internally generated funds which was reflected in their GEARING RATIOS and INTEREST COVER.

internal rate of return (IRR, investor's yield). The rate of interest which when used to discount a cash flow reduces the NET PRESENT VALUE to zero (*see* DISCOUNTED CASH FLOW). It effectively provides a break-even rate of return on an investment which can then be compared with the cost of capital for the investment. If the IRR exceeds the cost of capital the project is profitable. There are problems with its use, and the net present value is generally preferred. In economic theory IRR is the marginal efficiency of capital.

International Association of Independent Tanker Owners. *See* INTERTANKO.

International Bank for Reconstruction and Development. The full title for the World Bank.

International Chamber of Commerce. *See* INCOTERMS.

international company. *See* MULTINATIONAL.

International Energy Agency (IEA). After the FIRST OIL SHOCK, for a time oil-consuming countries felt the need for a counterpart to the Organization of Petroleum Exporting Countries (*see* OPEC). In the event, the nearest approach was the IEA, which was established in November 1974. IEA's 21 members are Australia, Austria, Belgium, Canada, Denmark, Germany, Greece, Ireland, Italy, Japan, Luxembourg, The Netherlands, New Zealand, Norway, Portugal, Spain, Sweden, Switzerland, Turkey, UK and USA; apart from Iceland, Finland and France all the members of the Organization for Economic Cooperation and Development (*see* OECD).

The IEA does not become involved in negotiations with oil-producing countries; its role in policy formation is purely advisory. It has, however, formulated an emergency plan to deal with any disruption in oil supplies. It has developed an information system on the oil market which has greatly improved knowledge of market events. In addition it has a considerable output of statistics and forecasts in most areas of the energy market, and it has had some influence on the energy policies of member governments. In its annual review — *Energy Policies and Programmes of IEA Countries* — it discusses recent trends in energy as well as the outlook and comments on the energy policy of each of its member governments. It also publishes surveys of the coal, natural gas and electricity markets, and reviews of research, development and demonstration programmes in IEA countries.

The basic objectives of the IEA are as follows: (a) cooperation among IEA participating countries to reduce excessive dependence on oil through energy conservation, development of alternative energy sources and energy research and development; (b) an information system on the international oil market as well as consultation with oil companies; (c) cooperation with oil-producing and other oil-consuming countries with a view to developing a stable international energy trade, as well as the rational management and use of world energy resources in the interest of all countries; (d) a plan to prepare participating countries against the risk of a major disruption of oil supplies and to share available oil in the event of an emergency.

international liquidity. The amount of money needed to finance international trade. The point of such money is that it must be generally acceptable. Any shortage of liquidity (i.e. lack of acceptable currency) could restrict the amount of international trade. Immediately following the FIRST OIL SHOCK, the RECYCLING problem was in essence the fear that OPEC would not recirculate their oil revenues thereby causing a shortage of international liquidity.

International Monetary Fund (IMF). A

fund that was set up by the Bretton Woods agreement in 1944 to oversee the international financial system.

International Petroleum Commission. *See* ANGLO-AMERICAN OIL AGREEMENT.

International Tanker Nominal Freight Scale Association Ltd. *See* INTASCALE.

international trade. The movement of goods and services between countries. The only difference between international trade and interregional trade is the existence of frontiers which give governments the power to intervene in the flow either directly by simple prohibition or by QUOTAS or indirectly by means of TARIFFS or regulations on handling, quality or technical specification. International trade occurs because of differences in costs of production (based upon COMPARATIVE ADVANTAGE) or because it increases welfare (*see* WELFARE ECONOMICS) by offering a wider range of goods and services. Since the 1950s crude oil has been the largest item in international trade in volume terms.

interruptible gas. Gas supplies (normally to large consumers) on the basis that at times of PEAK LOAD demand supplies may be interrupted to prevent loss to other consumers. This requires the consumer to have available a back-up source of gas, or an alternative source of power. Such supply contracts attract lower tariffs to induce the consumer to invest in the back-up system.

Interstate Oil Compact Commission. *See* PRORATIONING.

interstitial water. *See* CONNATE WATER.

Intertanko (International Association of Independent Tanker Owners). A trade association that was founded in 1971 to promote the interests of independent owners of tankers.

inventory. A term used in the USA for STOCKS.

investment. In terms of economics, expenditure on real CAPITAL. At the risk of gross oversimplification, its significance in economic theory is that investment can only be undertaken by sacrificing current consumption, but the act of investment provides future consumption. In more general usage, however, investment has come to mean the purchase of any ASSET or the undertaking of any commitment that involves an initial sacrifice followed later by benefits. Investment can be measured gross, which includes investment to replace used up capital (i.e. DEPRECIATION), or net, which measures only new additions to the capital stock (i.e. gross investment less depreciation). *See also* INVESTMENT APPRAISAL.

investment appraisal (project appraisal). The evaluation of the worth of prospective INVESTMENT projects. The evaluation may be of a single project (i.e. an ACCEPT OR REJECT DECISION) or of several projects from which one will be chosen (i.e. MUTUALLY EXCLUSIVE INVESTMENTS) in a situation of CAPITAL RATIONING. The techniques of appraising investment projects and allocating funds to them are sometimes referred to as CAPITAL BUDGETING.

The approach to investment appraisal most favoured by economists is DISCOUNTED CASH FLOW. There are, however, other methods which are often grouped under the heading TRADITIONAL METHODS OF INVESTMENT APPRAISAL. One such method is payback period which is a calculation of the period of time over which it is expected that the initial capital investment in a project will be recouped from profits (including DEPRECIATION). The shorter the payback period, the better the project is assumed to be. The main criticisms of this approach are that it ignores cash flows after the payback period and also it takes no account of the time pattern of cash flows within the payback period. Another method is ACCOUNTING RATE OF RETURN which is calculated from the ratio of average annual expected profit (net of depreciation and taxation) to either initial capital invested in a project or its BOOK VALUE. The return is usually expressed as the percentage of profit on capital. The principal disadvantage of this method is that it measures rate of return purely in terms of the quantities of profit and capital, making no allowance for the time value of money (*see* TIME PREFERENCE).

investor's yield. *See* INTERNAL RATE OF RETURN.

invisible hand. A concept in which the market mechanism automatically, by means of the PRICE MECHANISM, solves the economic problems of what to produce, how to produce it and for whom to produce it. The term was first introduced in Adam Smith's *The Wealth of Nations* published in 1776.

invisible trade. The import and export of services plus the flows of remittances, interest on loans and dividends on investments. *See also* BALANCE OF PAYMENTS.

I–O analysis. *See* INPUT–OUTPUT ANALYSIS.

IPAA (International Petroleum Association of America). An organization that provides important data on the US oil market.

i-paraffins. *See* ISO-PARAFFINS.

IPC (Iraq Petroleum Company). The successor to the Turkish Petroleum Company which had the concession for virtually all of Iraq with its affiliate company the Basrah Oil Company. The company, renamed IPC in 1925, was owned mainly by BP, SHELL, CFP, EXXON, MOBIL and the remaining 5 percent by Gulbenkian. In 1972, it was nationalized by the government of Iraq, although in 1961 LAW 80 removed all but the IPC's producing fields in terms of the acreage.

IPSA (Iraqi pipeline across Saudi Arabia). A pipeline built after the outbreak of the Iran–Iraq war.

IPSA I. The SPUR LINE which links IPSA into PETROLINE.

IPSA II. The pipeline built after IPSA I that runs parallel to PETROLINE.

Iranian Consortium. The group of oil companies that took over oil operations following the IRANIAN NATIONALIZATION and the overthrow of the government of Dr Mossadegh. The group consisted of BP (40 percent), Shell (14 percent), Mobil (7 percent), Exxon (7 percent), Texaco (7 percent), Gulf (7 percent), Socal (7 percent), CFP (6 percent) and IRICON (5 percent).

Iranian nationalization. The NATIONALIZATION in 1951 undertaken by the Iranian government under Dr Mossadegh. The MAJORS were able to use their market power to prevent Iran selling much of their potential production, and eventually the nationalization was thwarted when the US and UK intelligence services masterminded a coup which overthrew Mossadegh and replaced the Pahlavis in power. Dr Mossadegh was sentenced to death, but the sentence was later commuted to life imprisonment. The nationalization encouraged oil companies to be more flexible in the negotiations with governments. A further consequence was that the fate of Dr Mossadegh acted as a major deterrent to other governments until the early 1970s. *See also* IRANIAN CONSORTIUM.

Iraqi National Oil Company. *See* INOC.

Iraq Petroleum Company. *See* IPC.

IRICON. A group of nine US INDEPENDENTS which comprised 5 percent of the IRANIAN CONSORTIUM in 1955. It was effectively the first, but not the last, time such companies were able to get into the Middle East.

IRR. *See* INTERNAL RATE OF RETURN.

IS. *See* INTASCALE.

ISAG. *See* INDUSTRY SUPPLY ADVISORY GROUP.

isobutane. An ISOMER of BUTANE that is used to produce ISOOCTANE. *See also* CHEMISTRY.

isobutene (isobutylene). An ALKENE derived from the CRACKING of NAPHTHA. It is used to manufacture polyisobutylene and some butyl rubbers.

isobutylene. *See* ISOBUTENE.

isomer. A compound that has the same chemical composition as another, but with a different structural arrangement of the atoms within the molecule. This results in a distinctly different set of physical and chemical properties. *See also* CHEMISTRY.

isomerization. A refining process that improves the OCTANE RATING of motor spirit components. *See also* MOTOR SPIRIT UPGRADING.

isooctane. A reference fuel in the OCTANE

RATING test. It is also a component of high-quality motor and aviation fuel. *See also* QUALITY, AVIATION SPIRIT; QUALITY, MOTOR SPIRIT.

isopach map. A map used in geology that shows the thickness of a layer of sediment or sedimentary rock.

iso-paraffins (i-paraffins, branched-chain paraffins). A subdivision of the PARAFFIN series of hydrocarbons, according to the arrangement of the carbon atoms within the molecule; iso- denotes a branched-chain arrangement, as distinct from normal or straight-chain. *See also* CHEMISTRY.

issue. The floating of new shares in a company. It is used as a means to raise CAPITAL.

IUPAC. A naming system for petrochemicals developed by the International Union of Pure and Applied Chemistry. *See also* CHEMISTRY.

J

J. *See* JOULE.

jacket platform. A PLATFORM built entirely of steel.

jack-up rig. Mobile offshore RIG WITH LEGS THAT CAN BE RAISED FROM THE SEA BED IN ORDER TO RELOCATE THE RIG.

JET A. Aviation turbine kerosine (*see* FINISHED PRODUCTS, AVIATION FUELS).

Jet B. Aviation turbine gasoline (*see* FINISHED PRODUCTS, AVIATION FUELS).

jet bit. A drill BIT that has been modified to use a hydraulic jet to increase the drilling rate. *See also* DRILLING.

jet engine. An engine that converts fuel (*see* FINISHED PRODUCTS, AVIATION FUELS) and air into a fast-moving stream of hot gases which provides propulsion.

jet fuel. *See* FINISHED PRODUCTS, AVIATION FUELS.

jets. Nozzles (usually three) in the drill BIT that provide the DRILLING FLUID during drilling.

JOA. *See* JOINT OPERATING AGREEMENT.

jobber (resale agent). A middleman. The term developed in the context of the stock exchange, but it can also be used in the oil industry. A jobber may buy and sell products or may simply act as an 'introducer' of buyers and sellers. In the former case he earns a jobber margin on the difference between buying and selling prices; in the latter case he earns a commission on any subsequent sale.

jobber margin. *See* JOBBER.

joint bidding agreement. *See* JOINT VENTURES.

joint operating agreement (JOA). An agreement between the partners of a JOINT VENTURE. It lays down the partners' responsibilities.

joint product. *See* BY-PRODUCT.

joint production. A situation in which different products are produced from the same process. The most obvious example is oil refining. Its significance in economics is that it makes it very difficult to attribute costs to a specific product, which in turn makes it difficult to price the products.

joint ventures. A consortium that enters an agreement to operate in some licence area. The agreement determines the PRODUCTION SHARE of each company. In the UK, the preliminary agreement is known as a joint bidding agreement or an area of mutual interest (AMI) agreement. Later, the joint operating agreement (JOA) is negotiated which is the basic agreement between the companies in the joint venture governing their relationship under the terms of the licence. In the international oil industry, joint ventures have been the norm for OIL OPERATING COMPANIES. In the 1950s this was the result of two factors: (a) the intervention of the home governments of the MAJORS when the original concessions (creating the operating companies) were signed; (b) the need to regulate crude supply which meant that crude long companies should be partnered with CRUDE SHORT COMPANIES to minimize the size of the ARM'S LENGTH market. The resulting HORIZONTAL INTEGRATION of crude oil production was one of the key constraints on excess supply. During the late 1950s the term joint ventures began to be applied also to government–company agreements in which the governments had an EQUITY share in the operating company, although the foreign company took on its sole account the initial risk of exploration. During the late 1960s this type of arrangement tended to be superceded by the SERVICE CONTRACT arrangement.

joule (J). The basic unit of energy in the Système International. It is defined as the energy conveyed by one watt or power for one second. *See* APPENDIX.

JP4. Aviation turbine gasoline (*see* FINISHED PRODUCTS, AVIATION FUELS).

JP5. Aviation carrier turbine fuel (*see* FINISHED PRODUCTS, AVIATION FUELS).

jug hustler. A member of a seismic crew (*see* SEISMIC EXPLORATION) who operates the headphones used to pick up sound waves reflected from subsurface strata.

junk basket. A drilling tool used to recover equipment lost down a well during drilling. *See also* FISHING.

junked. Describing a well in which equipment has been lost and cannot be retrieved economically. Junking involves PLUGGING the well and then abandoning it.

K

kelly. The topmost section of the length of pipe used in DRILLING that fits into the DRILLING TABLE.

kerogen. The main organic component of OIL SHALES. On distillation at high temperature it yields oil and gas.

kerosene. *See* KEROSINE.

kerosine (burning oil, kerosene, paraffin (UK)). A MIDDLE DISTILLATE product used mainly for central heating and domestic heating stoves. *See* FINISHED PRODUCTS, KEROSINE.

key seating. The wearing of a groove in one side of a bore of a DEVIATION WELL by a DRILLING STRING. This can cause the DRILLING STRING to stick.

kick. A situation when the formation pressure in a well exceeds the pressure of the mud column causing a leakage of formation fluid into the well bore.

killing a well. Preventing the tendency of a well to flow naturally by filling the well bore with fluid of a suitable specific gravity, normally high-density DRILLING FLUID.

knock. A tendency to ignite spontaneously. Knocking is considered undesirable in the efficient combustion of MOTOR SPIRIT. *See also* QUALITY, MOTOR SPIRIT.

KOC (Kuwait Oil Company). An oil company created in 1934 as a JOINT VENTURE between BP amd GULF (using their modern names) and was solely responsible for the development of Kuwaiti oil. In 1975 the Kuwaiti government took over ownership of the company.

Kuwait Oil Company. *See* KOC.

L

lagged explanatory variables. *See* LAGS.

lagging indicators. *See* LEADING INDICATORS.

lags. It is rarely the case in reality that the full effect of an event is felt immediately. It is important when specifying an ECONOMETRIC model that the time lapse between the movement in one or more of the independent variables and the response of the dependent variable is incorporated into the model specification. The way in which this problem is normally tackled is by the introduction of lagged explanatory variables into the model. The assumption that changes take place over time forms the basis of the distributed lag model in which a series of lagged explanatory variables account for the time adjustment process. The introduction of a lag structure may cause problems in the estimation of the model, since a lengthy lag structure uses up degrees of freedom.

In attempts to forecast oil market movements, it is the lags which tend to be either forgotten or underestimated by many observers of the industry. Several examples serve to illustrate this.

(a) A change in the output of crude oil in the Gulf takes about six to eight weeks before it can translates itself into changes in the physical availability of products in north-western Europe; this ignores any possible STOCK changes which may be prompted by the change in output.

(b) A change in the price of crude oil may lead to a change in the behaviour of consumers either in their rate of usage of oil-using equipment or in the thermal efficiency of the oil-using equipment. In the case of the latter the lags before the full effect are felt can be very long indeed. For example, it can take several years to carry out the research and development needed to redesign the equipment, some time to retool the production lines and a very long time (depending on the expected life of the appliance) to turn over the particular stock of oil-using equipment. For example, much of the decline in the use of motor spirit that was observed in the early 1980s was the result of the price rises in 1973.

(c) A final example is the effect of a change in oil prices on the exploration for and development of new crude-producing capacity. The lags here can range from five to ten years.

Many believe that it is because lags are getting longer that forecasting in the oil and gas industries is becoming harder and more uncertain.

Lahee well classification. A long-established system of classifying wells when drilling is started and when it is completed either successfully or unsuccessfully. The system was developed for the USA by F.H. Lahee.

land. One of the economist's factors of production (*see* PRODUCTION FUNCTION). It includes not only the land surface, but also the NATURAL RESOURCES contained on or under the ground, or sea, such as petroleum. In practice, however, economics increasingly distinguishes between different types of natural resources.

landed cost (landed price). The actual cost of crude oil to a refiner taking into account all the costs incurrred to the refinery gate. It is equivalent to the CIF cost of crude oil plus the costs (including storage) involved in moving the oil from the tanker to the refinery.

landed price. *See* LANDED COST.

large-range tanker (LR tankers). The AFRA definition of LR tankers has varied over time as the average size of the world tanker fleet has changed. At present it covers vessels between 45 000 and 319 999 dwt. LR1 tankers are between 45 000 and 79 999, LR2 tankers are between 80 000 and 159 999, and LR3 tankers (also called VLCCs) are between 160 000 and 319 999 dwt.

last-in first-out. *See* LIFO.

latch on. To attach ELEVATORS to a section of DRILL PIPE. *See also* DRILLING.

Law 80. The law passed by Iraq in October 1961 which expropriated from the Iraq Petroleum Company all its non-producing acreage, leaving it only with producing fields. This was the first formal shot in a battle between the Iraqi government and the oil-producing companies that lasted into the early 1970s. It may also have encouraged other companies in the Middle East to relinquish acreage.

Law of Capture. The legal basis in the USA for the ownership of oil. In the original Supreme Court ruling, oil was likened to a wild animal, and therefore ownership accrued to whosoever's property the oil was on. However, because oil is a liquid (*see* CRUDE OIL CHARACTERISTICS) and can flow, irrespective of land boundaries, this means that so long as marginal costs (*see* INCREMENTAL COSTS) are met, it is worth pumping one's own oil (and that of one's neighbour) as rapidly as possible. The inevitable result is gross overproduction with probable damage to the fields (*see* PRODUCTION). To get round the consequences of the Law of Capture, in the 1930s, it was necessary to introduce PRORATIONING, which effectively allocated QUOTAS to producers.

law of diminishing returns. *See* DIMINISHING RETURNS.

law of the sea. Because of the absence of an obvious court jurisdiction the law of the sea is usually agreed in international conference. Of particular importance is the limit that can be claimed as territorial water, being an issue of crucial importance for offshore oil and gas activities.

lay barge. A BARGE specifically designed to lay underwater pipes.

lay days. The time allowed by charter for the loading or discharging of cargo.

LDF. *See* LIGHT DISTILLATE FEEDSTOCK.

lead alkyls. Organic compounds containing lead that are added to motor spirit to increase the OCTANE RATING, thereby improving the ANTI-KNOCK RATING of the fuel. The most commonly used lead alkyls are tetraethyl lead (TEL) and tetramethyl lead (TML). *See also* QUALITY, MOTOR SPIRIT.

lead-free motor spirit. Since the 1920s LEAD ALKYLS have been added to motor spirit to improve their performance. Their addition followed the development of more powerful internal combustion engines with higher compression ratios. However, the burning of leaded motor spirit results in the emission of metallic lead and lead oxides into the atmosphere which can cause health problems. During the 1960s as part of a growing environmental awareness, pressure was brought on governments to try to remove lead from motor spirit. The result was the spread of lead-free motor spirit which is more expensive than leaded motor spirit since more costly additives (e.g., MTBE) must be used to maintain the OCTANE RATING.

leading indicators. In a situation where there is a relationship between variables it is possible to predict the path of one variable by reference to what has happened to the other variable. A leading indicator is a variable that changes before the forecasted variable. For example, a change in tanker rates is reflected eventually by a change in oil supply and thus price. A lagging indicator changes after a change in the forecast variable and can therefore provide substantiation if the forecast variable is not easily measured. In this context, a coincident indicator is a variable which changes almost simultaneously with the forecast variable.

lead response. The increase in the octane number of motor spirit or aviation spirit as a result of the addition of a specific quantity of LEAD ALKYLS. *See also* QUALITY, MOTOR SPIRIT.

lead times. The period of time either from when a project was first conceived or from when first investment expenditures were made until the receipt of the first revenues. Capital projects, especially in the supply of oil and gas, often involve long periods of capital spending before any revenues are received. For example, an oil- or gas-field goes through several phases — EXPLORATION, APPRAISAL and DEVELOPMENT — before any oil or gas can be produced. Similarly, a major pipeline takes several years to plan and construct before any oil or gas can flow. Such projects are said to be front end-loaded, because of the heavy capital expenditures in the early years. It has been suggested that lead times in the oil and gas industry are getting longer which adds significantly to the problems of FORECASTING.

leak-off test. The application of pressure to the formation below the CASING seat to test the strength of CEMENTING and to determine the FRACTURE PRESSURE in the permeable zone just below the casing seat. *See also* DRILLING.

lean gas. A fuel gas with a low CALORIFIC VALUE.

LEAP. A model of the role of energy in an economy specially designed for use in THIRD WORLD countries.

lease commitments. A company's contracted hiring of equipment ranging from rigs to tankers. Since payment for the hire is fixed for a specific period the commitment effectively represents an addition to the company's debt.

lease sale. *See* BONUS BIDDING.

leverage ratio. The US equivalent of GEARING RATIO.

LIBOR (London Interbank Offer Rate). An average of the interest rates at which several major banks lend between themselves in the London money market. There are different LIBORs for different terms (e.g., overnight, one month, three months, etc.). LIBOR is a particularly important reference rate for short-term loans and roll-over syndicated medium-term credits. A loan agreement usually specifies a spread (premium) to be charged over LIBOR (traditionally three- and six-month LIBOR) and a list of reference banks to be used in calculating the LIBOR.

Libyan Producers Agreement. During the negotiations/bargaining over the Libyan POSTED PRICE in 1970 (*see* TRIPOLI AGREEMENT), the Libyan government successfully forced agreement by picking off the smaller and more vulnerable oil companies. To try to prevent a repeat, the oil companies operating in Libya signed the Libya Producers Agreement on 15 January 1971. It was agreed that if any producer faced threats of production cutbacks by Libya, the others would provide alternative supplies at TAX-PAID COST. The agreement, which was secret, required special dispensation from the US Justice Department to avoid problems arising from ANTI-TRUST LEGISLATION. The agreement failed to work, and its details were revealed three years later during Senate hearings.

licensing round. The process whereby the government invites bids for exploration acreage. It is used commonly in the context of the UK North Sea.

licensing systems. Where potential hydrocarbon reserves are publicly owned, governments, in order to regulate exploitation, issue licences (or leases) for tracts of land (or areas of the sea bed), facilitating exploration, development and extraction of resources. There are two broad systems of licence allocation.

(a) The DISCRETIONARY SYSTEM OF LICENSING is employed in most parts of the world. In the discretionary method applicants are assessed according to criteria formulated by the licensing authority — generally the central or regional government — and may comprise any condition the authority may see fit to include. The principle behind the discretionary system is that the government has the means to control the industry through the conditions and criteria of licence award and through its inspection of company work programmes. For example, a government may choose to alter an oil-field production profile if the government considers the market-determined rate of extraction is inconsistent with the socially optimum DEPLETION rate.

(b) The AUCTION SYSTEM OF LICENSING is most commonly used in the USA in cases where PROPERTY RIGHTS are not owned by the state or federal government. It involves companies entering some form of competitive bidding for the licence. The principle behind the auction system is that if it is a truly competitive auction the licence is awarded to the most efficient (i.e least cost) producer, and the owner captures all the ECONOMIC RENT.

LIFO (last-in first-out). An accounting technique for the valuation of STOCKS. In periods of rising prices such as the SECOND OIL SHOCK, LIFO tends to reduce the profit gain due to increased valuation of stock. There is no agreed convention over the use of LIFO or FIFO by the oil companies, and there is considerable variation between companies.

lifting. The amount of crude oil (or gas) taken by a company either by virtue of an equity share in the PRODUCTION operation or by virtue of simply purchasing from the producer. Once the oil has been lifted it becomes the property of the company.

light crude. A crude oil with a high API GRAVITY that yields a high proportion of the lighter, more valuable products after refining. For example, the Algerian crude, Hassi Messaoud, is one of the lightest crudes, with a gravity of 41.5°API and a distillate yield of over 60 percent. *See also* QUALITY, CRUDE OIL.

light distillate feedstock. A NAPHTHA feedstock used in the petrochemical industry, in steam cracking (*see* REFINING, SECONDARY PROCESSING), as a source of some of the major petrochemical raw materials. *See also* PETROCHEMICALS.

light distillates. *See* FINISHED PRODUCTS, SOLVENTS.

light ends (light fractions). The low BOILING POINT products (i.e. less than 300°C) which emerge from a refining process (*see* REFINING, PRIMARY PROCESSING). Examples are MOTOR SPIRIT and AVIATION SPIRIT. During the latter part of the 1970s many REFINERY CONFIGURATIONS were changed to allow the production of a higher proportion of light ends from the crude barrel because normally they have a higher market value than the heavy ends. The heavy ends or fractions are those products with a boiling point above 350°C, an example being FUEL OIL.

lightening. The practice of partially unloading a tanker so that it becomes LIGHT-LOADED which allows it to draw sufficient water to enter port. It is usually associated with long-haul crude carried in very large tankers. The unit costs of carrying crude in large tankers and then lightening are normally lower than those of shipping the crude oil in smaller tankers which could enter port without lightening.

light fractions. *See* LIGHT ENDS.

light fuel oil. A FUEL OIL with a lower viscosity and sulphur content than HEAVY FUEL OIL.

light-loaded. Describing a tanker carrying a load below capacity. A common reason for this is to allow the vessel to operate in waters that are shallow relative to the draught of the vessel. Light loading increases the unit transport costs on an exponential basis because total fixed costs are spread over a smaller volume. *See also* LIGHTENING.

Limits to Growth. *See* CLUB OF ROME.

linear programming (LP). A technique that deals with the formulation and solution of many problems concerned with the optimization of a variable subject to certain constraints. For example, LP may be used to maximize profits subject to a production QUOTA or input constraints. The variable to be maximized or minimized must be expressed in a linear form, and there may be only a finite number of linear constraints. LP can be applied where the standard methods of calculus fail, such situations being where the constraints are set out as inequalities.

An LP problem has three parts: (a) a finite number of linear inequalities or equations in a finite number of unknown variables; (b) sign constraints on some, or possibly all, of the unknowns; and (c) a linear function to be maximized or minimized. A feasible solution will satisfy the first two parts of the problem, and there may be several such solutions. An optimal solution will satisfy all three parts and the solution is usually unique. The most commonly applied method or algorithm for solving an LP problem is the simplex method.

LP methods are frequently applied to the field of energy modelling. The LP formulation of the transportation problem is particularly suitable for adaptation to energy system modelling, since it can deal with the capacity constraints characteristic of energy systems. LP models are not, however, well suited to forecasting purposes since they are optimizing models. But because they are optimizing models they are particularly attractive for planning and policy purposes. The ease with which sensitivity analysis can be incorporated into an LP model makes it a useful tool in evaluating alternative policies.

liquefaction. *See* LNG.

liquefied natural gas. *See* LNG.

liquefied petroleum gas. *See* LPG.

liquid assets. ASSETS that can be converted to cash. Cash itself is the most liquid of assets, and other assets decrease in liquidity as they become more difficult to convert to cash. Crude oil can be viewed as a relatively liquid asset because it can be easily sold, whereas an equity share in a concession is likely to be illiquid since it will take time to find a buyer, negotiate an agreement and complete the legal formalities.

liquidation. The termination of a company. The purpose is to convert the ASSETS of the company into cash in order to pay the company's creditors or provide cash for the shareholders/owners.

liquidity. The proximity (i.e. ease of conversion) of an ASSET to cash. *See also* LIQUID ASSETS.

liquid paraffin. A product in the class of white products (*see* BLACK PRODUCTS).

LN. *See* LUMINOMETER NUMBER.

LNG (liquefied natural gas). NATURAL GAS is transported from the gas-field in two principal forms: (a) in its gaseous state through high-pressure pipelines and (b) as a liquid in special sea-going tankers. The liquefaction process takes advantage of the fact that a given weight of liquid methane occupies only 1/442 of the volume of the same weight of the gas. The process necessary to liquefy the gas is complicated and costly as it involves reducing the temperature of the gas to –161°C. Initially the gas must be purified. Liquefaction involves a number of cooling stages. In the older cascade process the gas was first cooled in water, hence coastal sites were preferred, and then a series or cascade of compressors and heat exchangers with refrigerants (e.g., propane, ethene or methane) were employed, leading eventually to liquefaction. In more recent plant, precooling is often carried out using propane, and a mixture of refrigerants is used in later stages.

The storage and transport of LNG require highly secure and well-insulated vessels which add to the investment cost (*see* LNG/LPG CARRIERS). Sophisticated control equipment is also needed to prevent an unacceptable rate of reversion to gas (i.e. 'boil off'). Many tankers utilize boil off to maintain the low temperature of the gas in the ship's main turbine in order to save on fuel costs.

A method of regasifying the liquid is needed at the importing terminal. This consists of an evaporator with heat exchangers between salt water (the most convenient coolant) and the LNG to enable gas to be drawn off at a regulated rate.

Because of the high levels of capital investment required in LNG operations it is in the interest of the producer to fix the price on a long-term contract basis. Many of the initial contracts were arranged directly between producer governments and large gas utilities in the USA and Europe. As the price of oil has fallen since 1979, however, a spot market in LNG has developed, with prices tending to reflect competitive oil prices. It is likely that the trend away from fixed contracts will continue if large supplies of gas become available from such sources as the USSR and Qatar, and come on to a relatively stable market.

During the first half of the 1970s plans were laid in many countries for a significant expansion in LNG capacity, but the trade faced many problems.

(a) There was an inherent instability in a contract which had to set a price a number of years in advance of the first delivery because of the long LEAD TIMES associated with LNG projects. Thus, when the LNG plant came onstream, the price invariably had to be renegotiated. Since the seller had sunk very large amounts of capital into the construction of the plant this meant that the relative bargaining power had moved sharply in favour of the buyer. Hence the actual delivery of the LNG was often delayed as tortuous, and frequently bitter, renegotiations were carried out.

(b) Because of the very low temperatures involved, engineering tolerances in the LNG-handling plant were much less than in other hydrocarbon-processing plants, and many plants experienced technical problems sometimes involving long periods of shut down.

(c) There was growing concern over the safety of the LNG-handling plant, especially in the USA.

All these factors have led to a reduction in the attractiveness of the LNG route for gas sales, and many earlier plans have been cancelled.

LNG/LPG carriers. Ocean-going vessels specially constructed for the carriage of gas either at very low temperatures (LNG) or under pressure (LPG). The first LNG vessel was a converted oil tanker — *Methane Pioneer* — employed to take LNG from Louisiana to Canvey Island, UK. Early specialist LNG tankers were designed either with self-supporting aluminium tanks (the conch system used in the *Methane Princess* built in 1964 for the Algeria/UK trade) or nickel tanks (e.g., the Worms system as used in the *Jules Verne*, a 25 800 m^3 vessel built for Gaz de France in 1965). More recent tankers employ stainless steel membranes or spherical aluminium tanks. LNG carriers have high capital costs (approximately $200 million for a 125 000 m^3 vessel) and are also expensive to operate. They are generally tied into long-term (up to 20 years) contracts, although this has not prevented a substantial proportion (30 percent in mid-1985) from going into lay-up in depressed market conditions.

load on top system (lot system). A system, now widely used, for handling BALLAST water previously pumped into the sea causing pollution. The ballast water is pumped into slop tanks where it separates with the water below and oil floating on top. The water is then discharged and the oil remains. The next cargo is added to the oil remains for discharge at the unloading port.

loan capital. Fixed-interest borrowed funds of a company. *See also* DEBENTURES.

loans. The borrowing of a sum of money by one person or institution from another. Loans come in many different forms and may be secured or unsecured, interest-bearing or interest-free, short-term or long-term, and redeemable or irredeemable.

loan spread. The difference between the rate of interest paid by a bank to borrow money and the rate of interest paid to the bank by a borrower. In effect it measures the bank's profit margin on its role as financial intermediary.

local exchanges. An agreement between companies to exchange products in the same national market. It arises because different companies own refineries with different REFINERY CONFIGURATIONS or have refineries located in different regions. The prime function of such exchanges is to minimize transport costs.

location, refineries. *See* REFINERY LOCATION.

locational differential. An adjustment in the price of crude oil to take account of the distance of the crude from its market. The closer that a crude oil is located to the market the more valuable it is since, other things being equal, the lower will be the LANDED COST. The actual value of the locational differential will tend to vary as circumstances change. For example, the freight advantage held by Mediterranean crudes over Gulf crudes, because of the former's proximity to Europe, increased substantially in June 1967 following the closure of the Suez Canal.

logging. The evaluation of subsurface rock formations. *See also* DRILLING.

London Agreement. An agreement reached by OPEC in London in March 1983. The main points of this agreement were a $5 cut in price of crude oil and a production-sharing agreement. Its significance was that it was the first time OPEC had in operation a moderately successful agreement to restrain production, compared with attempts in the mid-1960s and in March 1982. The London Agreement itself was soon overtaken by events, but it served to bring OPEC back into the headlines of the western media.

London Policy Group. A group composed of the oil companies whose function was to provide the terms of reference for the teams who were to negotiate oil price agreements with OPEC in 1971. *See also* CRUDE OIL PRICES.

London Tanker Brokers Association. *See* AFRA.

long-run. *See* LONG-TERM.

long-term (long-run). Describing time horizons; however, such concepts are relative (*see* SHORT-TERM). In economics the definition is quite specific; long-run describes a period when all inputs are variable. In the oil industry, long-run is usually a period of over a year, but this is rather arbitrary and not always adhered to.

long-term capacity. *See* CAPACITY.

Lorenz curve. A means by which the distribution of a variable (e.g., income) can be measured. It effectively shows what percentage of the population has what percentage of income. A complete equal distribution is shown by a 45° straight line. The further away from the 45° line is the actual plotted curve the greater is the inequality of distribution.

lost circulation. A situation that occurs during DRILLING when less DRILLING FLUID returns from the hole than was originally pumped down. It can indicate that a potentially prospective formation has been encountered.

lost time. A period in which, due to unforeseen circumstances (e.g., bad weather or mechanical failure), work must cease. It is similar to downtime, although the latter includes anticipated loss of operation due to such factors as maintenance.

lot system. *See* LOAD ON TOP SYSTEM.

low absorbers. *See* ABSORPTIVE CAPACITY.

lower-tier crude. In the context of the USA pricing regulations in the 1970s this was another term for old oil. *Compare* NEW OIL.

low-grade fuel. A fuel with a low CALORIFIC VALUE.

low-grade heat. Heat that has no economic value.

LP. *See* LINEAR PROGRAMMING.

LPG (liquefied petroleum gas). A collection of petroleum HYDROCARBONS, of which PROPANE (C_3H_8) and BUTANE (C_4H_{10}) are the most important components. The LPG gases n-butane and isobutane, propene (C_3H_6) and the butenes (C_4H_8) are generally only used as FEEDSTOCK in the chemical industry, whereas propane and butane are used widely as fuels. They are produced from (a) natural gas production, (b) crude oil production or (c) crude oil refining. In the case of 'wet' natural gas-fields, small amounts of the heavier gases (LPGs and naphtha, i.e. C_3–C_7) may be present. In this case, the LPGs are removed prior to compression by a wide variety of methods including absorption by lean or refrigerant oil followed by fractionation of the light products from the heavier liquids. Where gas is present with crude oil, LPGs can be separated together with methane and other hydrocarbons. The third source of LPGs is from crude oil refining where each conversion process yields different amounts of LPGs. The widespread adoption of visbreaking distillation techniques (*see* VISCOSITY BREAKING) in order to achieve lighter products has improved both the value and quality of LPG yield. Also CATALYTIC REFORMING or CRACKING techniques affect the production of LPGs. After extraction the LPGs are sweetened (i.e. stripped of sulphur and other impurities), dried and bottled or piped under pressure for distribution.

LPG, like natural gas is free from impurities and easily controllable. It is used in all kinds of domestic, commercial and industrial situations as a 'premium fuel'. The fact that it can be supplied in containers of various sizes makes it attractive to fuel consumers. Its simple chemical composition also makes it suitable as the feedstock for many chemical processes such as steam reforming and petrochemical production.

The market for LPG is characterized on the demand side by two dominant buyers — the USA and Japan — who between them account for over 60 percent of world LPG use. On the supply side there is a large and non-homogeneous group of surplus LPG producers, including Saudi Arabia, Canada, Venezuela and, more recently, the North Sea, South-East Asia and Mexico. In view of the by-product nature of much LPG, pricing is largely dominated by the vagaries of the oil market: when demand for motor spirit increases LPG prices tend to weaken, but when oil demand in total falls due to rising oil prices LPG prices often tend to harden. The unpredictability of prices tends to deter investment in LPG facilities, including transport, so that it does not play such a large role in world energy products as its high quality might suggest.

LR tanker. *See* LARGE-RANGE TANKER.

lub oil feedstocks. The feedstocks from which LUBRICATING OILS are manufactured. They include VACUUM RESIDUE and VACUUM GAS OIL which are obtained from the VACUUM DISTILLATION process.

lubricating oils (lubs). A specialized range of products whose main function is to reduce friction and wear between metal surfaces, in all types of engines and machinery. *See* QUALITY, LUBRICATING OILS.

lubs. *See* LUBRICATING OILS.

luminometer number (LN). A product quality specification governing the combustion characteristics of AVIATION TURBINE KEROSINE. *See also* QUALITY, AVIATION FUELS.

lump-sum freight. A fixed freight rate whose level does not alter with the amount of cargo loaded.

lump-sum taxes. A system of taxation that is aimed at reducing the disincentive impact of taxation as it may be able to increase average rates but keep marginal rates low.

Lurgi process. A process that is used to manufacture gas from coal in which a mixture of oxygen and steam under pressure is employed.

luxury good. A luxury is that which is not necessary. What is necessary is, of course, dependent upon who you are, where you are and 'what' you are.

M

macro. Used by economists, with micro, to distinguish between the sectors of the economy being considered. Macroeconomics concerns itself with the economy as a whole at the national, regional or international level. Microeconomics in contrast is concerned with individual sectors or decision units or markets within the whole economy. The economics of the oil and gas industry involve micro issues such as crude oil pricing, integration, etc. Macro issues are also involved because of implications for output and balance of payments, etc.

macroeconomics. *See* MACRO.

magnesium aluminium silicate. *See* BENTONITE.

magnetometer survey. A method of geological survey in which sedimentary basins (hence possible reserves) are located and their size estimated by measuring the magnetic properties of the underlying rocks. *See also* GEOLOGY.

main products. A relatively small range of petroleum products which constitute the bulk of refinery output and which are used as sources of energy. They include MOTOR SPIRIT, AVIATION SPIRIT, AVIATION TURBINE KEROSINE, AVIATION TURBINE GASOLINE, KEROSINE, GAS OIL, DIESEL OIL and FUEL OIL. The term is used to distinguish them from the range of special products of much smaller volume, but more heterogeneous range, which are used in non-energy applications, examples being BITUMEN, WAXES, SOLVENTS, LUBRICATING OILS, etc. *See also* FINISHED PRODUCTS.

majors. A general term used to describe the seven largest oil companies that have dominated the international oil industry during the present century. They are BP, EXXON, GULF, MOBIL, SHELL, SOCAL and TEXACO. Sometimes, because of its intimate involvement with the other seven in the Middle East, CFP is also included as a major.

making a connection. The screwing of a length of DRILL PIPE onto a DRILLING STRING thereby lengthening the string. *See also* DRILLING.

making a trip. *See* ROUND TRIP.

managed exchange rate. *See* EXCHANGE RATES.

mandatory surrender. *See* RELINQUISHMENT.

manifold. A piping arrangement that divides a stream of oil or gas into two or more streams or allows different streams to be channelled into one.

manometer. An instrument that is used to measure gas pressures or pressure differentials in a system.

manufactured gas. A gaseous fuel that is made from a wide variety of substances including coal, oil and natural gas. The carbonization of coal produces a gas with a CALORIFIC VALUE of 450–550 Btu per cubic foot that is also known as town gas. Manufactured gas based on oil or natural gas has a higher energy content of between 850 and 1100 Btu per cubic foot. Much interest has been expressed in the underground gasification of coal using air and oxygen as the gasifying media, but there has been little commercial development so far.

marginal. A key concept in economics; it can roughly be translated into the 'last unit'.

For example, marginal cost or marginal revenue refers to the costs incurred or revenues gained, respectively, from production of the last unit of output. Generally it refers to small changes in a variable, and hence is amenable to the use of calculus. The concept carries much significance in economics because it is the unit over which the decision-maker has greatest control and therefore plays a key role in optimizing behaviour. Thus a decision-maker can decide whether to produce the next (last) unit or not (and incur the consequences of the decision).

In the economic analysis of the oil and gas industries, two key concepts are marginal cost and marginal revenue. In accounting terms this reflects the variable costs as opposed to the fixed costs. In the LONG RUN, when all costs are variable, the marginal cost would include the cost of installing new producing capacity including FINDING COSTS and DEVELOPMENT costs. The concept can be applied on a field, regional or global basis. In the SHORT RUN, because of the high capital intensity of oil and gas operations, the marginal cost is very low. In the long run, there is more disagreement about the size of the marginal cost due mainly to uncertainty over finding and development costs, with some arguing that they may be high reflecting more difficult operating conditions; this may be true at a field or regional level, globally it seems less likely to be true. Marginal revenue is the addition to total revenue from producing and selling the last barrel or cubic foot.

The significance of these two concepts in the economic analysis of the oil and gas industry arises because of their interrelationships. If the producer wishes to maximize his profit then he will produce as long as marginal revenue exceeds marginal costs. In the case of oil and gas, this is complicated by the fact that they are an EXHAUSTIBLE RESOURCE. Thus what is not produced today can be produced tomorrow. However, given that marginal cost in the short run is very low, in theory there is always pressure to produce more. If excess producing CAPACITY exists there is natural tendency in a purely competitive market to oversupply, thereby forcing down price. In theory under competitive conditions production will increase until the marginal revenue equals marginal cost. Thus marginal cost sets the competitive price. If the marginal revenue remains above marginal cost this indicates that there exist restraints on COMPETITION. In practice the mechanisms are far more complex than outlined above.

marginal barrel. *See* INCREMENTAL BARREL.

marginal costs. *See* MARGINAL.

marginal cost pricing (MC pricing). MC pricing means that an enterprise produces that output where price (P) equals MARGINAL cost (MC). This is what a profit-maximizing company operating in a perfectly competitive market (*see* COMPETITION) would choose to do. A theorem in WELFARE ECONOMICS says that if all companies in a market economy were perfectly competitive and setting P = MC, and if there were no instances of MARKET FAILURE (such as MONOPOLIES or EXTERNALITIES), so that all goods had appropriate market prices, the resulting allocation of resources would be efficient (or PARETO OPTIMAL). For, as long as price reflects society's marginal valuation of an extra unit of a good — its marginal social benefit (MSB) — and if marginal cost reflects the marginal SOCIAL COST (MSC) of an extra unit, then for any activity MSB = MSC. When this is so, the net social benefit of any activity, total social cost minus total social benefit, could not be increased either by raising or by lowering output, and the outcome is efficient.

This logic leads to the policy proposal that industries which the state either owns or can regulate should be instructed to operate at output levels where P = MC. MC pricing has been widely adopted in the pricing of electricity, for example. In cases where peak loads raise the marginal cost of supplying additional units of electricity, marginal cost pricing implies the use of higher PEAK LOAD tariffs.

In practice, however, the conditions necessary for the MC pricing rule to lead to efficient outcomes rarely hold. This is because there are 'deviant' sectors, which the authorities cannot or will not control, in which there is market failure. In these sectors there are either inappropriate price and cost signals or no prices at all. Here, the theory of the SECOND BEST suggests that MC pricing rules must be modified if a second-best Pareto optimal efficient solution is to be achieved.

marginal efficiency of capital. *See* INTERNAL RATE OF RETURN.

marginal revenue. *See* MARGINAL.

marine diesel. A range of DIESEL OILS used as fuel in marine transport. For many applications, a heavier fuel is suitable, and a proportion of RESIDUAL FUEL OIL can be included in marine diesel. *See* FINISHED PRODUCTS, DIESEL OIL.

marine riser. A pipe that connects a SUBSEA WELLHEAD or pipeline to the surface RIG.

marker crude. Oil is a DIFFERENTIATED PRODUCT in the sense that different crudes have different physical and chemical properties and also have different geographical locations near to or distant from the market. Therefore there is no set price for crude oil, but instead a structure of prices for crude oil exists. However, when crude prices are set by administrative decision, as was the case when the oil-producing companies set prices or OPEC appeared to set prices (*see* CRUDE OIL PRICES), it was too cumbersome to set all crude prices. Instead a benchmark or marker crude price was administratively decided, and other crudes were then priced according to a formula which allowed for differences in quality and location. Thus the marker crude was treated as the representative crude, and its price was viewed as 'the oil price'.

In August 1973 when Saudi Arabia was renegotiating its BUY-BACK PRICES with Aramco they insisted that the criterion to be used should be the actual sales price obtained by the national oil company — Petromin — when selling to third parties. This proposal was eventually accepted by the companies in September. The result was that this price for Arabian Light became the linchpin of the world's oil price structure; when OPEC announced a price change it was a change in the price of Arabian Light. Whether it was OPEC or Saudi Arabia who actually determined this price remains a matter of debate. Arabian Light was a sensible choice as marker since it was the largest-volume single crude oil in world trade with middle-of-the-road characteristics.

marker price. *See* MARKER CRUDE.

marketable natural gas. Gas from which some HYDROCARBONS and non-hydrocarbon compounds have been removed which means the gas can be used by the end user. In the USA it is sometimes called pipeline residue or sales gas.

market failure. Economic theory postulates that if a market is competitive there is an optimal allocation of resources in the sense that society is obtaining its desired output mix of good and services using the minimum resource inputs (*see* MARGINAL COST PRICING). This occurs by the operation of the PRICE MECHANISM. Thus no intervention is needed, merely the operation of the INVISIBLE HAND. In practice, however, two factors may operate to frustrate the achievement of this objective (i.e. market failure exists). (a) The existence of EXTERNALITIES which means that there are costs and revenues (benefits) that are not accounted for. These might range from pollution (an unaccounted cost) to regional development (an unaccounted benefit). (b) The market may not be competitive either because of a lack of information or because of the presence of monopolistic elements in the market (*see* MONOPOLY).

The presence of market failure provides for many the justification for the intervention of government in the working of the market, although the ECONOMIC THEORY OF POLITICS also points out the possibilities of 'government failure'. The oil and gas markets have always exhibited the characteristics associated with market failure — a prevalence of externalities, highly imperfect information and a non-competitive industry structure. This has in the past provided much of the justification for government involvement in the industries.

market period. The period in which the supply is fixed irrespective of price. It is used in economic theory in conjunction with the SHORT RUN and LONG RUN to explain price behaviour.

market price. The price ruling in the market place. It is usually used to imply a price determined by the interaction of supply and demand to distinguish it from a price set by administrative decision.

market share. The percentage of the total

market that is accounted for by a specific product or a specific supplying institution (i.e. a company or government). The use of such a concept implies that the product or institution faces competitive substitutes. Examples of the use of the term are the market share of gas, which is the percentage of total energy consumed accounted for by gas, and the market share of an oil-producing country. Normally, when the term is used it implies growing concern over the competitive position of suppliers and is often used in conjunction with COMPETITION and PRICE WARS in the context of a potential oversupply. In times of tight supply, producers are able to sell whatever they can produce, and the market share loses significance apart from its use as an expression of vulnerability of the consumer to supply disruptions. During the 1980s concern over market share grew, and in 1985 it became a key issue in the marketing strategy of OPEC.

market structure. The underlying characteristics of a market which in turn will set the degree of COMPETITION. These characteristics are the number and size of both buyers and sellers, the ease with which buyers or sellers may enter the market, whether it is a HOMOGENEOUS PRODUCT or a DIFFERENTIATED PRODUCT, the extent and nature of integration of the suppliers, etc., together with a number of other characteristics. Economists have identified four broad types of market structure: (a) perfect competition, which has a large number of buyers and sellers with a homogeneous product and PERFECT KNOWLEDGE; (b) monopolistic competition, which has a large number of buyers and sellers with a differentiated product; (c) OLIGOPOLY, which is characterized by a few sellers dominating the market; (d) MONOPOLY (*see* MONOPSONY), where there is a single seller (or buyer) and the product. Each type of market structure has certain characteristics and behavioural patterns, which provide the economist with an analytical framework with which to begin the analysis of a particular market. Conventionally the oil industry has been characterized by an oligopolistic structure, whereas the gas industry has been characterized as both oligopolistic and monopolistic (monopsonistic).

market value. *See* BOOK VALUE.

MARPOL. *See* BALLAST.

marshallian demand curve. *See* DEMAND.

marsh gas. METHANE generated by microbiological processes in shallow waters and mud.

maximum economic finding cost (MEFC). An estimate of rising development costs as an increasing amount of oil-in-place is depleted. It is a very important element in determining the MARGINAL cost of oil and gas production.

maximum sustainable capacity (maximum sustainable yield). The level of crude oil or gas production that can be maintained over a long period of time without reducing the amount of oil and gas that can ultimately be recovered from the field. It may be significantly below the maximum production from the field and is almost entirely a function of the geology of the field. *See also* CAPACITY.

maximum sustainable yield. *See* MAXIMUM SUSTAINABLE CAPACITY.

MC pricing. *See* MARGINAL COST PRICING.

median line. A line used to delineate national sectors of an offshore area where the territorial boundaries overlap. It is simply the points of equidistances between the shore lines of the two countries. In the absence of any other method or information it is viewed as an acceptable means of dividing ownership of the sub-sea minerals.

medium fuel oil. A FUEL OIL between light and heavy fuel oils with a sulphur content of less than 4 percent.

medium-range tanker. A category of tanker defined for AFRA purposes as being between 25 000 and 44 999 dwt. It is capable of carrying both crude oil and products. Because of its versatility as well as its ability to use almost all ports, medium-range tanker can command a premium over larger vessels. In mid 1985 the average freight rate for medium-range tankers was WS 101 compared with WS 28 for VLCCs.

MEES. (*Middle East Economic Survey*). One of the foremost trade papers covering the oil and gas industries. It was first published (weekly) in 1958 in Beirut, Lebanon, but in 1975 the publication was moved to Nicosia, Cyprus.

MEFS. *See* MAXIMUM ECONOMIC FINDING COST.

melting point. A product quality specification for waxes. It defines the temperature at which a wax starts to melt. *See also* QUALITY, WAXES.

mercaptans. A group of sulphur-containing organic compounds derived from hydrogen sulphide. Mercaptans, which are associated particularly with the lighter distillation fractions (*see* LIGHT ENDS), must be removed, or treated, to obtain finished products of acceptable quality. *See also* QUALITY, CRUDE OIL; REFINING, SECONDARY PROCESSING.

merger. The formation of one company when two or more separate companies join together. The difference from a takeover is that normally a merger is jointly agreed between two partners on a basis of rough equality, whereas a takover generally involves an unwilling partner. However the shades of distinction are not always clear, and many mergers are probably in truth takeovers. A merger of firms at the same stage of an industry (e.g., crude oil producing companies) leads to HORIZONTAL INTEGRATION, whereas a merger at different stages (e.g., a crude oil-producing company and a refinery) leads to VERTICAL INTEGRATION. Mergers involving companies in unrelated activities lead to CONGLOMERATE integration.

Merox process. *See* UOP MEROX PROCESS.

metals content. A quality characteristic of a crude oil. It is normal to include a reference to any traces of metals (mainly vanadium and nickel) present in crude oil, since these can have adverse effects at the refining or end use stage if not corrected. *See also* QUALITY, CRUDE OIL.

methane (CH_4). A light, odourless, flammable gas. It has the simplest structure and smallest molecule of any organic compound (*see* CHEMISTRY). It is the main constituent of natural gas and is the first member of the ALKANE series of paraffins. It can be used as a fuel or as a petrochemical FEEDSTOCK.

methanol (methyl alcohol). A colourless, volatile, flammable and toxic liquid produced by the STEAM REFORMING of hydrocarbons. Its main use is as a petrochemical FEEDSTOCK. During the 1970s interest in its use as a fuel grew considerably either alone for the generation of electricity or as an additive to motor spirit and coal slurry (*see* COAL LIQUEFACTION). The reason for this interest is that methanol can be produced from natural gas and is transported without the problems associated with LNG. It therefore appeared to present an alternative route for gas exports where pipelines were not feasible.

methyl alcohol. *See* METHANOL.

methylcyclopentadienyl manganese tricarbonyl. *See* MMT.

methyl tertiary-butyl ether. *See* MTBE.

micro. *See* MACRO.

microeconomics. *See* MACRO.

middle distillates. A general description of products in the boiling range of 160–360°C produced during REFINING. They include the kerosines, gas oils and light fuel oils. *See also* FINISHED PRODUCTS, DIESEL OIL; FINISHED PRODUCTS, GAS OIL; FINISHED OIL, KEROSINE.

Middle East Economic Survey. *See MEES.*

migration. The movement of oil from the source rock, in which it was originally formed, into the reservoir rock, in which it accumulates. *See also* GEOLOGY.

millidarcy. One-thousandth of a DARCY, the standard unit of permeability in a reservoir rock.

mineral fuels. A general term covering oil, condensates, gas, coal, tar solids and shales; the EXHAUSTIBLE RESOURCES as opposed to the RENEWABLE RESOURCES.

mineral oils. Any liquid product derived from MINERAL FUELS. The term is used to differentiate them from vegetable oils (e.g., olive oil, sunflower oil).

mineral rights. The legal ownership of sub-soil minerals. In most legal systems, except the USA, these belong to the state.

minimum exploration obligation. Once a company has been granted the right to explore for oil or gas in a particular acreage it is important from the point of view of the owner that exploration is actually undertaken. In the days of the concession agreement (*see* CONCESSIONS), it was not unusual for companies to take acreage with little or no intention of spending money on exploration simply to exclude competitors. Hence when JOINT VENTURE contracts and SERVICE CONTRACTS began to be signed in the 1950s they contained minimum exploration obligation terms which set out the minimum exploration that the company would be forced to undertake. The early agreements outlined these terms in the form of the number and/or depth of WILDCAT DRILLINGS. However, this was replaced by a financial commitment that gave the operating company greater flexibility. In situations where there was competition for acreage among the companies, the exploration commitment often became a key bargaining element.

minimum safeguard price. *See* FLOOR PRICE.

mixed base crude. A crude oil that contains both BITUMEN and PARAFFIN WAX.

mixing mud. The preparation of DRILLING FLUID by mixing fluids with clays and dry chemicals.

MMT (methylcyclopentadienyl manganese tricarbonyl). A non-lead additive that can be used to improve the OCTANE RATING of motor spirit instead of TEL, TML or MTBE. *Compare* LEAD ALKYLS. *See also* LEAD-FREE MOTOR SPIRIT.

MNC. Multinational corporation (*see* MULTINATIONAL).

MNE. Multinational enterprise (*see* MULTINATIONAL).

Mobil. The smallest of the US MAJOR oil companies.The company was a split off from the break up of the STANDARD OIL TRUST. In 1931 the company merged with the Vacuum Oil Company. Originally named Socony Mobil (i.e. Standard Oil Company of New York) and Socony–Vacuum, the link with Standard was dropped in 1966.

model. A simplification of a real-world process or system. It may be physical, graphical, verbal or mathematical. The aim of a model is to help to explain, predict and control some aspect of the real world. Because the real world is infinitely complex the model builder must maintain manageability while endeavouring to capture the essential elements of the process or system. Models range from the simple analysis of the direct relationship between two or more variables (e.g., oil demand and output) to a highly sophisticated analysis of the system as a whole with all its interacting relationships. An example of a large econometric model (*see* ECONOMETRICS) is the UK Treasury model which was designed to explain important behavioural relationships of the UK economy. Based on mainstream economic theory, it was first used to prepare the Treasury's forecasts in 1970 and at that time consisted of 40 equations and identities. It now has over 700 equations and identities.

One specific field of models that has received considerable attention over the past decade is that of energy models, stimulated by among other things the apparent end of an era of abundant low-cost energy. Energy models not only apply the economic theories of demand and supply to the energy sector, but are also concerned with aspects of government policy, technological development and environmental quality. There is no precise definition of an energy model since they differ considerably in their (a) formulation, (b) scope and (c) application.

formulation. Models vary widely in the techniques employed, among the most commonly used are statistical and econometric methodology, optimization where linear or non-linear programming techniques are generally applied, INPUT–OUTPUT ANALYSIS, simulation and SCENARIO building. Process analysis is also used as a descriptive model of resource allocation given conversion factors for different technologies.

scope. Energy models may be classified into: sectoral models, which focus on the

system and the economy as a whole are assessed (*see* BROOKHAVEN ENERGY REFERENCE SYSTEM; LEAP). These classifications are by no means mutually exclusive, and a model may be a combination of two or more groupings.

application. Energy models may be applied in the energy industry for production planning and investment appraisal, by governments for policy purposes and academics in an appraisal of both industry and government, as well as in the advancement of model-building methodologies. The models range from investigating the very short-run to highly speculative long-run models, forecasting more than 30 years ahead.

mogas. Motor gasoline (*see* MOTOR SPIRIT).

mole. *See* PIG.

monkey board. A high-level platform on the drilling DERRICK.

MON. *See* MOTOR OCTANE NUMBER.

monomer. A simple molecular unit such as ETHENE or STYRENE which when joined to other monomers forms polymers (e.g., styrene–butadiene rubber). Monomers are in effect the basic building blocks of petrochemicals.

Monopolies Commission. *See* MONOPOLY LEGISLATION.

monopoly. Literally, one seller (from the Greek *monos* meaning single and *polein* meaning to sell), where the seller (i.e. company, union, licensing body, trade association) is the only one from whom buyers can buy. A monopoly is thus a constraint on the relation between the demander and the supplier. Monopolies can only survive if they can prevent potential competitors from supplying the market. They can do this if there are limitations on entry to the industry such as: patent rights or other legal restrictions on the local area (a local monopoly); minimum efficient scale of output is so large in relation to market demand that only one firm can survive; and other barriers to entry which put the monopolist at a cost advantage compared with potential entrants. The monopolist is a 'price searcher', in contrast to the 'price taker' companies which operate in a situation of PERFECT COMPETITION. The monopolist faces the entire downward-sloping market DEMAND CURVE and can fix the price by choosing the appropriate quantity, or vice versa. Since increasing the quantity reduces the price on all units sold, MARGINAL revenue is less than price. A profit-maximizing monopolist (*see* PROFIT MAXIMIZATION) sets marginal revenue equal to marginal cost in order to find optimum output and price, with the result that price will be greater than marginal cost. An argument in WELFARE ECONOMICS suggests that if price indicates society's marginal valuation of the benefits from the output, then choosing an output where society's marginal benefit is above its marginal cost may be socially inefficient. Society might gain from raising output to the point where marginal benefit (price) equals marginal cost (i.e. MARGINAL COST PRICING). One way of measuring monopoly power is to find the ratio of a monopolist's cost to price.

A variant on monopoly is when companies in an oligopolistic market (*see* OLIGOPOLY) which, faced with uncertainty about each other's behaviour over price, band together to form a CARTEL whereby companies either agree on a common selling price or on a MARKET SHARE. Thus in theory the companies behave as a single seller.

The international oil industry has never been a monopoly in any sense of the word. However, from time to time, individual companies have found themselves in a monopoly position in specific markets either as a result of licence protection or control of refinery capacity. Such local monopolies are more common in the gas supply industries, although normally such monopolies are state-owned. This arises not only because of the ECONOMIES OF SCALE associated with gas distribution networks, but also because the problems of transporting gas create a 'protected' market for the controller of local supplies. *See also* MONOPOLY LEGISLATION.

monopoly legislation. There is always a strong probability that a private MONOPOLY will exploit its market power in order to earn what the public would view as excess profit. Hence governments pass legislation to investigate the possible existence of monopolies and to disband them if necessary or to make certain agreements illegal such as CARTEL agreements. In general two approaches with such legislation are possible. One approach, which was adopted by the ANTI-TRUST LEGISLATION in the USA, sets a market share above which companies may not go. The other approach, which has been adopted in the UK, also has a market share limit, but if a company exceeds this it is investigated by the Monopolies Commission to ascertain whether the company is behaving against the public interest. The former approach assumes that a market share greater than the limit is automatically bad, whereas in the latter this is not necessarily so.

monopsony. A market situation in which there is only a single buyer. This may occur as a result of legislation. It may also result from the nature of the market (e.g., a refinery close to an oil-field where the export prospects are very poor). An example of a monopsony is British Gas, which is the only organization entitled to buy gas production in the UK.

Monte Carlo method. A method used in operations research to assist in the decision-making process. It involves creating a model which is then run many times in order to derive the probabilities of an outcome in situations where statistical or mathematical derivation of probability is impossible. *See also* RISK.

moon pool. The hole in a drill ship through which the DRILLING STRING operates.

motor gasoline. *See* MOTOR SPIRIT.

motor octane number (MON). A product quality specification for MOTOR SPIRIT defining its combustion properties. MON indicates a fuel's performance under more severe operating conditions than the ROAD OCTANE NUMBER. *Compare* RESEARCH OCTANE NUMBER. *See also* QUALITY, MOTOR SPIRIT.

motor spirit (motor gasoline, mogas, petrol (UK)). The fuel consumed in the majority of private cars. It undergoes extensive secondary processing, which results in its having a high added value. *See also* FINISHED PRODUCTS, MOTOR SPIRIT; QUALITY, MOTOR SPIRIT.

motor spirit upgrading. The distillation fraction (approximately 30–200°C) from which motor spirit is manufactured, which includes the straight-run components of gasoline, benzine and naphtha, is of insufficiently high quality to be used directly as finished motor spirit. The main feature that has to be improved is its OCTANE RATING.

alkylation. This, like catalytic polymerization (*see below*) works on the principle of joining smaller molecules to form molecules within the motor spirit range (in this case isooctanes). Both alkylation and catalytic polymerization are useful processes for converting C_3/C_4 hydrocarbons, which might otherwise be in surplus, to motor spirit.

catalytic cracking. A process for producing motor spirit components of high OCTANE RATING, but its essential function is as a yield shift conversion plant (i.e. it converts heavier fractions of low value into lighter higher-value ones). There is a range of conversion plants designed to operate directly on the straight-run light fractions and improve their quality without bringing about any significant yield shift. Rather than breaking up the molecules, therefore, they restructure them from hydrocarbon types with low octane number to those of a higher number.

catalytic polymerization. A less common process for producing motor spirit components is based on the chemical principle of POLYMERIZATION. This can be regarded as the opposite of cracking (which breaks up large molecules into smaller ones) in that it joins two smaller molecules together to form a larger one. By combining two C_x molecules together to form a C_{2x} molecule, it effectively converts gaseous hydrocarbons into liquid hydrocarbons suitable for use in motor spirit.

catalytic reforming (cat reforming). The major process for producing high-quality motor spirit components. It takes as feedstock most of the lighter straight-run materials except light motor spirit (i.e. boiling range of approximately 70–90°C). These straight-run components have an OCTANE NUMBER of about 40, and this is increased to 95 or 100 in the course of the process. Several different chemical reactions take place during the process, but the net effect is to convert the paraffins and naphthenes in the straight-run material, which have low octane numbers, into aromatics, which have much higher octane numbers. The process is carried out at high temperatures in the presence of a catalyst. Some of the chemical reactions taking place involve the removal of hydrogen (i.e. dehydrogenation) from the feedstock, and the cat reforming plant is therefore a useful source of the hydrogen required for HYDROFINING.

isomerization. A process that improves the octane number of the lighter straight-run motor spirit fractions (30–70°C) that are less effectively dealt with by catalytic reforming (*see above*). The straight-chain paraffins found in the straight-run material have low octane numbers, whereas the corresponding branched-chain isomers (*see* CHEMISTRY) have high octane numbers. If the atoms within the molecule of, for example, pentane (C_5H_{12}) can be arranged into those of isopentane (also C_5H_{12}, but with a different arrangement of the atoms) then a considerable gain in octane number results. This is effectively what the isomerization process does, operating at a relatively low temperature in the presence of a CATALYST.

mousehole. A hole in the drilling floor in which the KELLY joint and stands of pipe are stored when making a connection. *See also* DRILLING.

movement matrix. A table that relates oil or gas consumption with origins of supplies. The rows identify consuming areas, whereas the columns identify supplying areas. Hence a row gives exports by destination and a column imports by source.

moving average. *See* MARGINAL.

MR. Marginal revenue (*see* MARGINAL).

MSV. *See* MULTISERVICE VEHICLE.

MTBE (methyl tertiary-butyl ether). A lead-free octane improver for motor spirit (*see* OCTANE RATING). As governments legislate to reduce the lead content of gasoline the use of MTBE is likely to expand considerably. *See also* LEAD-FREE MOTOR SPIRIT.

mud. *See* DRILLING FLUID.

mud pits. Storage tanks through which the DRILLING FLUID is circulated. *See also* DRILLING.

mud pumps (slush pumps). Pumps that are used to circulate the DRILLING FLUID. *See also* DRILLING.

mud ring. A ring of solid material left on the wall of the well while drilling a porous formation. It is due to water being absorbed by the formation.

mud weight. The density of the DRILLING FLUID. It is normally measured in pounds per gallon and can be varied by altering the proportion of solids in the fluid.

Mullion rates (spot rates). An index of spot freight rates supplied by the London tanker brokers Harley Mullion Ltd and published in the *Petroleum Economist*. It is a composite average of dirty (i.e. crude and heavy fuel oil) single-voyage rates.

multigrade oil. A lubricating oil whose viscosity changes as the operating temperatures varies from relatively thin when cold to relatively thick when hot. The change is achieved by the addition of a chemical polymer — a viscosity index improver.

multinational (multinational corporation, MNC; multinational enterprise, MNE; international company; transnational corporation, TNC). A company that operates in additional countries to its country of origin. A widely accepted definition is a company that produces at least 25 percent of its output outside of the country of origin. The implication of key importance is that a subsidiary or AFFILIATE of such a company will take decisions based upon the interests of its parent's global strategy; this may bring it into conflict with its host government. The expansion of MNCs after World War II has resulted in increasing attention being paid to MNCs by economists in terms of their behaviour, market power and impact.

The oil companies provide an excellent example of MNCs with their differently located production, refining and distribution facilities. Other issues associated with MNCs that are revelent to multinational oil companies include relations with the host government, political influence, TRANSFER PRICE policy and PROFIT MARGINS.

Multinational Corporations Subcommittee. *See* CHURCH COMMITTEE.

multiple regression. *See* REGRESSION ANALYSIS.

multiservice vessel (MSV). A support ship used in offshore operations to provide a range of services such as construction, fabrication, diving, emergency, etc.

mutually exclusive investments. Alternative investment projects for achieving the same objective; only one can be chosen. For example, the transport of crude oil from the oil-field to the refinery might be feasible (and profitable) by road tanker, sea tanker or pipeline. The three alternatives are mutually exclusive since if one is chosen the other two will normally become redundant. *See also* INVESTMENT APPRAISAL.

N

naphtha. A LIGHT DISTILLATE FEEDSTOCK used as a petrochemical feedstock. *See also* PETROCHEMICALS.

naphthenates. *See* NAPHTHENIC ACIDS.

naphthenes (cycloparaffins). HYDROCARBONS that contain at least one closed ring of carbon atoms. The most important naphthene is cyclohexane which is used as a petrochemical FEEDSTOCK. *See also* CHEMISTRY.

naphthenic acids. Organic acids found in some crude oils. If the concentration is sufficient (1 percent or greater) they are removed and used to produce naphthenates employed as paint driers and as wood or textile preservatives.

naphthenic crudes. *See* PARAFFINIC CRUDES.

national income. A measure of the total payments made to the factors of production employed in the production of output in an economy in a specified period such as a year. As such it is a proxy measure for the output of the nationals of an economy and is an alternative measure of gross national product (*see* GROSS DOMESTIC PRODUCT).

National Iranian Oil Company. *See* NIOC.

nationalization. The taking-over by the state of all or part of the activities of a private company. It is different from PARTICIPATION where the state negotiates a share in the company, although often the distinction is blurred because it is impossible in many cases to identify how far the private company is free to negotiate the take-over. Given the prevalence of foreign companies in oil and gas operations in many countries nationalization is a frequent occurrence. Prior to 1970 there were two major nationalizations — in Mexico in 1938 and in Iran in 1951 — both of which provoked extremely hostile reactions from the oil companies and their parent governments. These reactions forced both countries into a humiliating settlement, which caused other countries contemplating such a move to draw back from actual nationalization except in a number of relatively insignificant cases. They also reminded the host governments very sharply of their limited freedom of action in their domestic hydrocarbon sector, which led to growing resentment in many of the producing countries. The fears of the consequences of nationalization were eventually laid to rest in the early 1970s when the governments of Algeria, Libya and Iraq successfully nationalized their oil-fields. Thereafter, nationalization became much more common. *See also* IRANIAN NATIONALIZATION.

national product. *See* DOMESTIC PRODUCT.

natural drive (natural recovery). The spontaneous flow of oil without artificial assistance when the natural pressure in a reservoir is sufficiently high. *See also* PRODUCTION, NATURAL DRIVE MECHANISMS.

natural gas (casing head gas). A group of gases that are found either in association with crude oil (i.e. associated gas) or in independent reservoirs (i.e. non-associated gas). Where gas occurs in crude oil reservoirs it is found either dissolved in the crude oil (solution gas) or in a separate layer above the oil (the gas cap). Natural gas comprises the following gases — METHANE, ETHANE, liquefied petroleum gases (LPG) and heavier compounds — in proportions that vary according to the geographical location and degree of association with crude oil. The non-associated gas is mainly methane, whereas associated gas contains a much greater proportion by volume of ethane and the LPGs, which are collectively called NATURAL GAS LIQUIDS. Hence non-associated gas is often called dry gas and associated gas wet gas. Sometimes, however, liquids occur in non-associated fields and in this case they are referred to as condensates. These differences are illustrated in Table N.1 for four selected gases.

Table N.1.

	Associated		Non-associated	
	N. Sea Forties	Abu Dhabi Zakum	Algeria Hassi R'mel	N. Sea W. Sole
Composition				
Methane	44.5	76.0	83.5	94.4
Ethane	13.3	11.4	7.0	3.1
LPGs	31.9	7.6	2.8	0.7
Pentanes +	8.4	1.3	0.4	0.2
Sulphur+impurities	1.9	3.7	6.3	1.6
Classification				
Wet	√	√		
Dry			√	√
Significant condensates			√	

Gas reserves are less well known than those of oil. In the past gas was less attractive commercially than oil, due both to relatively higher oil prices and to the regulated state of many gas markets. World proved reserves have been estimated by British Petroleum to be 3400 trillion cubic feet (81 600 mtoe) at end 1984, only a little lower than proved oil reserves of 96 000 mtoe. Although the USSR (with 42.7 percent of the total reserves) and Iran (with 14.7 percent of the total reserves) have the majority of reserves. Significant amounts are also to be found in the USA, Qatar, Saudi Arabia, Canada, Mexico, Norway and The Netherlands. The main markets exist between the gas-deficit areas of Western Europe and Japan. and the surplus gas producers of the USSR, North Africa, South Asia, Australia and Latin America.

Natural gas can be used either as a fuel or as a chemical feedstock for the production of fertilizers and PETROCHEMICALS. Associated gas is only suitable for the former use as its supply depends on the output of crude oil, and hence does not have the reliability required for the production of chemicals. This has led many countries such as Saudi Arabia to develop non-associated gas-fields in recent years to supply their new chemical industries.

Trade in natural gas expanded rapidly in the 1970s, although because of high transportation costs traded gas represents only a small part of total world consumption (4.2 out of 50 trillion cubic feet in 1977). Of the transport methods available, pipelines are the most important (4 trillion cubic feet). These are largely confined to overland trade, although some undersea pipelines (e.g., between Algeria and Western Europe) are now in operation. Most long-distance transport requires liquefaction of the gas (*see* LNG) and the use of special cryogenic tankers (*see* LNG/LPG CARRIERS). These take advantage of the vast reduction in volume which occurs when LNG is frozen at atmospheric pressure or above.

Natural gas prices are influenced by three factors: (a) capital costs; (b) degree of association with crude oil; and (c) regulation of consumer prices. Capital costs both of processing and transportation are extremely high. On the processing side, gas treatment requires pipelines, gas separation and purification plants. On the transport side, the additional requirements are for liquefaction, storage, tanker and regasification plant. This may be contrasted with the simple refinery requirements of oil processing. Where gas is produced in association with oil it must either be flared or sold at market prices. Erratic movements in prices may be introduced (e.g., by rapid changes in crude oil output levels) causing shortages or gluts of natural gas. Markets in the USA and Western Europe have until recently been highly regulated. This has both retarded supply and stimulated demand, leading either to rationing (as in the UK where natural gas is denied to power station consumers) or to excessive public investment in gas supplies. On the producer side attempts have been made to apply OPEC-style pricing policies to natural gas (*see* CRUDE OIL PRICES). These have until recently proved to be largely unsuccessful in the face of falling world oil prices.

natural gas liquids (NGL). The higher-molecular-weight hydrocarbons in NATURAL GAS which are liquid at moderately low temperature or under moderate pressure. The two main components are LPG (propane and butane) and NATURAL GASOLINE (pentane).

natural gas prices. *See* NATURAL GAS.

natural gasoline (casing head gasoline). The part of NATURAL GAS LIQUIDS that is liquid at ambient temperature and atmospheric pressure. It is composed mainly of pentane (C_5H_{12}).

natural monopoly. Economic activities that have ECONOMIES OF SCALE over very large output ranges. As output increases, average costs continue to fall which implies that greater EFFICIENCY is obtained by allowing only one firm to supply (hence minimizing costs), normally under some form of state control. The gas supply industry is frequently held up as an example of a natural monopoly. Equally some have argued that oil exhibits similar pressures towards a single supplier, but this is based on the misconception that the industry faces decreasing costs as output rises.

natural recovery. *See* NATURAL DRIVE.

natural resources. The products of nature used as inputs into economic activity. Natural resources are classified either as EXHAUSTIBLE RESOURCES or RENEWABLE RESOURCES, although renewable resources (e.g., forestry or fish stocks) can in practice become exhausted if over-used. Oil and gas are both natural resources.

NCE. *See* NON-COMMERCIAL ENERGY.

NCW. Non-communist world (*see* WOCA).

needle coke. *See* PETROLEUM COKE.

Nelson indices. Cost indices for refinery construction and operation published in the *Oil and Gas Journal*.

netback. *See* NET PRODUCT WORTH.

net book value. The acquisition cost of a company's assets (*see* BOOK VALUE) less DEPRECIATION. Net book value has been widely used as the basis for COMPENSATION in cases of NATIONALIZATION.

net cash flow. *See* CASH FLOW.

net pay. *See* NET THICKNESS.

net present value (NPV). The sum of the difference between cash inflows and outflows of a project discounted to their PRESENT VALUE. In essence it is the current value of expected profits from a project. If the NPV is positive, in theory, the project is worth undertaking (i.e. it is profitable). NPV is a widely used criterion in INVESTMENT APPRAISAL. *See also* DISCOUNTED CASH FLOW.

net product worth (NPW, netback). A means of valuing crude oil. It is the value of refined products (from a barrel of oil) sold at the refinery gate (i.e. GROSS PRODUCT WORTH) less the costs of transporting and refining the crude oil. This means of pricing crude oil became widely used in September 1985 when Saudi Arabia began to sell increasing amounts of crude oil on this basis. The NPW of the crude oil varies depending upon which geographical product markets are taken as well as which product prices (i.e. SPOT PRICES or TERM PRICES) are used, although normally spot prices are used.

net profit. *See* PROFIT.

net profit bidding (percentage profit rate bidding). A means of allocating acreage when the acreage is put up for auction to the highest bidder. It involves offering different percentage shares of the profits to the owner of the acreage. *See also* AUCTION SYSTEM OF LICENSING.

net thickness (net pay). The thickness of RESERVOIR ROCK actually containing producible oil or gas. *See also* GEOLOGY.

net-to-gross ratio. The ratio of the net volume of rock that contains producible oil to the total amount of reservoir rock. *See also* GEOLOGY.

network purification. The getting rid of unprofitable elements of a company's operations.

neutral tax. *See* TAXATION NEUTRALITY.

new buildings. Tankers under construction; gross additions to the tanker fleet.

new issues. Shares and securities offered by

firms in order to raise long-term capital when initially offered for sale.

new oil. After the FIRST OIL SHOCK the US government introduced even greater controls over oil pricing. A base production control level was established from the monthly production achieved in 1972 (STRIPPER WELLS excluded); this was known as old oil. Production above these levels or from new fields was known as new oil. The old oil was subject to a price ceiling set initially at $4.25 per barrel, whereas new oil was exempt. Since this limited the ability of producers to secure WINDFALL PROFITS from the rise in international prices it led the industry to mount a serious, and eventually successful, campaign for the decontrol of oil prices.

New York Mercantile Exchange. *See* NYMEX.

NGL. *See* NATURAL GAS LIQUIDS.

NIOC (National Iranian Oil Company).The national oil compoany which was first created in 1951 to take over the operations of the then nationalized BP. *See also* IRANIAN NATIONALIZATION.

nodding donkey. A pump used to bring crude oil to the surface for a well in which there is insufficient pressure to force the crude oil to the surface. Nodding donkeys are normally associated with STRIPPER WELLS. The cost of running them represents a high proportion of the higher operating costs of the stripper wells. When the price of crude oil fell in 1985 and later, many expected that a significant number of these high-cost wells would be SHUT-IN because of their high cost. However, they had overlooked the fact that lower oil prices made the nodding donkeys cheaper to run, thereby reducing operating costs.

nomenclature of hydrocarbons. *See* CHEMISTRY.

nominal. Money values expressed in money of the day (i.e. unadjusted for changes in purchasing power arising from inflation). Nominal prices are not useful when comparisons over time are made. *See also* REAL.

non-associated gas. NATURAL GAS produced from gas-fields as opposed to gas produced from oil-fields (*see* ASSOCIATED GAS). Non-associated gas is more attractive to the producer since its depletion profile can be chosen independently to that of crude oil.

non-commercial energy (NCE). A term used to describe fuels, such as wood and animal and vegetable residues, that are used extensively as energy sources in THIRD WORLD countries. It therefore distinguishes these fuels from COMMERCIAL ENERGY. The use of the term is extremely misleading since much of what is called non-commercial energy is in fact commercial in the sense that it is actively traded within the Third World. The use of the term implies that non-commercial energy is both of marginal importance and incapable of influence by government policy. Both impressions are, however, totally false. NCE is better thought of as a subset of TRADITIONAL ENERGY.

non-communist world. *See* WOCA.

non-excludable public good. *See* PUBLIC GOOD.

non-majors. *See* INDEPENDENTS.

non-price competition. In an oligopolistic market (*see* OLIGOPOLY) there is always fear that price reductions, in order to attract customers, will trigger a PRICE WAR with competitors. Therefore companies tend to indulge in non-price competition which can range from simple advertising to the providing of free gifts with the purchases. The motor spirit market in particular has always been characterized by such competition.

non-renewable resources. *See* EXHAUSTIBLE RESOURCES.

Norbec. Created in 1984, it became a crude oil trading unit for Saudi Arabia. Its significance was that it was responsible for the very substantial floating storage of crude oil which the Saudis developed in 1984–85. As a result it was in a position to influence rapidly the spot market since the crude oil was on the market, and hence the usual time lag was absent between production and price change. *See also* CRUDE OIL PRICES.

normal good. Sometimes if a consumer's income increases he will buy less of a good whereas common sense would suggest that more would be bought. This indicates that the good is viewed as a poor substitute for what is really wanted, but due to poverty the obtaining of the desired good is constrained Such a good is called an inferior good as opposed to a normal good or a superior good; demand for a normal or luxury good would increase following a rise in income. An example of an inferior good in industrial countries is kerosine which might be used for cooking, lighting and heating although gas or electricity would be preferred.

normal-paraffins. (n-paraffins, straight-chain paraffins). A subdivision of the PARAFFIN series of HYDROCARBONS, according to the arrangement of the carbon atoms within the molecule; normal denotes a straight-chain arrangement. *See also* CHEMISTRY.

normal profit. The minimum amount of profit a firm must receive to induce it to stay in any particular line of business. Any profit received above normal profit is called super-normal profit. The implication is that governments can therefore tax away super-normal profits without damaging the firm or causing it to leave. The distinction has always been of great importance in the oil and gas industry. Governments (and the public) frequently accuse oil and gas companies of earning super-normal profits. However, the problem is that there is no objective means of quantifying normal profit, except to remove super-normal profit until the firm packs up and does something else, which may defeat the object of the exercise. This uncertainty over the levels of normal profit is compounded in situations of high risk such as oil and gas exploration because the risk premium will increase the level of normal profit. However, perceptions of what the risk is (or was) varies in the eye of the beholder and hence so too do perceived levels of normal profit.

Norwegian trench. An area of very deep water off the coast of Norway. In the 1970s its existence exercised a significant influence over the development of the North Sea oilfields because of the inability to build pipelines across it. Thus Norwegian fields in the northern sectors had to seek alternative means of moving the crude oil.

notional output. *See* APQ.

n-paraffins. *See* NORMAL-PARAFFINS.

NPV. *See* NET PRESENT VALUE.

NPW. *See* NET PRODUCT WORTH.

Number 2 fuel. In the USA LIGHT FUEL OIL used for domestic and small-scale commercial central heating. *See also* FINISHED PRODUCTS, GAS OIL.

numeraire. A unit of account or standard measurement for the aggregation of different things. The most widely used numeraire is money.

Nymex. The New York Mercantile Exchange where much of the crude oil futures market (*see* FORWARDS MARKET) operates.

O

OAPEC (Organization of Arab Petroleum Exporting Countries).

OAPEC's Creation

Following the defeat of the Arabs by the Israelis in June 1967, several Arab countries began to examine the use of oil as a possible weapon in their struggle against Israel. Their ideas surfaced at the conference of Arab Ministers of Finance, Economy and Oil held in Baghdad in August 1967. Iraq and Algeria led a movement to use a form of oil embargo to counter the military defeat. Saudi Arabia, Kuwait and Libya, which was still a monarchy, were against these proposals and felt that the embargo (which had actually operated for a short time after June 1967) would hurt the Arabs more than anyone else. Instead they proposed a plan to maximize oil revenues in order to provide financial assistance for the front-line states.

In view of the difference of opinion among the Arab countries, the Saudi Arabia, Kuwait and Libya decided that the coordination of their policies would be far more effective outside of the Arab League. As a result, after a meeting in Beirut on 9 January 1968 they formed OAPEC. In the prevailing political context, this was an organization set up by the 'traditionalist' states to protect their interests against action by the 'radical' states. In fact, the founding agreement was designed specifically to exclude Algeria and Egypt by stating that the members' principle and basic source of national income should be petroleum.

Between its foundation in 1968 and the end of 1973, the most significant development within OAPEC was the increase in the membership. In September 1969, the Libyan revolution replaced the monarchy, thus shattering OAPEC's traditionalist base and opening the way for a much wider and broader-based membership. In May 1970 Algeria, Abu Dhabi, Dubai, Qatar and Bahrain joined the organization. Algeria was able to join because oil had supplanted wine as its main source of national income in 1969. In May 1972 Dubai and Abu Dhabi withdrew from OAPEC as individual members and combined their membership as constituents of the United Arab Emirates.

In December 1971 the founding agreement was altered fundamentally with the criterion for membership becoming that oil should form 'an important source' of national income. This change permitted Syria and Egypt to join OAPEC in 1972 and 1973, respectively. Since by 1972 all the Arab oil producers had joined the Organization or applied to join, Iraq decided to follow suit rather than risk isolation. Thus by 1973 the membership of OAPEC had increased from three to ten and included all the oil producers in the Arab world.

The First Oil Shock

The FIRST OIL SHOCK acted as a significant watershed for OAPEC. Firstly, because it came into public prominence; mistakenly it was viewed as being responsible for the October price hike (*see* CRUDE OIL PRICES) and the ARAB OIL EMBARGO.

The source of the error is easy to understand. On the morning of 17 October 1973 the OPEC Gulf States met in Kuwait under the chairmanship of Iran to discuss the oil price increase. They were offered the use of the OAPEC building, and the announcement of the price increase was made from OAPEC headquarters. But because Iran chaired the meeting it could in no way be considered an OAPEC meeting. In the afternoon the Arab ministers of oil, who had been invited by Kuwait by direct representation and via the Arab League, met at the Sheraton Hotel in the capacity of ministers of oil and not as the OAPEC Council of Ministers, and it was from there that the oil embargo was announced. Despite the fact that the official communiqué was headed quite clearly 'the Arab Ministers of Petroleum' some of the press assumed it was an OAPEC decision.

This error was reinforced by two further developments. On 8 December there was an OAPEC Council of Ministers meeting in Kuwait in the morning at which the annual budget and other OAPEC issues were discussed. In the afternoon the same people met again to clarify some aspects of the embargo, but this time as Arab ministers of oil. The whole issue became totally confused when the Arab League placed a full-page advertisement in some Western newspapers explaining the embargo and linking it to OAPEC with the words 'OAPEC had decided ...'.

The error was important since OAPEC desperately tried to distance itself from pricing decisions and politics on the grounds that their intrusion would damage the real objectives of the organization. These were set out in Article 2 of the founding agreement.

OAPEC's Objectives

Article 2 stated that: 'The principle objective of the organization is the cooperation of the members in various forms of economic activity in the petroleum industry, the realization of the closest ties among them in this field, the determination of ways and means for safeguarding the legitimate interests of its members individually and collectively; the unification of efforts to ensure the flow of petroleum to its consumer markets on equitable and reasonable terms, and the creation of a suitable climate for capital and expertise invested in the petroleum industry in the member countries'.

The Secretariat interpreted this article under five different headings: (a) coordination of the oil policies of its members; (b) harmonization of the legal systems of member countries to the extent needed for OAPEC to carry out its activities; (c) assistance to members in the exchange of information and expertise and the provision of training to improve the latter; (d) the promotion of cooperation among members in working out solutions to problems facing them in the oil industry; (e) the use of members' resources and common potentialities to establish joint projects in various stages of the petroleum industry.

The boost given to the hydrocarbon sector in the Arab world by the First Oil Shock correspondingly accentuated the potential role of OAPEC. In essence two areas were crucial in the 1970s — joint ventures and coordination. The Arab countries individually sought increased control over their oil and gas sectors. This, in turn, implied the provision of various support services associated with the different stages of the industry. OAPEC saw that these services could be provided by means of Arab JOINT VENTURE companies which began to emerge in the 1970s. The need for coordination arose because the DOWNSTREAM plans of many of the oil producers were in danger of aggravating the already significant overcapacity in refining and petrochemicals. In the event this tended to be less successful, although publication of many of the plans forced the postponement and cancellation of many of the proposed projects.

OAPEC's Decline

The real and genuine progress and contribution of OAPEC were inhibited in the 1980s for two reasons. (a) The Camp David Accords between Egypt and Israel split the Arab world, which in turn severely weakened the ability of OAPEC to make or influence decisions. (b) The falling oil revenues meant that OAPEC was starved of funds, a process reinforced by the creation of the Gulf Cooperation Council which for the richer members gained first claim on funds for inter-Arab organizations. As a result many of the staff taken on in the 1970s left, and the role of the organization was effectively reduced to that of an information centre.

obligatory well. An exploration well which a company is obliged to drill as part of the conditions for being granted the operating licence. Obligatory wells are a variant of MINIMUM EXPLORATION OBLIGATIONS.

OBO carrier (ore/bulk/oil carrier). A COMBINED CARRIER capable of carrying oil, bulk products or ores in a three-way voyage.

observation well. A well drilled to permit observation of the behaviour of a reservoir (e.g., fluid or pressure levels) as production continues. *See also* DRILLING.

ocean leakage. Losses incurred in the ocean transportation of crude oil and oil products.

It has been estimated that 0.5 percent may be lost in this way.

OCN. *See* MOTOR OCTANE NUMBER.

octane number. A measure of the OCTANE RATING. *See also* MOTOR OCTANE NUMBER; RESEARCH OCTANE NUMBER; ROAD OCTANE NUMBER.

octane rating. A product quality specification indicating the tendency of MOTOR SPIRIT to ignite spontaneously (i.e. knock). *See also* QUALITY, MOTOR SPIRIT.

odorant. A compound (e.g., MERCAPTANS) that is added to natural gas and natural gas liquids, which are normally odourless, to impart a smell and thus assist in their detection in the case of leakage.

OECD (Organization for Economic Cooperation and Development). An organization based in Paris. OECD is often used as a convenient shorthand for the industrialized countries.

official price. The publicly stated price at which sellers of crude oil or gas (normally government sellers) will sell. It is usually signified by the GOS.

offshore. Until the 1920s oil and gas were produced only where facilities could be placed on land (i.e. onshore). However, in the 1920s companies began to develop the ability to explore and develop capacity in swamp regions such as Louisiana and in lakes such as Erie and Maracaibo. It was not until the late 1940s that such facilities began to be developed at sea (i.e. offshore) in the Gulf of Mexico. As offshore technology developed it effectively opened up vast areas of new potential acreage. The potential acreage expanded as the technology allowed operations in increasingly deeper and more distant waters. This was an important factor in the increase in excess producing capacity which was experienced from the 1950s.

offtake facilities. The installation of equipment, such as pipelines, storage and pumping facilities and tanker-loading facilities, at the development stage of an oil-field project.

offtake nominations. In JOINT VENTURE oil and gas operations the owners of the company usually are entitled to take a share of the oil or gas produced based on their EQUITY holding. The offtake nomination is the amount of their entitlement which they wish to LIFT in the coming month, quarter or year. If they want less than their entitlement they are described as underlifters and if they want more, overlifters.

offtake obligations. In a JOINT VENTURE, each participant must make arrangements to lift the quantity of each product allocated to it so as not to prejudice the interests of other participants.

offtake provisions. In the UK, detailed offtake provisions are decided upon after the offtake system has been agreed in the ANNEX B. The arrangements include, for instance, forecasts of production and lifting entitlements (*see* OFFTAKE NOMINATIONS; OFFTAKE OBLIGATIONS).

OGSP. Official government selling price (*see* GOS; OFFICIAL PRICE).

oil basin. A geological structure that may contain oil or gas. *See also* GEOLOGY.

oil equivalent. The amount of heat contained in 1 tonne of crude oil. The oil equivalent can be used to compare or aggregate the heat content of other fuels. Hence different fuels are compared to oil. Because crude oils vary in composition it is an imprecise measure, and the actual conversion figure should always be stated. For example, British Petroleum's *Statistical Review of World Energy* assumes that 1 tonne of oil is equivalent to 1.5 tonnes of coal or 1111 cubic metres of natural gas. *See also* APPENDIX.

oiler. A small fuel oil-carrying ship.

oil-field. *See* FIELD.

oil-in-place. An estimation of the total amount of oil in a reservoir whether it is recoverable or not. *See also* RESERVES.

oil operating company. An oil-producing company that is a JOINT VENTURE. The administrative format of such companies can vary enormously, ranging from independent corporations in which parent companies own an EQUITY share (such as ARAMCO was) to simple field operators who made crude oil available to the parents (such as the IRANIAN CONSORTIUM). Whichever format is used, the operating company is responsible for all activities associated with the production of crude oil. This is different from some joint venture arrangements in which a production licence is granted to a joint venture company, but one of the partners takes on the responsibility for producing the crude oil.

The history of the international oil industry shows such operating companies were common and important. They were common because, at the time that many large concessions were granted, the parent governments intervened to secure participation for their oil companies. In addition, in cases of very large discoveries such as Saudi Arabia, the original operator took in partners in order to minimize the impact of increased supply on competition. Such companies were important since they provided HORIZONTAL INTEGRATION at the crude-producing stage of the industry which led to joint control of crude oil supply. Thus, for example, the main operating companies of the Middle East were almost exclusively owned by various combinations of the MAJORS.

oil pledging (royalty bidding). The offering of a proportion of the oil to the owner of an acreage when companies are competing for the acreage by auction. *See also* AUCTION SYSTEM OF LICENSING.

oil region. An area containing a number of OIL-FIELDS and OIL BASINS in close proximity.

oil reservoir. A rock formation in which oil has accumulated. *See also* GEOLOGY.

oil seepage. The slow release of oil when the CAP ROCK of a reservoir is broken. Eventually gas or oil may reach the land surface.

oil shale. A sedimentary rock that contains a high proportion of organic matter (largely KEROGEN). It yields oil when destructively distilled, although the percentage yield varies considerably, but averages around 10 per cent. In theory, the world contains large quantities of oil 'reserves' in shale. However, its production requires the handling of massive quantities of shale and almost as much spent shale after the oil has been extracted. This causes significant environmental problems. In addition, given present technology, the cost of producing shale oil is roughly twice that needed to produce North Sea crude.

oil shocks. Sudden large changes in the price of crude oil. The term grew in popularity after 1978 when the terms FIRST OIL SHOCK and SECOND OIL SHOCK were used to distinguish between the price increases of 1973 and 1979–80, respectively. Subsequently the term THIRD OIL SHOCK has been used to identify the period of falling prices since 1982. The term grew in popularity as the mass media discovered energy as an issue, but it is increasingly used by oil industry analysts as a convenient piece of jargon.

oil spill. An unintended release of oil into the environment. In the popular mind they are normally associated with tanker disasters at sea such as the *Torrey Canyon* in 1967 and the *Amoco Cadiz* in 1978. More damaging oil spills have accompanied well BLOW-OUTS. Thus although the *Torrey Canyon* released 95 000 tonnes of crude oil and the *Amoco Cadiz* 220 000 tonnes, the Ixtoc-1 blow-out off the coast of Mexico in 1979 released about 400 000 tonnes of crude oil.

oil string. *See* DRILLING.

oil trap. A geological structure that prevents the further migration of oil or gas hence creating an oil-field. *See also* GEOLOGY.

oil/water contact. The boundary surface between an accumulation of oil and an underlying accumulation of water. *See also* GEOLOGY.

oil well production. The volume of oil that can be produced from an oil well and, less explicitly, at what cost. The range of production can vary enormously from STRIPPER WELLS which may produce less than 20 barrels a day to some wells that can produce in excess of 10 000 barrels a day. There is an inverse relationship between production volume and costs.

OLADE (Organizacion de Latinamerica de Energía). A group of 22 Latin American

Energía). A group of 22 Latin American countries formed in 1972 to cooperate in areas of energy.

old oil. *See* NEW OIL.

olefins. *See* ALKENES.

oligopoly. A MARKET STRUCTURE that exhibits a high degree of concentration (*see* CONCENTRATION RATIOS), hence supply is dominated by relatively few firms. The importance lies in the nature of the pricing/output decisions of the firm, since before making such decisions a single firm must consider how its rivals will respond to that decision. Therefore the market is characterized by relatively high degrees of uncertainty. This can often result in PRICE WARS when a cut in price by one firm aimed at increasing its market share can provoke hostile reactions from its rivals. To avoid the consequence of this uncertainty there is a natural tendency for oligopolistic firms to collude. Such collusion can range from a loose agreement to exchange information to a full-scale CARTEL which sets levels of market share. Because such collusion is regarded as against the consumer interest or in economic terms produces a misallocation of resources most countries use MONOPOLY LEGISLATION to prevent such agreements. Economic theory has developed a number of variants of the oligopoly model such as duopoly (two firms dominating) and the dominant firm oligopoly (one firm leads on price/output decisions and others follow).

The international oil industry has always been regarded as an oligopolistic market. Many attempts have been made to explain the industry's behaviour using oligopoly as a starting point. Up to the 1970s the MAJOR companies dominated all aspects of the industry, although during the 1960s this was somewhat weakened as a result of the entry of the INDEPENDENTS into the international scene and the creation of the national oil companies. During the 1970s it might be argued that this concentration of the whole industry was weakened as the producing governments took over the operating companies. In addition, the HORIZONTAL INTEGRATION and VERTICAL INTEGRATION, which had characterized the industry, disappeared. However, in effect at the crude-producing stage the oligopoly of the companies was replaced by an oligopoly of the producing countries. The countries (most of which were OPEC members) proceeded to exhibit classic oligopolistic behaviour — uncertainty arising from the absence of information, price wars and attempts at collusion. Meanwhile the oligopolistic structure of the companies at the refining and distribution stage of the industry remained intact.

OMV. The Austrian national oil company.

onshore. *See* OFFSHORE.

OO carrier. Ore/oil carrier (*see* COMBINED CARRIER).

OPEC (Organization of Petroleum Exporting Countries).

OPEC's Creation

The organization was formed in September 1960 in response to pricing decisions taken unilaterally by the oil companies. As a consequence of the Gulf BASING POINT PRICING system the international price of oil had been directly linked to the US domestic oil price (*see* CRUDE OIL PRICES). Following World War II, the oil companies tried to maintain this link, and when in 1957 US domestic prices increased (following the Suez crisis) the oil companies increased POSTED PRICES to match. However, the price link was becoming increasingly untenable as relative production costs in the USA and the Middle East diverged and as those companies with large-scale access to Gulf supplies looked to increase their market share. In 1959 and again in 1960, the oil companies cut posted prices in the Middle East and Venezuela without consulting the governments. Such action, apart from reducing the producing government's revenue, highlighted the extent to which such governments were hostage to the companies.

Prompted by Venezuela, who as a relatively high-cost producer stood to lose much from further price cuts, a conference was held in Baghdad in September 1960. At the conference Venezuela, Saudi Arabia, Iran, Iraq and Kuwait signed the founding agreement to create OPEC with the explicit purpose of restoring the cuts in posted price and requesting more consultation from the oil companies.

During the 1960s membership of OPEC expanded, with Abu Dhabi, Algeria, Indonesia, Libya and Qatar joining, and a Secretariat was created based in Vienna. The impact of OPEC, however, was rather limited. Its main function during the 1960s was to act as a forum in which the less important terms of the CONCESSIONS could be discussed and to some extent renegotiated, most notably the issue of royalty EXPENSING.

OPEC's Changing Role in the 1970s

The role of OPEC began to change in the late 1960s and early 1970s as it sought to extend its 'negotiating net' to the issue of PARTICIPATION and crude oil pricing. At the same time its precise role in events also became increasingly controversial with the debate revolving around the question of how far OPEC as an organization initiated and directed and how far OPEC as an organization simply reacted and followed, although individual members may have initiated and directed. Certainly OPEC assisted the oil producers by allowing an easy and effective exchange of information and at certain crucial times provided support via bargaining solidarity. However, the participation issue became largely a Gulf affair and in pricing the initial breakthrough came from Libya. Although OPEC was central in the TEHERAN AGREEMENT and the TRIPOLI AGREEMENT it was again the Gulf countries that initiated the October 1973 price increase. However, the December 1973 price increase was a full OPEC affair, albeit orchestrated by the Shah of Iran. Many minor changes to the fiscal arrangements were also settled in the OPEC forum. Whatever the reality, in this period OPEC came to the attention of the world, and since this was courtesy of the Western media much of the perception was grossly distorted. OPEC became fixed in the Western mind as a collection of grossly rich Arab Sheiks. This, in turn, began to influence consumer government attitudes to the international oil scene.

The role of OPEC in the period 1974–79 was equally controversial as economists began to describe the organization as a CARTEL pointing to the apparent price-setting role of the ministerial meetings. However, it is possible to argue that it was in fact Saudi Arabia which set the marker price (*see* MARKER CRUDE) within the OPEC forum, whereas the DIFFERENTIAL PRICES were set by the market place. If this view is accepted, the absence of formal production control means that the organization could not be characterized as a cartel. During the period between the two oil shocks (*see* FIRST OIL SHOCK; SECOND OIL SHOCK) OPEC began to set up mechanisms to provide aid to the THIRD WORLD, culminating with the creation of the OPEC Fund in 1976. The giving of aid, both individually and collectively, although at times clumsy and rather selective was in general extraordinarily generous.

During the Second Oil Shock most observers would agree that the role of OPEC as an organization was minimal. Meetings were held and price structures were agreed, but these were ignored by most members as they followed the chaos and panic of the market. OPEC was very much a bystander as prices simply responded to the short-term whims of the market place. However, it was OPEC that eventually restored a relatively orderly price structure with its meeting of October 1981 which set the marker price at $34 per barrel.

OPEC Unequivocally Moves Centre Stage

Subsequently, OPEC played a crucial role in the oil market. Much of the VERTICAL INTEGRATION of the industry was effectively destroyed during the Second Oil Shock as the governments began to take full responsibility for marketing the crude oil. Effectively this left only Saudi Arabia as SWING PRODUCER to act as the mechanism to control the excess crude-producing capacity. Because the oil demand had fallen as a result of the two oil shocks and non-OPEC oil production had increased for the same reason, excess capacity was rising rapidly. The Saudi swing role was not strong enough to control this, and prices began to weaken.

The creation of an additional control mechanism was urgently needed if prices were not to collapse. In this context, OPEC stepped in to try to recreate the horizontal integration of the 1950s and 1960s, which the oil companies had so successfully used to defend prices. They did this by trying to introduce production controls by allocating a QUOTA to each member. This was the first time OPEC had attempted such explicit production controls since an abortive attempt in the mid-1960s. Whatever the debate of the 1970s, OPEC was now definitely trying to behave as a cartel.

The first attempt at controlling prices came in March 1982, but discipline over quotas very rapidly collapsed, and the downward pressure on the price structure re-emerged.

A year later OPEC tried again, and after an extremely long meeting produced the LONDON AGREEMENT which reimposed production quotas and at the same time cut $5 per barrel off the marker price.

The Divisions

Much about this meeting epitomized what has bound and divided OPEC as an organization. Their common ground is derived from the fact that the members of OPEC are all Third World countries, although the smaller Gulf States do not fit the per capita income criterion for Third World status. The common dependence on oil revenues is the other binding feature. In 1981, probably the year of peak dependence, around 95 percent of OPEC's exports were in the form of crude oil, gas or oil products. The final binding element of the organization is the history (since 1960) of meeting, disagreeing, compromising and meeting again. The sources of divisions are legion. They are divided geographically in terms of location and size (both area and population). They are divided economically in terms of their degree of poverty, their overall economic base and their economic potential. They are divided politically in terms of ideology, objectives and decision-making structures. They are divided socio-culturally in terms of religion and history. The final division, and possibly the one of greatest operational significance, is a division of time horizon. In the mid-1980s, of the 13 members (Venezuela, Saudi Arabia, Iran, Iraq, Kuwait, Algeria, Indonesia, Libya, Qatar, United Arab Emirates, Ecuador, Gabon, Nigeria) six have reserves that would last more than 50 years, whereas five have reserves of less than 30 years. Thus, although short-term pressures make current revenue maximization a general goal, the effect of current price on future oil demand is of more interest to some members than to others.

The Problems Facing OPEC

Faced with high probability of a price collapse OPEC has managed to frighten itself into burying the members' differences and these to agreement on a division of the production, as seen with the London Agreement of March 1983. Two problems then emerged and dominated events until September 1985 and after December 1986 to the time of writing. The first was the loss of information arising from the destruction of much of the vertically integrated structure referred to earlier. When OPEC met to set the overall level of their output there was no way of knowing what supply was needed to balance the market. This was of crucial importance since OPEC decided to set both price and production. To do this it is necessary to have exact information since the price and production decision is constrained by the EXOGENOUS existence of a DEMAND CURVE. Normally cartels set either price or production, and then accept the other. The second problem was that of members accepting their quota in the meeting, but then going home and pumping more than their quotas in the face of pressures to obtain either revenue or ASSOCIATED GAS.

Both factors, in the 1983–85 period, tended to maintain the downward pressure on prices. However, OPEC decisions were made viable by the existence of Saudi Arabia's swing role (*see above*). While the Saudis were willing to swing error and cheating could be absorbed. Without it the OPEC control mechanism was too clumsy and inflexible to survive. The problem was that cheating meant that Saudi production fell lower. By summer of 1985, production in Saudi Arabia fell so low that the government felt no longer able to continue, and the swing role was explicitly rejected. At the December 1985 meeting it was announced (largely under the impetus of Saudi Arabia and Kuwait) that in the future the organization would seek its 'fair market share'. Although in economics this concept has no meaning, the market interpreted it as a declaration of a price war and the oil price collapsed.

In effect OPEC unwittingly had fallen over the edge of the cliff from which they had been pulling back since 1982. In free-fall the enormity of what had happened became apparent. Under enormous pressure from all sides (and especially the USA which could not live with such low oil prices) the Saudis reversed their policy, sacked Sheikh Yamani and invoked the calling of the special OPEC committees which had been created to maintain events and encourage a collective OPEC response. At the December 1986 meeting a new production agreement was hammered out based around an oil price (of a basket of crudes) of $18.

operating companies. *See* OIL OPERATING COMPANIES.

operating costs. The costs of an operation.

Operating costs vary with output and are simply the costs of operating an existing activity. They are distinct from fixed costs (or sunk costs) which are incurred whether the operation is producing or not. In effect they are short-term MARGINAL costs and are therefore crucial in the decision over whether to produce or not and how much to produce. In the case of both oil and gas there is a range of operating costs depending upon the geological structure. However, since oil and gas are highly capital-intensive (which represents fixed costs) the absolute range of operating costs (which are a relatively small percentage of total PRODUCTION COSTS) is much smaller than is popularly believed.

operation contract. *See* SERVICE CONTRACT.

operator. The company that undertakes the actual work to find, develop and produce an oil- or gas-field. It may either be a CONSORTIUM company (an OIL OPERATING COMPANY) or a company that is a consortium member.

opportunity cost. When an economist thinks of costs, he thinks in terms of the idea that the resources used to produce one good could have been used to produce something else. This thought process derives from what might be called the basic economic problem. Society and its members have unlimited wants and since the resources used to satisfy these wants are limited, they are insufficient to satisfy all wants. Therefore, the basic economic problem is to choose which wants to satisfy and which to leave unsatisfied. For example, if a consumer with a choice of two goods — A and B — and an income sufficient to buy only one chooses good A, then the opportunity cost of the choice is good B. The opportunity cost of a barrel of oil or cubic metre of gas consumed today is that it cannot be consumed tomorrow. The use of the term is aimed at emphasizing the need to ensure that resources are allocated in an EFFICIENT way, thereby ensuring that the maximum number of 'wants' which can be satisfied (given the resource base) are in fact satisfied. The term can be applied to an individual, company or economy, or indeed on a world basis.

opportunity cost of capital. All actions have OPPORTUNITY COSTS because, resources being limited, when one action is taken some other opportunity is necessarily foregone. Thus undertaking a particular investment project implies foregoing the opportunity to undertake some other project. The rate of return on the foregone investment is the opportunity cost of capital (*see* DISCOUNTED CASH FLOW). For example, an oil company investing in an oil-field in the North Sea foregoes the opportunity to use the resources in the Middle East and therefore gives up the return it could have expected on the alternative investment project. The opportunity cost of capital, which is invariably difficult to estimate in practice, is the appropriate measure of the DISCOUNT RATE. A less appropriate measure, but one which is frequently used, is the WEIGHTED AVERAGE COST OF CAPITAL. In a company with some EQUITY CAPITAL and some DEBT (fixed-interest borrowing), for example, the weights used would be the shares of equity and debt in total capital. They would be applied to the interest costs of each form of capital to obtain a weighted average.

optimal. Literally meaning the best possible. It is used in economics to refer to the most efficient resource allocation. A commonly used version in economics is PARETO OPTIMAL.

option. An agreement between a buyer and seller which allows the holder to buy (or sell) at a set price within a given time period. An option to sell is a put option and one to buy is a call option. An option is different from a contract used in the futures or FORWARD MARKET since the holder of an option is free to choose whether to buy or sell, whereas a signatory to a forward contract must buy or sell. In effect options are rather like insurance in the sense that in return for a 'premium' the buyer obtains a policy (the options contract) which allows him to buy (i.e. call) or sell (i.e. put) a crude futures contract for a period of time at a given price (the strike price). An example may help to illustrate the mechanism.

A buyer of a '$18 April put' might pay a $2 per barrel premium in return for the right to buy a short futures position before April. If prices then fell to $15 the buyer could exercise his option forcing the seller to bear the cost of acquiring a short futures position at the $18 strike price. Thus the option buyer gains $1 a barrel (i.e. the $3 difference between strike price and the actual price less the $2 premium) whereas the option seller loses $1 (i.e. the $3 drop in price plus the $2 premium).

ordinary shares. Shares that only pay a dividend after holders of DEBENTURES and preference shares (*see* BONDS) have been paid. *See also* EQUITIES.

ore/oil carrier. *See* COMBINED CARRIER.

organic trade. The trade of a company that is based on TERM CONTRACTS as opposed to INORGANIC TRADE which refers to one-off transactions. Prior to the SECOND OIL SHOCK most crude oil was traded on an organic basis; subsequently increasing amounts were traded inorganically. Gas transactions are predominantly organic because of the fixed facilities required for transport, pipeline or LNG.

Organizacion de Latinamerica de Energía. *See* OLADE.

Organization for Economic Cooperation and Development. *See* OECD.

Organization of Arab Petroleum Exporting Countries. *See* OAPEC.

Organization of Petroleum Exporting Countries. *See* OPEC.

original-in-place reserves. The original amount of oil or gas in a RESERVOIR. It is equal to OIL-IN-PLACE plus past production.

other things being equal. *See* PARTIAL EQUILIBRIUM ANALYSIS.

outage costs. The costs associated with a cutoff of the supply of an input (normally unforeseen). For example, a refinery's outage cost refers to the consequences of being unable to obtain sufficient crude to run the refinery units at full capacity. If outage costs are high, as they are with refineries because OPERATING COSTS rise exponentially with below-capacity operation due to high FIXED COSTS, then security of supply of the input becomes a matter of crucial concern. High outage costs are frequently an encouragement to VERTICAL INTEGRATION into UPSTREAM operations to try to ensure supply security. This has been the case in the oil industry.

outstep well. *See* APPRAISAL WELL.

overlifter. *See* OFFTAKE NOMINATIONS.

overshot. A drilling tool used to recover equipment lost down a well during DRILLING.

owned crude (equity crude). The crude oil to which a company is entitled by virtue of its equity share in an oil-producing company. The term is somewhat misleading since in most legal systems ownership of the crude is vested in the state.

P

packed column. A DISTILLATION COLUMN or absorption tower filled with small objects (packing) that force rising vapours to mix.

packer. A device that can be expanded against the well bore or casing to isolate annular sections.

packing. *See* PACKER.

paper barrel. *See* WET BARREL.

paraffin. A term used in the UK for high-grade KEROSINE burned in lamps and portable space heaters. *See also* FINISHED PRODUCTS, KEROSINE.

paraffin-based crudes. *See* PARAFFINIC CRUDES.

paraffinic crudes (paraffin-based crudes). Crude oil with a high content of paraffins (*see* ALKANES). The term is sometimes used as a classification of crudes together with the term naphthenic crudes, which are crudes high in NAPHTHENES.

paraffins. *See* ALKANES.

paraffin wax. WAXES normally derived as a by-product of the production of lubricating oils. Paraffin waxes are composed largely of NORMAL-PARAFFINS.

paramarginal reserves. Oil and gas resources which at current prices are not worth developing and/or producing, but only just. The implication is that an increase in current price would bring such reserves into the market.

parameter. A term frequently used in a mathematical context to refer to a constant quantity in an algebraic expression. For example, in the expression

$$Y = aX + bZ$$

X, Y and Z are variables, whereas a and b are parameters. By definition the parameters are always constant, but only for a particular situation or condition. Parameters may vary if the situation changes. The concept is widely used in economics.

parcel carrier. *See* TANKER TYPES.

Pareto improvement. *See* PARETO OPTIMAL.

Pareto optimal. A concept that underlies much of modern WELFARE ECONOMICS. A situation is Pareto optimal if it is impossible by reallocating resources to make somebody feel better off except by making someone else feel worse off. This provides a definition of efficiency: when a resource allocation is not Pareto optimal, society is wasting resources through unnecessarily foregoing an opportunity for improvement. Conditions for Pareto optimality have been specified and provide the basis for a 'first best' efficient reference solution in welfare economics. For example, it can be shown that under stringent assumptions a perfectly competitive economy (*see* COMPETITION) will be Pareto optimal, whereas one that is imperfectly competitive or generates EXTERNALITIES will not. Pareto optimality offers a criterion — the Pareto improvement — for evaluating alternative economic situations and suggesting policies. However, it is limited because many policy options (e.g., building a gas pipeline) involve both gainers and losers, and hence cannot be ranked by this criterion. A more complete criterion, but one that is more demanding in its assumptions, is used in COST–BENEFIT ANALYSIS and is based on the

COMPENSATION PRINCIPLE: a 'potential Pareto improvement' occurs if the gainers from a new situation could compensate the losers and retain a surplus. The terms are named after Vilfredo Pareto (1848–1923) an Italian economist and sociologist.

partial equilibrium analysis. A tool of analysis commonly used by economists to analyze specific markets. It consists of examining the behaviour of the variables in the market such as supply and demand and their effects (e.g., on price), ignoring by holding constant via the other things being equal assumption other markets that may well have linkages with the market under scrutiny. For example, the effects on price of a sharp fall in oil supply will be examined assuming that the resultant price change does not affect the general level of economic activity, and hence the demand for oil. The main justification for such an approach is that in complex situations it enables the economist to focus, for exposition purposes, on individual aspects of the situation in order to highlight important issues. This can be valuable as an aid to understanding, but is not a reflection of the way the real world operates (nor is it meant to be) unless a very short period of time is considered. Because it is simple and requires little data, partial equilibrium analysis is the starting point used in most introductory economics courses. The alternative and much more complex GENERAL EQUILIBRIUM ANALYSIS, which tries to account for different interactions, does not occur until a relatively advanced stage of study and normally on a fairly mathematical basis. This means that many people view economists as being naive and unrealistic, since they assume that partial equilibrium analysis is the way economists actually view the world. It also means that many who do not understand the purpose of the tool use it anyway, thereby producing meaningless and misleading results which further discredit the methodology. *See also* EQUILIBRIUM.

participation. The government's involvement, usually by means of ownership of EQUITY, in the activities of private companies. In the international oil industry the term has a long history and an important role. In many cases, when the first oil CONCESSIONS were being granted, the host governments were promised the possibility of participation when the OIL OPERATING COMPANIES offered shares to the public. The problem was that no share issue ever appeared, and therefore the host governments had no *de jure* say in the way that the operating companies actually operated. This was viewed by the governments as an infringement of their sovereignty and caused growing resentment against the oil companies, especially where oil dominated the local economy. This resentment, which emerged also in non-oil mineral sectors, was reflected in a series of United Nations resolutions on the issue of 'permanent sovereignty' over natural resources. The first resolution was passed in 1952.

During the 1950s and 1960s, in a context of growing opposition to 'imperialism', pressures on governments to nationalize the oil companies grew. In the Arab world this reached a peak in June 1967 as both the USA and the UK were popularly implicated in the Israeli victory. For the most part, however, Arab governments were opposed to such a move partly because of the memory of the IRANIAN NATIONALIZATION of 1951 and partly because the crude oil market was weak, and hence selling crude oil was unattractive. However, an alternative was required to mute the growing clamours for nationalization. This emerged tentatively in 1967, more explicitly in 1968 and in detail in 1969, when the Saudi Arabian Oil Minister Sheikh Zaki Yamani outlined his ideas on participation. Yamani argued that crude oil prices were weakening because the major oil companies' control of supply was being weakened by the rise of the INDEPENDENT oil companies and the producing countries' national oil companies. Thus to take over the crude oil production by nationalization would only aggravate the market weakness. Instead he proposed that producing governments should take on equity share not only in the oil operating companies, but also in the refinery operations of their parents. This would retain the VERTICAL INTEGRATION of the industry which was crucial to defend the price structure.

Yamani's ideas caught on rapidly. In July 1968, OPEC endorsed a policy of 'reasonable' participation, and in July 1971, a further OPEC resolution called on members to take 'immediate steps' to implement participation. The strategy for the Middle East countries was to hold joint negotiations between the companies and governments to reach a broad agreement which would form the basis of more specific negotiations between individual governments and companies. These negotiations were begun in May 1972, and the General Agreement On Participation was reached in October, although Iran negotiated separately and Iraq did not sign leaving Saudi Arabia, Kuwait, Abu Dhabi and Qatar. The Agreement gave the governments an initial 25 percent stake (in the oil operating companies only) rising to 51 percent by 1982. The agreement was extremely complex, but since it was soon overtaken by events the details are only of academic interest. Gradually, the signatories demanded and obtained a greater participation share until the governments secured 100 percent of the operating companies, the exception being Abu Dhabi with 60 percent. The rapid dating of the October agreement was the result of several factors. Crude oil prices had strengthened, and some countries had nationalized without disastrous results. Libya had negotiated a similar deal, but with much more favourable terms such as an immediate 51 percent equity. The key consequence of these developments was the destruction of the HORIZONTAL INTEGRATION of the international crude supply industry (*see* CRUDE OIL PRICES).

Participation re-emerged in 1976 when the UK government obtained a 51 percent share in the oil-fields of the North Sea.

participation oil. The crude oil entitlement of governments by virtue of their equity share in an OIL OPERATING COMPANY. For example, the BNOC, established by a Labour government on 1 January 1976 and later privatized by Conservative government, was granted a 51 percent equity share in all new licences (*see* LICENSING SYSTEMS). In the sixth licensing round, companies could offer BNOC a greater than 51 percent share. Participation oil refers to the 51 percent of oil produced under then existing licences which BNOC was entitled to purchase at the market price. *See also* DISCRETIONARY SYSTEM OF LICENSING.

payback period (payout period). The period of time it takes for an investor to recover his initial investment from the stream of revenues. It was widely used as a criterion to evaluate investment projects (*see* INVESTMENT APPRAISAL): the shorter the payback period the better the investment. Unfortunately, it suffers from the fact that it ignores any revenue benefits after the payback date. Furthermore, it can only be used to assess which project should be done, but it cannot establish if any project at all should be undertaken. Hence its use in investment appraisal has declined, and it would never be used in isolation from other criteria. It can have its uses, especially in conditions where political uncertainty is high because the company may wish to protect itself against eventual nationalization and thus needs to know how soon it can recover its initial outlay.

payout period. *See* PAYBACK PERIOD.

pay zone. The stratum of rock in which the oil and gas is found. *See also* GEOLOGY.

peak load. The daily demand for gas has two components: a base load and a peak load. The base load is the amount demanded relatively consistently throughout the day, whereas the peak load is the higher demand at certain times of the day when demand rises sharply. This division can create problems in supply because in order to meet the peak load demand it may be necessary to build capacity which for much of the time may remain unused. Although the concept normally relates to daily demand fluctuations the analysis can equally be applied when demand is seasonal (*see* SEASONALITY).

peak sharing. The practice of introducing extra supplies of gas to meet a PEAK LOAD demand.

penetration. A product quality specification for BITUMEN. *See also* QUALITY, BITUMEN.

pentanes. *See* NATURAL GASOLINE.

percentage profit rate bids. *See* NET PRODUCT BIDDING.

percussion drilling. A type of DRILLING in which the hole is made by continuous raising and lowering of the BIT.

perfect competition. *See* COMPETITION.

perfect knowledge. An economist's assumption required for the existence of perfect competition (*see* COMPETITION). It means that all necessary market information on prices being asked and offered is immediately and costlessly available to all the market participants.

perforating gun. A tool used to perforate the formation allowing the oil to flow. *See also* DRILLING; PRODUCTION.

perforating the casing. An operation that may be carried out during DRILLING operations during which the casing, or lining, of the well is pierced in order to allow oil to flow into the well for testing. *See also* PRODUCTION.

performance numbers (PN). A product quality specification used for aviation fuels. *See also* QUALITY, AVIATION FUELS.

permeable. A geological term for rock that permits fluid, and hence oil, to pass through it. *See also* GEOLOGY.

perpetuity. An ANNUITY that is paid for an infinite period of time. The PRESENT VALUE of a perpetuity (P) is P/r, where r is the rate of discount. *See also* DISCOUNTED CASH FLOW.

Persian Gulf. *See* GULF, THE.

petrochemicals. Substances (e.g., plastics, fertilizers, synthetic fibres, etc.) that have been manufactured by chemical means using petroleum compounds as raw materials. The processes by which petrochemicals are manufactured belong within the chemical industry and hence are outside the scope of this book. Petrochemicals are discussed here mainly in terms of the FEEDSTOCK requirements which they present to the oil industry.

A wide range of raw materials is required for the manufacture of chemicals, plastics, synthetic fibres, etc. These raw materials are defined in terms of individual hydrocarbon compounds (e.g., ETHENE or PROPENE) which may not be directly available from refineries in the required quantities. The basic raw material requirement therefore has to be translated into a requirement for a feedstock (e.g., naphtha) which can be processed to give the required individual raw materials.

Many of the raw materials used in the petrochemical industry are UNSATURATED HYDROCARBONS, since these are much more chemically reactive than the SATURATED HYDROCARBONS, combining readily with other substances to form new compounds (*see* CHEMISTRY). A key stage in the process of converting available petroleum feedstocks into petrochemical raw materials is therefore the formation of unsaturated hydrocarbons from saturated hydrocarbons by the removal of atoms of hydrogen. The most important route by which this is carried is by STEAM CRACKING, using principally petroleum gases or NAPHTHA as feedstock. One example is the conversion of ethane (C_2H_6) to ethene (ethylene).

Another example is the conversion of propane (C_3H_8) to propene (propylene, C_3H_6). The desired end products of this process are mainly ethene, propene, butene and butadiene, and these can be isolated for use in a wide range of products. These products are the major petrochemical raw materials and are often referred to as the lower olefins (i.e. olefins — unsaturated hydrocarbons with one or more double bonds — with a small number of carbon atoms and hence at the lower end of the olefin series).

A more severe version of the steam cracking process can be used to obtain ACETYLENE (C_2H_2), essentially by removing more hydrogen.

The high cost of steam cracking, together with the availability of alternative manufacturing routes to many of the acetylene derivatives has led to the decline in importance of acetylene as a petrochemical raw material.

Another important group of petrochemical raw materials are the AROMATIC COMPOUNDS, principally BENZENE, TOLUENE and the XYLENES, collectively known as BTX. These are produced by removing hydrogen in a process which is virtually identical to that used in CATALYTIC REFORMING.

Two routes of less importance in the preparation of raw materials for the petrochemical industry involve the introduction of other chemical elements (e.g., chlorine, nitrogen, oxygen) into the hydrocarbon feedstock.

petrochemical solvents. *See* SOLVENTS.

petrocurrency. A phenomenon that emerged after the FIRST OIL SHOCK. The exchange rate of the oil-producing countries began to be influenced by expectations with respect to oil prices. After 1973, many oil producers' exchange rates became overvalued as the price of oil, and hence oil revenues, rose. Linked to the idea of a petrocurrency is the Dutch disease. This followed observations concerning the developments in the economy of The Netherlands after the large gas discoveries of 1960s; overvaluation of the exchange rate made imports cheaper and exports more expensive. This, in turn, was thought to be responsible for the decline of tradeable goods output (e.g., manufactured goods) and the rise in the output of non-tradeables (e.g., public services), giving rise to what was seen as a process of deindustrialization.

petrodollars. Dollars owned by a national of an oil-producing country. The term came into prominence after the FIRST OIL SHOCK. As oil is normally priced in dollars, the OPEC surpluses which arose because of the inability of the OPEC countries to spend their revenues due to their low ABSORPTIVE CAPACITY were in the form of dollars. The implications of the holding of these dollars for the world's financial system became known as the recycling problem.

petrol (gasoline (US)). The name used by the general public in the UK for MOTOR SPIRIT. *See also* FINISHED PRODUCTS, MOTOR SPIRIT.

petroleum. The generic name for CRUDE OIL, NATURAL GAS LIQUIDS, NATURAL GAS and their products.

Petroleum and Gas Revenue Tax (PGRT). *See* PETROLEUM REVENUE TAX.

petroleum coke. The coke that is produced by the carbonization of high-molecular-weight hydrocarbons and is a residual from THERMAL CRACKING. The three most common types are honeycomb coke used to produce electrodes and in aluminium refining, needle coke used in electronic arc furnaces to produce steel and SPONGE COKE often blended with coal coke. *See also* COKING.

petroleum ether. A special boiling-point spirit (a petroleum solvent) of high volatility and a narrow distillation range. *See also* FINISHED PRODUCTS, SOLVENTS.

petroleum fractions. Divisions of crude oil usually based on a BOILING POINT range which emerges from a DISTILLATION process. *See also* REFINING, PRIMARY PROCESSING.

Petroleum Industry Advisory Committee. *See* PIAC.

Petroleum Weekly News. *See PIW.*

petroleum resins. Solid or semi-solid resins derived from the distillation of special crudes or lubricating oil extracts. They are used as substitutes for natural resins.

Petroleum Revenue Tax (PRT). A specific oil tax central to the UK oil taxation system. It was initially proposed in the 1974 Oil Taxation Bill and implemented, in a somewhat modified form due to industry pressure, in the 1975 Oil Taxation Act. PRT was introduced at a flat rate of 45 percent payable on the landed value of gross revenues less several deductions. The taxable unit is the oilfield, and the tax is levied on six-month chargeable periods. The deductions include OPERATING COSTS, capital costs, Supplementary Petroleum Duty (SPD) payments (when applicable), ROYALTY payments and the equivalent value of the PRT Oil Allowance of 0.5 million long tons per six-month period, subject to a cumulative total of 10 million long tons per field (a measure intended to assist smaller fields). In addition, capital expenditure was uplifted (*see* UPLIFT), so that an amount in excess of the actual capital spent may be offset against PRT, by 75 percent for the period up to PRT payback (i.e. the time when PRT profits have been earned), after which only 100 percent of capital expenditure may be deducted. This was intended as a relief against the high-capital front end loading (*see* FRONT END-LOADED) in the offshore industry and because capital expenditure is calculated in historical terms with no allowance for interest payments. This calculation provides the untapered mainstream PRT liability.

The tapering and safeguard provision provides an annual limit on PRT liability in order to protect oil-field returns from falling to unacceptably low levels. The annual limit is 80 percent of the landed value of gross revenues less royalty payments, spd payments (when applicable), operating costs and 30 percent of accumulated capital expenditure. If the untapered PRT liability exceeds the tapered liability the PRT annual limit applies; otherwise it is ignored.

In the UK offshore taxation system other taxes apply in conjunction with PRT, and the overall system has changed many times since its inception in 1975. Initially there were three levels of taxation: PRT, royalties and corporation tax (CT).

Royalties are a barrelage tax — a direct levy on production — calculated for the licence area, but payable for each field. In 1975 royalty payments were set at a rate of 12½ percent of gross revenues. Royalties are assessed six-monthly and paid two months after the end of each chargeable period. Prior to the Fifth Licensing Round, gross revenues were calculated at the wellhead thus excluded transport costs, whereas after the Fifth Licensing Round, the landed value of gross revenues applied. The Secretary of State for Energy retains the discretionary power to refund all or part of the royalty payments of a field.

CT is levied on the profits of oil companies operating in the North Sea. A RING FENCE around a company's oil extraction activities prohibits losses made outside the North Sea being deducted from North Sea CT profits. CT profits are the landed value of gross revenues less capital costs, operating costs, interest on debt, royalty payments and a deduction for PRT payments. In 1975 the CT rate was 52 percent.

In August 1978 the UK government proposed oil taxation changes. The rate of PRT was to be increased to 60 percent. From July 1978 the PRT capital uplift was to be reduced to 135 percent from 175 percent. The oil allowance in PRT was reduced to 0.5 million tonnes per year subject to a cumulative total of 5 million tonnes per field.

From January 1980 the rate of PRT was increased to 70 percent, and in March 1980 it was announced that advance payments of PRT would be introduced. The advance payment was made two months into each chargeable period and was calculated as being 15 percent of the higher of the two previous period's PRT liabilities.

In March 1981, Supplementary Petroleum Duty (SPD), a fourth tax applying to the North Sea oil extraction industry, was announced. SPD was a 20 percent charge on the landed value of gross revenues less an allowance equivalent to the value of 0.5 million tonnes of oil per six-month period. SPD was deductable against PRT profits and CT profits. From January 1982 the PRT capital uplift was limited to the period up to PRT payback and the tapering and safeguard provision was limited to a period half as long as the period to PRT payback once PRT payback is reached.

In the March 1982 Budget it was announced that as from January 1983 SPD would be abolished and replaced by Advance Petroleum Revenue Tax (APRT). APRT was wholly creditable against later mainstream PRT liability. APRT was 20 percent of the landed value of gross revenues after a deduction for the equivalent value of 0.5 million tonnes of oil. It was calculated six monthly and, as announced in June 1982, applied for 10 chargeable periods. At the end of the tenth period all outstanding APRT was to be refunded immediately. APRT was intended to bring forward government tax revenues. From July 1983 a system of spreading PRT payments replaced advance payments of PRT.

In March 1983 it was announced that a distinction for tax purposes would be drawn between fields granted development consent before and after April 1982. For those fields granted development approval (i.e. their ANNEX B submission had been ratified by the Department of Energy) after April 1982 the PRT oil allowance was doubled to 1 million tonnes each year and 10 million tonnes cumulatively per field, and the 12½ percent royalty payment was abolished. For all fields, restrictions on PRT relief for shared assets such as pipelines were eliminated, and all future exploration and appraisal expenditure could be treated as an allowable expense for PRT relief. In addition, for all fields, APRT would be phased out, and from January 1987 would be zero.

The 1984 Budget gave notice of CT changes. The CT first-year capital allowance of 100 percent would be reduced in stages, and after April 1986 would be 25 percent. The rate of CT was also to be reduced, in stages, and from the 1986–87 fiscal year would be 35 percent.

petroleum spirit. Another name for BENZINE. In the USA it is also used to describe white spirit (*see* FINISHED PRODUCTS, SOLVENTS).

Petroline. The trans-peninsular crude oil and gas pipeline line that links the eastern producing fields of Saudi Arabia (Abqaiq) to the port of Yanbo on the Red Sea. It began operation in 1981, and its capacity has been upgraded to 3.05 million barrels per day.

Petromin. The Saudi Arabian national oil company. It was created in 1962, but its role has always been somewhat unclear with the existence of ARAMCO, the Ministry of Petroleum and later with the Saudi Basic Industries Corporation, which has responsibility for petrochemicals.

petrosulphur compounds. Hydrocarbon chemicals that contain sulphur. Examples include MERCAPTANS.

PFD. Primary flash distillate; the total DISTILLATE from primary distillation. *See also* REFINING.

PFO. Polymer fuel oil; the bottom oil resulting when ETHENE is cracked. It is also known as steam-cracked fuel oil (SCFO). *See als* REFINING, SECONDARY PROCESSING.

PGRT. Petroleum and Gas Revenue Tax (*see* PETROLEUM REVENUE TAX).

phantom freight rate. In Gulf BASING POINT PRICING the sum used to equalize landed prices (*see* LANDED COSTS) in any market with the landed prices from the Gulf of Mexico. *See also* CRUDE OIL PRICES.

phase-down clause (phase-out option). A clause in a crude oil supply contract that allows the buyer to reduce LIFTING — usually at a specified rate in a given period — if the buyer does not like the price being charged by the seller.

phase-in crude. The marketing arrangement for crude oil in the GENERAL AGREEMENT ON PARTICIPATION. This agreement gave the governments a 25 percent equity share in the OIL OPERATING COMPANIES, and hence an entitlement to 25 percent of the crude oil produced. However, many of the governments lacked the marketing expertise or infrastructure to sell the crude. Thus phase-in crude was the part of the government's entitlement which the companies would temporarily lift until the marketing network was developed by the government. *See also* BUY-BACK CRUDE.

phase-out option. *See* PHASE-DOWN CLAUSE.

PIAC (Petroleum Industry Advisory Committee). A committee that provides advice and information on the oil and gas industry to the UK government.

pig (go-devil, mole). A piece of equipment that is used to clean or monitor the interior of a pipeline or to differentiate the throughput of different crudes or products during BATCHING. The pig is inserted into the pipeline and is carried along by the flow of the oil or gas.

pigouvian taxes. *See* 'POLLUTER PAYS' PRINCIPLE.

pinch-out trap. A trap where the reservoir rock holding the oil or gas gets thinner as it nears an impervious area and eventually disappears. *See also* GEOLOGY.

pipe hooks. Equipment that is used to store PIPES used in DRILLING.

pipe-laying barge. *See* LAY BARGE.

pipeline residue. *See* MARKETABLE NATURAL GAS.

pipe racks. Equipment that is used to store PIPES used in DRILLING.

pipes. Sections of the DRILLING STRING around 10 metres in length. *See also* DRILLING.

pipe tongs. Equipment that is used to screw and unscrew sections of the DRILLING STRING (*see* PIPES).

pitch. The black residue (normally solid) produced from coal distillation. Pitch should not be used to describe petroleum products. *Compare* BITUMEN.

PIW. (*Petroleum Intelligence Weekly*). A much respected trade newspaper published weekly.

plateau level. The peak production level

from an oil- or gas-field. Maximum production declines as the reserves are depleted; how long the plateau is maintained can be influenced by the application of SECONDARY RECOVERY and enhanced oil recovery. *See also* PRODUCTION, ENHANCED OIL RECOVERY; PRODUCTION, NATURAL DRIVE MECHANISMS; PRODUCTION, SECONDARY RECOVERY MECHANISMS.

plate column. A DISTILLATION COLUMN used in refineries that contains a number of equally spaced, perforated horizontal plates. *See also* PACKED COLUMN.

platform. An OFFSHORE structure used for drilling and/or production.

platforming. A refining process that uses a platinum-based CATALYST which includes fluorine or chlorine on an alumina base. It is used in the UOP catalytic reforming process. *See also* MOTOR SPIRIT UPGRADING.

Platts. A publisher of newsletters and oil prices. In the crude oil and products SPOT MARKET there is no official central collection of the prices at which contracts are exchanged. Instead spot prices are produced by specialist agencies who in effect 'trawl' for information on prices on a daily basis, and they are usually published as a range with high and low values being quoted. Platts is the most famous of these agencies, and Platts prices are frequently used as the basis for a contract price.

plugging. The filling of a well with concrete once production has ceased and the well is to be abandoned.

PMS. *See* PREMIUM MOTOR SPIRIT.

PN. Performance number (*see* QUALITY, MOTOR SPIRIT).

PNA analysis. A method of classifying crude oils in order to identify the dominant type of hydrocarbon present; PNA refers to the three main hydrocarbon types — PARAFFINS, NAPHTHENES and AROMATICS. *See also* QUALITY, CRUDE OIL.

point elasticity. *See* ELASTICITY.

Point Six Rule. The relationship, established by W.L. Nelson, between the capacity of a refinery and its cost of construction which arises as a result of an aspect of the ECONOMIES OF SCALE. Because oil moves in three-dimensional space, although investment is proportional to surface area output is determined by volume. The relationship can be expressed in the simple formula

$$c_2/c_1 = f(Q_2/Q_1)$$

Where Q_1 and Q_2 refer to two different capacity refineries and c_1 and c_2 refer to their respective costs. The exponent f has a value of about 0.6, hence the name. Similar principles apply to chemical plants, where the exponent varies between 0.45 for chlorine and caustic soda to 0.83 for methanol.

polluter pays principle. This states that polluters should meet the full social costs (*see* COST–BENEFIT ANALYSIS) of their activities, including the costs of any EXTERNALITIES which would otherwise be borne by someone else. If the external damage is 'internalized' in this way, the gap between the polluter's private costs and social costs will be closed. Consequently, even if the polluter passes on some of the extra costs the private decision about how much to pollute will be taken as if it were based on the cost to society of an extra unit of pollution. The instruments that can be used to achieve this internalization include pollution taxes (i.e. pigouvian taxes), direct controls and even the auctioning by the state of 'pollution rights' (i.e. permits to discharge specified amounts of pollutants). Application of the polluter pays principle, in full or in part, has been recommended by the OECD for its member countries.

polymer. A chemical compound formed by the joining together of MONOMERS to form complex, long-chain molecules (e.g., plastics, rubbers).

polymer fuel oil. *See* PFO.

polymer injection. A form of enhanced oil recovery that uses chemicals to improve oil flow. *See also* PRODUCTION, ENHANCED OIL RECOVERY.

polymerization. The formation of chemical compounds (i.e. POLYMERS) by the joining together of MONOMERS. *See also* CATALYTIC POLYMERIZATION.

pool. A single separate RESERVOIR of oil or gas. The key point of a pool is that it is under the influence of a single pressure system. *See also* FIELD.

pool association. *See* ACHNACARRY AGREEMENT.

pore pressure. The natural pressure contained within a formation. *See also* GEOLOGY.

porosity. A measure of the amount of 'gaps' in a rock; it is usually expressed as the volume of gaps to total volume. *See also* POROUS.

porous. A geological term describing rock that contains pores which permit the retention of fluid, and hence of oil. *See also* GEOLOGY; PRODUCTION.

portfolio. The total ASSETS held by an individual or institution. The important element of a portfolio, apart from its size, is its structure: for example, the proportion of the assets that are quickly convertible to cash (i.e. liquid assets), the proportion that are high-risk or low-risk, etc. In petroleum economics, the idea can be used to express views on production levels. Oil in the ground is an asset to the country or company. Its production and sale reduce the oil reserve component of the asset portfolio, replacing it with currency from the oil sale or whatever investment is made with the currency.

There are numerous theories in economics and finance to explain the choice of a portfolio, and such theories were used in the 1970s to determine if the oil-producing countries were behaving in their own best interests. No clear conclusions emerged.

possible reserves. Oil and gas RESERVES that have a less than 50 percent probability of being produced given the existing economic and technical conditions. *Compare* PROBABLE RESERVES.

posted price, crude. Posted prices for crude oil were first used by the STANDARD OIL Trust when it 'advertised' the prices at which it would buy crude by posting notices on the wellheads. This practice continued in the USA, but posted prices reached prominence internationally when the oil companies agreed in the late 1940s and early 1950s to a fifty–fifty profit split with the producing governments. To assess the profits on crude oil production in a country required the calculation of a profit figure by subtracting cost from revenue. However, this generated an immediate problem; because of the high level of VERTICAL INTEGRATION in the international industry most traded crude oil was moving through inter-AFFILIATE channels. In effect, there was no open market for crude oil and hence no price. With no price, revenue (and hence profit) could not be calculated. To overcome this problem the companies set posted prices for the crude oil to be used as the basis for profit calculation. These prices were theoretically supposed to represent the price at which the company would be willing to sell to all-comers. This was designed to prevent companies setting a low price in order to reduce their tax obligation.

Initially posted prices were supposed to approximate to market reality had there been a market, and in general they did. However, the companies attempted to retain the link between US domestic prices and international prices inherent in the GULF PLUS pricing system, and in 1957 the companies increased posted price. This was untenable given the market conditions, and in early 1959 and again in mid-1960 the companies cut posted prices. This led to a cut in the revenue of the oil-producing government about which they had not been consulted; the companies had insisted that the setting of posted prices was their prerogative alone. Not unnaturally the governments objected to the actions of the companies, and they created OPEC in order to restore the cuts in posted price.

Despite the fact that OPEC failed to reverse the cuts during the 1960s, they did manage to provide sufficient threat to prevent any further erosion of posted prices, although realized prices of crude continued to decline. After 1960 the link between posted and market prices was broken, and the posted price become a tax reference price used to compute revenues, profits and tax liability. As a consequence, when in the early 1970s producer governments began to take over the OIL OPERATING COMPANIES the posted price lost its importance, and now it only remains relevant in the few situations (e.g., Abu Dhabi, Libya and Nigeria) where foreign companies still retain an equity interest. *See also* CRUDE OIL PRICES.

posted price, products. The prices that the sellers of products 'advertise' as the wholesale prices at which they would be willing to supply products at specific locations.

pour point. The lowest temperature at which an oil or product will flow (i.e. pour). It is generally taken as the solidifying temperature plus 5°F. *See also* QUALITY, CRUDE OIL; QUALITY, DISTILLATE AND RESIDUAL FUELS.

power kerosine (vapourizing oil). A volatile kerosine that is used as a fuel in some spark-ignited engines such as used in tractors. *See also* FINISHED PRODUCTS, KEROSINE.

predatory pricing. *See* DUMPING.

preference shares. The holders of preference shares are paid out before the holders of ORDINARY SHARES, but after the holders of DEBENTURES. They often carry a fixed rate of dividend and limited voting rights. *Compare* BONDS.

preliminary assay. The carrying out of a detailed analysis on a small quantity of crude oil from an oil-field. Such an assay gives a general indication of crude type, but it does not give sufficient information for designing a refinery or establishing details of product quality. To achieve this, a TBP distillation (*see* QUALITY, CRUDE OIL) is carried out.

premium motor spirit (PMS). A MOTOR SPIRIT with a RESEARCH OCTANE NUMBER (RON) of 97–99 as opposed to REGULAR MOTOR SPIRIT (RMS) with an RON of about 90. The distinction is largely used for pricing purposes with wholesale motor spirit prices (posted or spot) being divided into PMS and RMS prices.

present value. A sum of money expected in the future is worth less at the present time than on its future repayment date. This is because the value now (i.e. the present value) could be invested at a rate of interest to produce more in the future. For example, if the rate of interest were 10 percent then £110 promised in one year would have a present value of £100; this is because if the £100 were invested at 10 percent in one year it would become £110. The present value concept is an integral element in DISCOUNTED CASH FLOW analysis which involves trying to compare money spent today (e.g., on a capital project) with the stream of revenue from the project expected in the future. To make the two comparable the future stream of revenue must be converted into its present value. *See also* ANNUITY.

pressure distillate. The untreated distillate product of THERMAL CRACKING. *See also* REFINING, SECONDARY PROCESSING.

price control. The intervention by a government in order to control prices. Such control may take the form of the direct setting of price ranging from a minimum price (i.e. FLOOR PRICE) to a maximum price (i.e. CEILING PRICE), or the control may be of a more indirect nature.

price cutting. *See* PRICE WAR.

price deregulation. The process that occurred in the USA between 1975 and 1981 whereby the US government removed controls over domestic oil prices.

price discrimination. *See* DISCRIMINATORY PRICING.

price effect. Changes in the amount of a good or service bought as a result of a price change, holding other determining variables constant. Price changes are different to changes that arise from a variation in income (i.e. income effect, *see* SUBSTITUTION EFFECT). *See also* DEMAND.

price elasticity of demand. A measure of the responsiveness of changes in the quantity demanded holding other determining variables constant. ELASTICITY is measured by the elasticity coefficient. If it is less than one, demand is characterized as non-responsive (i.e. inelastic); if greater than one demand is characterized as responsive (i.e. elastic).

price elasticity of supply. A measure of the responsiveness of changes in the quantity supplied holding other determining variables constant. *See also* ELASTICITY.

price index. The price of a good or collection of goods (e.g., oil products) expressed as a percentage of their value at some base period. If a crude oil price index is 120 with the base year 1980 = 100 this means that since 1980 the price has increased by 20 percent. Such indices can be expressed in both NOMINAL or REAL terms.

price maker. *See* PRICE TAKER.

price mechanism. The basic problem with which economics concerns itself is that society has an unlimited appetite for goods and services, but the resources required to satisfy this appetite cannot meet all the goods and services wanted. Because resources are scarce relative to 'wants' society must somehow choose which wants will be satisfied and which will remain unsatisfied. This fundamental problem can be subdivided into three problems which require solution: (a) what will be produced; (b) how they will be produced (i.e. which combinations of resources should be used); and (c) how the output will be distributed. Economics identifies three archetypal solutions. The traditional solution used in simple societies solves the problems by a process of precedence: what is produced is that which has always been produced, it is produced in the way in which it has always been produced and is distributed to those who in the past have received the goods and services. In the command solution a central body decides on the output mix, directs resources to produce that mix and then allocates the output on a rationing basis. The final solution is the price or market mechanism: what is produced is the most profitable as measured by the volume and price of output (i.e. total revenue) less the volume and price of inputs (i.e. total costs). This output is produced by means of the lowest cost method, which is determined by the price of inputs and the technology. The good or service is then distributed on the basis of ability to pay. This is a function of income which is determined by the ownership of factors of production together with their price. The economic problem is solved therefore by an automatic means which Adam Smith (1723–90) described as the 'invisible hand'. The price mechanism can apply at the MACRO level (i.e. for the whole economy) or at the micro level (i.e. for a particular sector such as the oil and gas industry).

The three solutions are not mutually exclusive, and most economies and sectors use all three elements in the allocation of resource. In the international oil and gas industry it has always been the price mechanism which has dominated the resources allocation. This is not to say that both supply and demand, and price have not been manipulated by the various actors in the market which inject significant elements of command into the decision-making process.

price ring. A type of CARTEL in which there is agreement on prices within a market.

price setter. *See* PRICE TAKER.

price shading. The reducing of a price by a small amount in order to increase sales. In the case of oil, there are many mechanisms by which a price may be shaded, ranging from secret discounts to BARTER DEALS. Another common method is to change the credit terms on which the oil is sold.

prices in real terms. *See* CONSTANT PRICES.

price structure. In those cases where a commodity such as oil or gas is not a HOMOGENEOUS PRODUCT and thus has different characteristics (e.g., different chemical characteristics or geographical location), there is no such thing as 'a price'. Instead there exists a structure of prices. Thus the price structure of oil, for example, can change in the sense that all the prices can rise or fall (i.e. the whole structure moves up or down). Alternatively, the price structure can change in the sense that DIFFERENTIAL PRICES within the structure can alter. An example is the widening or narrowing of the range between heavy and light crude oils.

price taker. A supplier can influence price by virtue of the impact of his supply decisions on the total market supply. If his supply decisions can significantly alter market supply, as is the case, for example, with Saudi Arabia, then he is a price maker. However, if his supply decisions have a limited impact on total supply he is a price taker in the sense that the market price is outside of his control or influence; he simply accepts the price as given.

The distinction between price makers and price takers is often not as clear cut as the definition suggests. For example, it is usually argued that the UK as an oil exporter is a price taker because the UK sector of the North Sea has supplied a relatively small proportion of internationally traded oil. However, in a market where perceptions and expectations are important, the decisions of some small producers such as the UK can influence the market and give them the status of a price maker, at least in a partial sense.

price war. In an oligopolistic market (*see* OLIGOPOLY) which is characterized by a few sellers dominating the market, price decisions are constrained by the possible response of rival price setters. If one supplier decides to increase its market share by cutting prices, other suppliers may respond. This may cause the original price cutter to cut prices further, which in turn provides further price cutting from the rivals. The result is a price war in which price cutting is used to increase market share and possibly to drive competitors out of business or force competitors to collude. It is these latter points that differentiate price wars from price competition. If the rivals are strong the final result is usually a similar share of the market, but at a lower price. Although this may be a satisfactory situation for the consumer, it is not for the suppliers. In general price wars are avoided either by tacit understanding or by some more formal collusion such as a CARTEL.

Prior to the ACHNACARRY AGREEMENT there were periodic price wars in the international oil industry. A price war also followed the decision by OPEC in December 1985 to pursue a 'fair market share' strategy (*see* CRUDE OIL PRICES).

primary distillation unit. The basic distillation unit in a refinery. It undertakes the initial stage in the processing of crude by heating and drawing off the various DISTILLATION CUTS. When the capacity of a refinery is stated the figure normally refers to the primary distillation units, with secondary refinery capacity being specified by the type of unit. *See also* REFINING, PRIMARY PROCESSING.

primary energy. An energy resource that requires extraction, refining or processing to provide a useable fuel. Examples of primary energy include crude oil, gas, coal, nuclear energy and hydroelectricity.

primary products. A term used by economists for unprocessed raw materials and agricultural output. Both oil and natural gas are primary products.

primary recovery. The production of oil and gas by natural drive mechanisms. *See also* PRODUCTION, NATURAL DRIVE MECHANISMS.

principal. The amount of money owed on a debt before interest payments are added.

private benefits. The benefits of production that a producer gains. It is used to differentiate from social benefit. *See also* COST–BENEFIT ANALYSIS; PRIVATE COSTS.

private brand distributors. A term used in the USA to describe a wholesaler (and possibly retailer) who buys gasoline (i.e. MOTOR SPIRIT) to sell under his own brand name.

private costs. The costs of production that a producer must pay himself. It is used to differentiate from social costs or EXTERNALITIES which fall to society in general, but do not enter the company's accounting balance sheet. For example, unregulated air pollution from a refinery is a social cost, but if legislation is passed to reduce noxious emissions the cost of controlling the emissions becomes a private cost (i.e. it is internalized). *See also* COST–BENEFIT ANALYSIS.

private placement. A share issue offered to institutional investors by a financial intermediary acting for the issuing company. This is used instead of issuing the shares on the stock exchange where a proportion must be made available to the general public. Private placement tends to be a cheaper way of raising finance than stock exchange issue.

private sector. *See* PUBLIC SECTOR.

probable reserves. Oil or gas RESERVES that have a better than 50 percent probability of being produced given the existing economics (including the price) and the technology. *Compare* POSSIBLE RESERVES.

processing capacity. *See* CAPACITY.

processing contract. An agreement by the refiner to refine oil for an oil producer (i.e. a company or a government) and supply the resulting products to the producer.

process train. In LNG production, each unit of production capacity is known as a process train. For example, the CAMEL LNG plant at Arzew in Algeria was constructed with three process trains each of 0.7×10^9 m^3 per annum capacity, yielding a total plant capacity of 2.1×10^9 m^3 per annum. Three separate refrigerants are used — propane, ethene and methane.

producer gas. A low-grade fuel gas produced by blowing air through heated coke.

producing horizon. The rock from which oil and/or gas is produced. *See also* GEOLOGY.

product barrel. The products (normally measured in value terms) produced from refining a barrel of crude oil. Because of different characteristics of crude oils and different refinery processes, there is considerable variation in the value. It is equivalent to the GROSS PRODUCT WORTH.

product differentiation. A method by which one producer distinguishes his good from that of a rival's product. Real differentiation exists where there is an intrinsic difference between the goods. Imaginary differentiation exists where the products are fundamentally the same, but the consumer is persuaded (e.g., by advertising) that there is a difference. The implication is that the producer is not a price taker (i.e. the producer has a degree of influence over the product price) and is faced by a downward-sloping demand curve for the good.

production. When technical and economic appraisals of a reservoir have been carried out, and the decision has been taken to produce from a particular well the first step is perforating the casing. This involves lowering a device — a perforating gun — down the CASING until it is level with the producing formation, when the gun then is fired. The special bullets used fly out sideways to form openings through which the oil can flow from the reservoir into the well.

The technical and economic feasibility of producing oil or gas from a particular reservoir depends on (a) the characteristics of the rock formation and (b) on the properties of the oil or gas.

A petroleum reservoir consists of a section, or stratum, of POROUS rock, which contains pores, or spaces, within which the oil is retained, rather than consisting of an underground pool of oil (*see* GEOLOGY). The proportion of the reservoir rock taken up by pores defines its degree of porosity. The porosity of a particular formation therefore determines the amount of oil it contains. The rate at which the oil can be recovered can be influenced by the permeability of the rock (i.e. the extent to which the individual pores are interconnected), thus permitting the oil to flow freely out of the reservoir at the production stage.

Crude oil normally occurs together with salt water and gas. The gas is often dissolved to some degree in the oil and therefore can influence its VISCOSITY; the more dissolved gas contained in a crude oil the less viscous (i.e. the thinner) the oil and the more readily it flows out of the reservoir rock.

In order that the flow of oil may be maintained the pressure in the reservoir must be higher than that in the bottom of the well; in most reservoirs this is the case initially. However, oil production techniques used in a particular field are likely to vary with its age as well as with the particular characteristics of the formation.

There are fields (most notably the major Middle East fields) where oil will flow unassisted for years. There comes a point, however, when the pressure in the reservoir has fallen to such a level that the pressure differential between reservoir and well is no longer sufficient to sustain the natural drive mechanism (*see* PRODUCTION, NATURAL DRIVE MECHANISMS). Then, unless artificial means of maintaining reservoir pressure are employed, no more oil can be produced. Such artificial means of maintaining reservoir pressure are known as secondary recovery techniques (*see* PRODUCTION, SECONDARY RECOVERY MECHANISMS).

In the early days of oil production, primary and secondary recovery derived their names from the fact that they tended to be used sequentially. Nowadays, however, it is common for secondary recovery techniques to be introduced at an early stage of a field's development. For this reason, primary and secondary recovery are now more commonly referred to collectively as conventional oil recovery techniques, whereas tertiary recovery is known as enhanced oil recovery (*see* PRODUCTION, ENHANCED OIL RECOVERY).

production, conventional oil recovery techniques. *See* PRODUCTION, NATURAL DRIVE MECHANISMS; PRODUCTION, SECONDARY RECOVERY MECHANISMS.

production, enhanced oil recovery (production, EOR; production, tertiary recovery). Although the employment of secondary recovery techniques (*see* PRODUCTION, SECONDARY RECOVERY MECHANISM) considerably improves oil recovery up to an estimated 50 percent on average, further approaches are being evolved to boost recovery still further. These involve a range of technologies falling into two broad categories: (a) those involving thermal techniques (i.e. the application of heat in some form) and (b) those using solvents or chemicals.

All enhanced oil recovery techniques are expensive, either by virtue of their energy costs (e.g., steam generation) or because of the intrinsic costs of the chemicals and solvents used, and as a result their introduction may not always be economically justifiable.

thermal techniques. This approach involves either injecting heat into, or generating heat in, the reservoir. The underlying principle is that when the oil is heated it becomes less viscous and therefore flows more easily and is more easily displaced from the rock pores. Most commonly steam or hot water is injected into the reservoir under pressure in order to heat the oil. This process is known as steam injection or steam flooding. The alternative process of generating heat in the reservoir itself (i.e. *in situ* combustion or fireflood extraction) requires the injection of air into the reservoir in order to facilitate the controlled burning of some of the oil.

chemical and solvent techniques. Chemicals and solvents are used in conjunction with water or gas injection (*see* PRODUCTION, SECONDARY RECOVERY MECHANISMS) with the object of making the injection medium more efficient. For example, polymer injection consists of the addition of polymers to the water injected. This has the effect of thickening the consistency of the water so that it displaces the oil more evenly. One of the risks of water injection is that water sometimes breaks through the oil it is supposed to be displacing; this is called fingering. If this happens at several points then pockets of oil can be isolated by water and left behind, becoming virtually irrecoverable.

production, natural drive mechanisms (production, primary recovery). The amount of oil which can be recovered using only primary recovery methods varies with the type of formation, but on average worldwide it is estimated to be around 25 percent of the oil initially in place. The reason why the yield from primary recovery is so low is that natural production mechanisms depend on the difference in pressure between the reservoir and the well — the pressure in the reservoir must be greater than that at the bottom of the well — but with the production of oil and gas the pressure in the reservoir gradually falls. In order to understand the basic principles of these, it is necessary to be aware of the typical arrangement of oil, gas and water in the most common type of reservoir structure.

The simplest reservoir structure is the ANTICLINAL TRAP, or dome-shaped structure, which is typical of many Middle East reservoirs. The lower part of the porous rock is occupied by salt water — the aquifer. Above it lies the crude oil which has gas dissolved in it (i.e. associated gas), and above this lies the free gas, called the gas cap. The presence of each of these in the reservoir represents a source of potential energy, and each gives rise to a different form of natural production mechanism as described below.

At the primary recovery stage when oil is flowing spontaneously as a result of the reservoir's potential energy, the oil flow will merely need to be controlled at the surface by a choke — an orifice on the flow line — at the wellhead. At some stage it may be necessary to increase the flowing capacity of a well by improving the permeability of the reservoir rock, which means enlarging the channels through which the oil can flow. This can be done for limestone formations by acid treatment (acidization), in which hydrochloric acid is pumped into the formation to dissolve the limestone thus creating larger flow passages for the oil.

water drive. In reservoirs where the aquifer, or water-bearing part of the structure, is as porous and permeable as the oil-bearing part the potential exists for water to flow freely into the oil-bearing part. When oil is produced, its removal from the reservoir causes an overall reduction in pressure, which results in the water that underlies the oil extending through the pores of the formation. This displaces the oil upwards, which helps to maintain the reservoir pressure. To obtain the maximum benefit from this form of natural recovery mechanism, it is desirable to control the rate at which oil is produced. If the rate of oil production is too high, reservoir pressure will decline too rapidly to enable all the potentially recoverable oil to be produced, and water, rather than oil, will start to be produced (a phenomenon known as water coning).

In reservoirs where the aquifer is not very porous or permeable, there is little, if any, contact between the oil and water, and hence there is no scope for a water drive mechanism. In such cases, natural recovery mechanisms depend on gas (*see below*).

solution gas drive. In this case, it is the gas that is dissolved in the crude oil which ultimately provides the recovery mechanism. As the oil is produced, the reservoir pressure falls and continues to do so until the oil in the reservoir reaches a certain point — the saturation pressure — at which the gas starts to come out of solution. The gas that has escaped then expands, and as it does so it displaces oil from the pores of the rock, and eventually gas and oil together pass into the producing wells. As with water drive (*see above*), the oil production rate has to be regulated to avoid too sharp a reduction of reservoir pressure, and hence the loss of some potentially recoverable oil.

gas cap drive. In this natural recovery mechanism, it is the free gas which occupies the space above the oil — the gas cap — that provides the energy for recovery. The reduction in the reservoir pressure that follows the production of oil causes the gas cap to expand to fill the space vacated by the oil. This results in oil being further displaced downwards into the producing well. As with the other natural drive mechanisms (*see above*) the rate of oil production must be regulated in order to control the decline in reservoir pressure and ensure optimum oil recovery.

production, primary recovery mechanisms. *See* PRODUCTION, NATURAL RECOVERY MECHANISMS.

production, secondary recovery mechanisms. These mechanisms involve the injection of water or gas into the reservoir, which reinforces the potential reservoir energy and displaces the function of the naturally occurring water or gas (*see* PRODUCTION, NATURAL RECOVERY MECHANISMS), forcing out oil which would otherwise not be recovered.

water injection (water flooding). This is one of the most commonly used secondary recovery techniques. It involves injecting water under pressure into the reservoir by means of INJECTION WELLS, drilled below the oil/water level.

gas lift (gas injection). If the water injection technique is not suitable for a particular formation and if sufficient quantities of gas are available, gas injection may be used, in which case gas is injected into the reservoir using wells drilled into the gas cap. Using a series of valves, pressure in the tubing is gradually reduced, which encourages oil to start flowing because the required pressure differential between reservoir and well is re-established.

sub-surface pumping. This involves lowering a pump into the well which is operated from the surface to keep the oil flowing.

production bonus. A lump sum payment paid by the licensed operator of the field to the government (or owner) when certain specified levels of production are reached. It is used as a means of making the original bid for the acreage seem more attractive. *See also* BONUS PAYMENTS.

production capacity. *See* CAPACITY.

production ceiling. The maximum production possible from a field (or producing country or region). This upper limit may be set by constraints imposed by technical or capacity measures, or it may be the result of a production-sharing agreement. Alternatively, it may be set by a government as part of its depletion policy.

production costs. The SHORT-TERM costs of producing crude oil or gas from a field consist of the fixed costs of exploration and development together with the cost of the capital equipment needed to produce, handle and transport the crude to the point of sale. In addition there are the variable costs associated with actually operating the installed equipment. The variable cost is the MARGINAL cost of production, and in most fields is small both in absolute terms and relative to the fixed costs. It is often argued that North Sea variable costs are significantly higher than those of other other areas; several dollars per barrel compared with less than 50 cents per barrel for onshore Middle East fields. In practice many apparent variable costs are in fact over periods of up to a year fixed costs. For example, labour — normally regarded as a variable cost — is generally under contract. In the long run, all production costs are variable because the decision makers are free to make decisions about new exploration and development. In this situation, there are real and significant differences in per barrel costs reflecting geology and the harshness of the geography.

production decline curve. *See* DECLINE RATE.

production functions. A mathematical expression of the relationship between the inputs into the production process, often referred to as factors of production, and the quantity produced for a given technology. In economics all factors of production are classified as land, which may be interpreted to include energy and other raw materials, labour or capital.

Traditionally economic theory has concerned itself with the relationship between output (Q), labour (L) and capital (K) with the production function f being represented as

$$Q = f(K, L)$$

More recently, however, the production function has been extended to include energy (E) and raw materials (M), thus

$$Q = f(K, L, E, M)$$

The explicit functional form of the production function relies on the treatment of the important characteristics of the production process. One of the most important characteristics is the nature of the returns to scale exhibited. Returns to scale refer to changes in output resulting from increases in all factors of production in the same proportion; they may be constant, increasing or decreasing. Also of interest is the relationship between the factors of production, in particular the degree of substitutability (i.e. elasticity of substitution) of any two of the inputs. The treatment of TECHNICAL PROGRESS in the production function has also been a feature of considerable importance in both the theoretical and empirical application of production functions.

One of the most widely used production functions for empirical work is the Cobb–Douglas (C–D) production function which takes the form

$$Q = A\,L_{\alpha}\,K_{\beta}$$

where A, α and β are fixed parameters. The C–D production function is popular because it is relatively straightforward to estimate using ECONOMETRIC techniques, and the parameters are easily given economic interpretation. The sum of α and β denotes the degree of returns to scale, whereas independently they represent the elasticity of output with respect to labour and capital, respectively, and the factor shares. Unfortunately the C-D production function has a number of limitations that make it an unrealistic approximation of a real life production process. Since the exponents are constant the implications of the specified process are that the input elasticity and the factor shares will be constant, although they might in fact be expected to vary with output or input levels. Returns to scale are fixed for all output levels; again an unlikely feature of the production process. Finally the specification imposes a unit elasticity of substitution between the factors of production. This is a restriction which is unlikely to be justified in most circumstances.

A less restrictive family of production functions, of which the C–D function is a special case, is the constant elasticity of substitution production function (CES). Rather than restricting the elasticity of substitution to be unity the CES production function only imposes a constant elasticity of substitution on the factors of production.

Production functions have been applied extensively in the analysis of industrial energy demand. Extensions of the production function theory allow factor demand functions to be derived and estimated. By incorporating energy into the production function as a factor of production one can derive a demand function for energy and analyze the interrelationships between energy and the other factors of production, as well as the impact of relative prices and technical progress.

production licence. A licence granted by the state that allows a company to produce oil and gas from a specified acreage.

production phase. The period during which a field actually produces oil and/or gas. Its length varies considerably between fields and is a function of the geology of the field and the extent of investment in secondary recovery (*see* PRODUCTION, SECONDARY RECOVERY MECHANISMS) and enhanced oil recovery techniques (*see* PRODUCTION, ENHANCED OIL RECOVERY).

production platform. An OFFSHORE structure from which DEVELOPMENT WELLS are drilled and production is controlled.

production share. The division of output from a field when the field is owned by a JOINT VENTURE. It may refer either to the theoretical share based upon EQUITY holding or the actual share based on LIFTING.

production sharing. A type of agreement that emerged in the 1960s, pioneered by Indonesia. It combines elements of the JOINT VENTURE and the SERVICE CONTRACT. The company undertakes exploration at its sole risk, but on COMMERCIAL DISCOVERY the company acts purely as a contractor for the government, recovering its initial outlay from production. The company's crude oil entitlement (for the contract services) is not subject to taxation and therefore enables the company to gain fully from any increase in the price of crude.

production string (production tubing). The pipe from the RESERVOIR to the surface through which the oil or gas flows.

production tubing. *See* PRODUCTION STRING.

production well. A well drilled into a proven reservoir.

productivity. The rate of output from a given unit of input, usually expressed as a ratio of output to input. The concept is most commonly used in the context of labour and capital inputs, but since the FIRST OIL SHOCK interest in energy productivity — which is a measure of conservation — has grown.

product quality giveaway. The supplying of a product that is over the minimum specification when a refiner has to meet stringent quality controls and faces a penalty cost if the product is below specification. It provides a margin of error and, in effect, the extra quality is provided free. Such a practice is most commonly found in the OCTANE RATING of motor spirit.

product quality specification. The quality characteristics that a product should have in order to be classified as a certain type of product. The tolerances in the specifications tend to vary significantly between different products, usually being dependent on their end use. For example, jet fuels require much stricter specifications than fuel oils. *See also* QUALITY.

products carrier. Oil tankers that specialize in the ocean transport of oil products. They vary in size up to 70 000 tons, but the majority are between 30 000 and 39 999 tons. Traditionally product tankers have traded along short routes between refineries and market centres. The development of refineries in the Middle East is likely, however, to increase the demand for the larger vessels to supply Europe and the Far East. Product tankers switch readily between clean products and dirty products or crude oil, and therefore rates are affected by the state of the overall oil market. Technological developments include the use of double-skinned tanks and engines with improved fuel efficiency.

product spectrum. The range of products sold by a company. For an oil company, because of the existence of so many specialist products, the range can be very large indeed.

product worth. *See* GROSS PRODUCT WORTH.

product yield. The main products (expressed as a percentage of the crude volume) obtained from REFINING a barrel of oil.

profit. In an accountancy context, gross profit is total sales revenue less costs, as represented by money outlays on wages, rents, fuel, raw materials, etc. Net profit is gross profit less interest on loans and DEPRECIATION. In an economic context, costs have to be adjusted to take account of the OPPORTUNITY COST of factor inputs owned by the company. Economists also draw a distinction between NORMAL PROFIT, which is the minimum an entrepreneur must receive for him to remain in a particular line of activity, and SUPER-NORMAL PROFIT.

profit margin. A financial ratio that gives PROFIT, before interest or tax is deducted, as a percentage of TURNOVER.

profit maximization. In simple economic analysis, the assumed objective of all firms. As the analysis becomes more complex and tries to approximate more closely to the way in which firms do behave the assumption is widened and other objectives of the firm enter as possibilities. Thus a firm may seek to maximize turnover or growth. In modern corporations where control and ownership are split, as is the case in most oil companies, management may have further objectives that relate to maximizing their security or enjoyment of the job or indeed maximizing their opportunity for a quiet life.

progressive royalty rates. ROYALTY that is fixed on a sliding scale, depending on the size of the field; a large field attracts a larger royalty. Such a system was adopted by The Netherlands in 1967 and by Norway in 1972. The system has attractions because it is supposed to be more favourable to a small field. Although this is true other things are equal, but other things are seldom equal, and the profitability depends equally on the field's location (i.e. deep or shallow water, the need for transit pipelines, etc.). There is always a danger that the royalty may provide a disincentive to the optimal depletion of the fields if size alone is the criterion.

progressive profits tax. *See* PROGRESSIVITY.

progressivity. A progressive profits tax implies that as profits increase the proportion taken in tax increases at a more than proportionate rate. For this to occur, the marginal rate of tax must be above the average rate.

project appraisal. *See* INVESTMENT APPRAISAL.

Project Independence. The response of the US administration to the FIRST OIL SHOCK, culminating in the *Project Independence Report* produced in November 1974. Its broad aim was to reduce the dependence of the USA on imported oil.

projection. The process of taking historical data and using it as the basis for future predictions. Projections can be performed by methodologies of varying sophistication, ranging from simple EXTRAPOLATION to sophisticated econometric techniques (*see* ECONOMETRICS). *See also* FORECASTING.

prompt barrel. *See* WET BARREL.

propane. A hydrocarbon gas with the formula C_3H_8 that is liquid under pressure at ambient temperatures. Together with BUTANE propane is commonly known as LPG, although the latter requires higher pressure to keep it liquid.

propene (propylene). A flammable, gaseous olefin (*see* ALKENES) with the formula C_3H_6 that is important as a raw material in PETROCHEMICALS.

property rights. The ownership, control and use of resources, or goods and services, as constrained by legal or contractual limits. MINERAL RIGHTS to oil and gas are less extensive, referring to certain legally defined activities associated with the exploitation of reserves. For example, where a tract is rented or leased, the licence conditions and tax regulations lay down the rights of the licensee.

propylene. *See* PROPENE.

prorationing. An attempt to control the supply of oil or gas by agreement or legislation in order to prevent oversupply in the market. Prorationing began in the USA where the LAW OF CAPTURE encouraged overproduction. Initially it was organized on a state-by-state basis, the most famous controlling institution being the TEXAS RAILROAD COMMISSION. In 1934, the system was expanded to allow for interstate control by the creation of the Interstate Oil Compact Commission. In effect OPEC during the 1980s attempted to introduce a prorationing system with its QUOTA agreements.

protection. An attempt to limit the import of a good by the imposition of physical quantity limits (i.e. QUOTAS) or an indirect tax (i.e. TARIFF). Such measures are normally adopted to protect domestic producers either for economic, political or strategic self-sufficiency reasons.

proven field. A FIELD that contains PROVEN RESERVES of oil or gas.

proven reserves. RESERVES of oil and gas that will be produced with reasonable certainty given existing economic and technical conditions. *See* POSSIBLE RESERVES; PROBABLE RESERVES.

PRT. *See* PETROLEUM REVENUE TAX.

public good. A good for which consumption by one person does not deplete the amount available to others in the society. Once provided to anyone, it can, therefore, be supplied to everyone else at no extra cost. Such goods are called non-rival in consumption (e.g., lighthouse light, radio and television broadcasts, clean air). Public goods may also be non-excludable in that once the good is supplied to anyone nobody else can be excluded from access (e.g., clean air, mosquito abatement, some forms of national defence). Non-excludable public goods can also be EXTERNALITIES. If someone provides a good, others will benefit without having to pay. Private firms estimate demand only for those who are willing to pay. Thus non-excludable public goods are likely to be under-supplied by private firms and may require government provision, financed out of taxation. Private goods, by contrast, are normally both rival and excludable.

public sector. Describing those economic activities under the ownership and/or control of the state. The public sector includes the departments, ministries or agencies of the central (or regional) government together with state-owned corporations. This contrasts with the private sector, where ownership and control is vested in private individuals. In most THIRD WORLD countries and many industrialized countries the oil and gas sector to one extent or another is part of the public sector. In theory at least, the significance of the distinction is that the public sector has different objectives (e.g., maximizing society's welfare) compared with the private sector's PROFIT MAXIMIZATION objective. In reality the distinction between objectives is neither simple nor clear cut as can be seen in the ECONOMIC THEORY OF POLITICS.

pump price. A commonly used expression that denotes the final price to the consumer of oil products, most commonly motor spirit.

purchasing power. An attempt to describe what income or revenue can actually buy. It is income adjusted for inflation and also for changes in the exchange rate. In the context of oil and gas the concept is commonly used to refer to the revenues of oil-exporting countries. With constant unit revenues for oil, for example, high inflation and a depreciating dollar (in which oil is priced) means the purchasing power of the unit revenue will fall sharply. This is particularly important when a country's revenue requirements are considered as part of the analysis of how much oil the country will be willing, or forced by revenue needs, to supply.

purging. The cleaning of oil- and gas-handling equipment (e.g., pipelines, processing units or storage tanks) to remove unwanted gases or materials such as water.

put option. *See* OPTION.

pygas. *See* PYROLYSIS GASOLINE.

pyrolysis gasoline (pygas, steam-cracked gasoline). A by-product from ethene crackers. *See also* REFINING, SECONDARY PROCESSING.

Q

quad. A measure of energy equal to 1000 million million (i.e. 10^{15}) British thermal units (Btu). It is normally used in the context of national power supplies.

quality. The quality of a product, in the petroleum context, can be defined as its suitability for a specific purpose. To find out whether a particular product performs satisfactorily, without carrying out elaborate simulations of its performance in end use appliances, it is necessary to translate desirable performance characteristics into a set of properties that can be specified and tested in quantitative terms. Such a set of properties is called a product quality specification. As petroleum products are used for hundreds of different purposes, with correspondingly different quality requirements, the subject of quality is a highly complex and technical one. For general purposes, it is sufficient to understand what the most commonly quoted quality specifications indicate about the performance of a product. For example, a residual fuel oil is described as having a kinematic viscosity of 370 centistokes at 50°C, with a sulphur content of 3.5 percent by weight and a pour point of 9° C. This simply translates into a more straightforward description, namely a heavy black fuel oil more solid than liquid, with a pungent smell when burned, which must be stored in heated tanks if it is not going to clog up the flow lines to the boiler.

Quality specifications differ widely from one product to another. There are some properties that are relevant to certain products only; for example, octane rating for motor spirit and aviation fuels. Other properties (e.g., sulphur content) are relevant to most, or all products, although in varying degrees. *See also* QUALITY, AVIATION SPIRIT; QUALITY, AVIATION JET FUELS; QUALITY, BITUMEN; QUALITY, CRUDE OIL; QUALITY, DIESEL OIL; QUALITY, DISTILLATE AND RESIDUAL FUELS; QUALITY, LUBRICATING GREASES; QUALITY, LUBRICATING OILS; QUALITY, MOTOR SPIRIT; QUALITY, WAXES; QUALITY, WIDE-CUT GASOLINE.

quality, aviation spirit. The main quality requirements of aviation spirit are similar to those of motor spirit (*see* QUALITY, MOTOR SPIRIT), although in most cases they are more restrictive because of the higher power outputs of aviation engines (*see* FINISHED PRODUCTS, AVIATION FUELS).

anti-knock rating. As with motor spirit, aviation spirit must be free from any tendency to KNOCK. The definition of anti-knock or OCTANE RATING is rather more complex than for motor spirit, since wider variations of engine performance are required of aviation than of automotive engines. For added power at take-off, a higher proportion of fuel in the fuel–air mixture is required, whereas cruising after take-off can be maintained with a much lower proportion of fuel in the mixture. Both these types of conditions — known as rich-mixture and lean-mixture, respectively — give rise to different levels of anti-knock performance, and both must be defined. It is therefore the practice to quote a two-fold octane rating for aviation spirit (e.g., 100/130, which defines the fuel's performance in rich-mixture and lean-mixture conditions, respectively). Octane ratings above 100 — achieved as with motor spirit by the addition of lead additives (e.g., LEAD ALKYLS) — are normally referred to as performance numbers.

volatility. In aviation spirit volatility is controlled in a similar way to motor spirit, but controls are more restrictive to allow for greater variations in temperature and engine performance.

specific gravity. The density of aviation spirit is defined for use in fuel load calculations. Tank capacity normally limits the volume of fuel that can be carried, and usually a higher-density fuel oil will be preferred, since this gives greater power per unit of volume (i.e. gallon or litre).

maximum freezing point values. These are laid down for aviation fuels in order to safeguard performance at high altitude and thus low temperatures. The maximum freezing point defines the point at which the fuel starts to solidify and hence clog fuel lines. In the case of aviation spirit its value is –60°C.

other properties. Storage stability and corrosivity of aviation spirit are measured and controlled basically as for motor spirit.

quality, aviation jet fuels. Fuels for aviation gas turbines (i.e. aviation turbine gasoline and aviation turbine kerosine) conform to quite different requirements from those used in aviation piston engines (*see* QUALITY, AVIATION SPIRIT). The actual combustion process is a continuous one, at constant pressure, which means that the problems associated with knocking which can occur in a piston engine do not arise. The most important characteristics of a jet fuel are its energy content and its combustion characteristics.

calorific value. The energy content is expressed in terms of calorific value, which defines the amount of heat in a given unit of fuel. It is expressed in units such as British thermal units per pound (Btu/lb) or kilocalories per tonne (kcal/t). Together with the specific gravity of the fuel, which gives the weight per unit volume, it is thus possible to define the energy content of any given fuel on a weight or volume basis.

combustion characteristics. These characteristics refer simply to the type of flame produced when the fuel is burnt. They are related to hydrocarbon type, with AROMATICS producing a very smoky and hence undesirable flame. One method of controlling the type of flame produced is to place a limit on total aromatics content. The main quality specification governing combustion characteristics of jet fuel is related to its smoke point. This was originally derived for illuminating kerosine and defines the maximum height of flame (in millimetres), which can be achieved in a wick-fed lamp under prescribed conditions before the flame starts to smoke. Smoke point correlates fairly well with combustion performance for aviation kerosine fuels, but in the case of wide-cut gasoline fuels (*see* QUALITY, WIDE-CUT GASOLINE) it has to be related to volatility also, in which case it is called a SMOKE VOLATILITY INDEX (SVI). A modified version of the smoke point test, which takes account of the radiant heat characteristics, gives rise to a further specification called LUMINOMETER NUMBER (LN).

Aviation kerosine fuels have much lower volatility requirements than aviation spirit, and in the absence of a more restrictive volatility limit, it is necessary to quote a flash point. This defines the temperature at which the fuel can be stored and handled without serious fire hazard.

quality, aviation turbine gasoline. *See* QUALITY, AVIATION JET FUELS.

quality, aviation turbine kerosine. *See* QUALITY, AVIATION JET FUELS.

quality, bitumen. Bitumens have a complex chemical composition, which can differ widely according to the feedstock and the manufacturing method. Its three main constituents are asphaltenes, resins and mineral oil, and the variations in the proportions of each in a particular grade of bitumen can affect its suitability for specific applications.

Bitumen is used mainly for road construction, roofing felts, damp-proof courses and other specialist impregnating applications, for which quality requirements differ widely. For most applications, however, two main tests are used — penetration and softening point — which can be supplemented with other more specialist tests for specific applications.

penetration. This is an empirical test which defines the depth to which a standard needle penetrates a bitumen sample under defined conditions. It is a measure of the hardness or consistency of the material, and is measured in units of penetration (Pen) at 25°C, ranging from 0 (very hard) to around 500.

softening point. This is measured by heating the bitumen under controlled conditions and noting the temperature at which a defined change in consistency (i.e. softening) occurs.

quality, crude oil. Crude oil quality is essentially a derivative of product quality, because with the exception of those properties relevant to its ease of handling, the characteristics of a crude oil are of interest purely in relation to their predictive value as to the quantity and quality of the products that it will yield.

In addition to defining specific characteristics (*see below*) more detailed information about the crude oil is obtained by means of an assay. An assay is essentially a detailed analysis which involves distilling the crude oil under specified conditions to establish the percentage weight of the oil that is distilled at given temperatures. The various percentages distilled within specified temperatures correspond roughly to the percentages of different straight-cut products and are collectively termed the product yield (*see* REFINING, PRIMARY PROCESSING). It is possible to conduct a preliminary assay on small quantities of crude oil from field samples. This provides a general indication of the crude oil type, but it does not give sufficient data for the purposes of designing refinery plant or establishing detailed product quality. For such purposes, true boiling point (TBP) distillation is carried out. In this, the crude is distilled, and the quantity of oil distilling at different temperatures — the product yield — is noted. The relationship between temperature and yield is plotted to give a TBP curve, which is unique for each crude oil. Each fraction is then analyzed to establish its quality.

Despite the influence of hydrocarbon type (*see below*) on crude quality, it is not very common to characterize crude oils in this way outside the refining context. By far the most commonly quoted quality characteristics in the industry are those of gravity and sulphur content. Quality characteristics of lesser importance, but which help to establish a crude's ease of handling are viscosity and pour point. There are, in addition, other quality characteristics of minor importance (metals content, asphaltene content, carbon residue content, other contaminants).

hydrocarbon type. There are wide variations between different crude oils in terms of composition and properties. The differences arise from the fact that crude oils consist of mixtures of the various types, or series, of hydrocarbons (*see* CHEMISTRY), and the relative proportions of each determine a crude oil's unique characteristics. Attempts to classify crude oils have not been entirely successful, largely because of the complex composition of crude oils. The most obvious method of classification is to designate a crude oil according to the dominant hydrocarbon type present in its make-up (e.g., paraffinic, naphthenic, etc.). This type of analysis is known as PNA analysis, after the main hydrocarbon types present (i.e. paraffins, naphthenes and aromatics). Although this method is commonly used (e.g., Venezuelan crudes are referred to as naphthenic, Libyan as paraffinic, etc.), it is not entirely satisfactory because many crude oils have a different dominant hydrocarbon in different parts of the distillation range (*see* boiling range). To overcome this 'crossover' effect, the US Bureau of Mines derived a method that quotes preponderant hydrocarbon types in two specified fractions of the distillation range, giving rise to crude oil characterizations of the type naphthenic–naphthenic, naphthenic–paraffinic, etc. Additional information is often provided by introducing other characteristics, such as high- and low-wax content (i.e. pour point, *see below*).

specific gravity (SG). This is a measure of the weight per unit volume of a substance. The basic unit of volume used in relation to crude oil is the barrel, which is equivalent to 42 US standard gallons or 35 imperial gallons. A barrel of Kuwait crude weighs more than a barrel of Algerian crude, thus Kuwait crude has the higher specific gravity. Most of the major crude oils have specific gravities in the range 0.8–0.9; the higher the specific gravity the heavier the crude. However, crude oil gravities are rarely quoted in this form. The common unit of measurement for crude oil gravity is that of API gravity, which was devised by the American Petroleum Institute. This is an arbitrary scale of measurement that is related inversely to specific gravity as follows

$$API = (141.5/SG \text{ at } 60°F) - 131.5$$

API values are expressed in degrees API (°API). The values derived on this scale range from approximately 30–45°API for most of the world's major crudes. On the API scale, the higher the API value the lighter the crude. Thus the Algerian crude Hassi Messaoud, which is one of the lightest crudes, has a higher API gravity (41.5°API) than Kuwait, which is one of the heaviest (31.1°API). The gravity of a crude is significant for two reasons.

(a) The API gravity of a crude is related to its value, since a light crude (i.e. one with a high API gravity) yields a larger percentage of the lighter, more valuable products after refining than a heavy crude of low API gravity. For example, Algerian crudes, when refined, yield more than 60 percent light and middle distillate products, compared with Kuwait crude, which yields only 40–45 percent distillate products, and this is reflected in the fact that they can command a much higher price. Such a price differential, based on potential product yield, is called a gravity differential or premium.

(b) At certain stages of the industry it is conventional to use volume measurements (e.g., barrels a day) and at others weight (e.g., millions of tonnes per annum). For example, RESERVES are expressed in billions of barrels, and crude oil production rates and pipeline throughputs are expressed in barrels per day. At the transportation stage, tanker capacities and cargoes are expressed in tons. Individual refinery plants are designated in barrels per day, although the total crude distillation capacity of a refinery is often referred to in millions of tonnes per annum. Conversion from volume to weight, and vice versa, is thus fundamental to many calculations in the industry and accurate conversions require a knowledge of the specific gravity of the crude or product in question. *See also* APPENDIX.

sulphur content. There are wide variations in the sulphur content of crude oils, which is expressed as a percentage of total weight. Most of the world's major crude oils fall within the range 0.3–3.0 percent. The sulphur present in crude oil, after refining, is distributed throughout the product range, but is most concentrated in the residue or fuel oil. For many end uses, the presence of sulphur in products is regarded as highly undesirable, since it gives rise to atmospheric pollution and can cause plant corrosion. Refining processes exist for removing or counteracting the effects of sulphur in distillates, but as yet the desulphurization of residue (or fuel) is largely unknown outside Japan (*see* REFINING, SECONDARY PROCESSING). The only way in which fuel oil with a low sulphur content can be obtained, therefore, is from the residue of low-sulphur crudes. For this reason, low-sulphur crudes command much higher prices than high-sulphur crudes and, on a sliding scale of value, medium-sulphur crudes also show a significant price differential over high-sulphur crudes, reflecting their lower sulphur content. Such a price differential is called a sulphur premium. Middle Eastern crudes tend to be high- or medium-sulphur, with most of the African crudes (e.g., Nigerian, Libyan, Algerian), North Sea and North American crudes being of low sulphur content.

viscosity. Viscosity can be regarded as the thickness, or stickiness of an oil, being a measure of its resistance to flowing freely; the higher the viscosity of a crude oil the less easily it flows. This can be an important factor influencing the design of pumping facilities and pipelines to transport the oil from the oil-field to the terminal. As a rule, the heavier the crude oil, the higher the viscosity. The unit most commonly used for measuring viscosity is the centistoke (cSt), which is a very small unit of quantity per second, since most tests of viscosity involve measuring the time it takes for a sample of the oil to flow through a standard narrow space. Because viscosity is affected by temperature — it improves on heating — it is necessary to quote viscosity (usually called kinematic viscosity) at a standard temperature (normally 38°C). Typical crude oil viscosities are 1.4 cSt at 38°C for a light North African crude and 9.6 cSt at 38°C for a Middle East crude.

pour point. At low temperatures, the wax present in crude oil can start to solidify, which makes normal pumping and pipeline operations impossible. It is therefore necessary to know the precise temperature at which this occurs for each crude: such a temperature is called the pour point, expressed in degrees centigrade. Like viscosity, pour point improves on heating, so very high pour point crudes (e.g., Libyan Sarir) require heated storage and tankers to maintain them in a fluid state at ambient north-western European temperatures.

wax content. The wax content of a crude oil is related to pour point (*see above*).

carbon residue. This provides an indication of the percentage residue yield of a crude oil.

asphaltenes content. This as also provides an indication of the percentage residue yield of a crude oil.

metals content. Metals present in crude oil can have adverse effects both at the refining stage (e.g., by poisoning CATALYSTS) and in some end use applications of fuel oils.

Traces of vanadium and nickel are sometimes present.

other contaminants. Salt, water and sediment can also be present in crude oil. They are routinely tested for.

quality, diesel oil. The range in size and power of diesel engines is extremely wide varying from small, high-speed automotive engines to the large, low-speed engines used for marine propulsion and for electric power generation. Quality requirements of diesel fuels therefore differ markedly, depending on the engine size, load, speed range, etc., with the result that a number of different grades of diesel fuels exist. In general, the larger the engine and the lower the speed, the less restrictive the quality specification. In fact, the heaviest grade of diesel fuel — marine diesel — includes a proportion of residual fuel oil.

The combustion process in a diesel-powered engine differs markedly from that of the motor spirit-powered engine. In the latter, a mixture of fuel and air is ignited by a spark from a sparking plug. In the diesel engine, air alone is compressed, which raises it to a high temperature. Only then is the fuel introduced into the combustion chamber where, after a short delay while the fuel droplets absorb the heat from the compressed air, the mixture ignites spontaneously. Thus, although spontaneous ignition is regarded as undesirable in the motor spirit engine, it is the fundamental principle on which the operation of the diesel engine is based.

The most important properties of a diesel fuel are its ability to produce rapid spontaneous combustion and its ability to flow at low temperatures without crystallizing or solidifying. Numerical means of measuring, and hence specifying and testing these two properties are given by the cetane number and the cloud point, respectively.

cetane number. Efficient spontaneous combustion is related to the length of the interval during the combustion process between the injection of the fuel and the spontaneous ignition of the fuel/air mixture. This interval is referred to as the ignition delay, and the shorter the ignition delay the more efficient the combustion process. The cetane number is essentially a numerical description of the length of ignition of a diesel fuel. The test method used for rating a fuel for the cetane number adopts a similar principle to that used in deriving octane ratings for motor spirit (*see* QUALITY, MOTOR SPIRIT). A CFR (Cooperative Fuel Research Committee) diesel engine is used and a reference fuel, in this case cetane, is selected which has a particularly good ignition delay and is arbitrarily assigned a value of 100. A second reference fuel with poor ignition delay is assigned a zero or very low value. The limits of the possible range of cetane numbers thus having been defined, the individual fuel being tested is run through the test engine and its ignition delay is 'matched' with that of a blend of the two reference fuels in different proportions. The fuel being tested is then given a cetane number which is based on the cetane content of the blend that matches its performance. For example, if the fuel has an ignition delay which is the same as that of a blend with 70 percent cetane and 30 percent of a zero-cetane number fuel, it is given a cetane number of 70. As with octane ratings, the higher the cetane number the better the fuel. The cetane number of diesel can theoretically be improved by the addition of AMYL NITRATE, although in practice this has been little used.

As an alternative to this rather time-consuming empirical method, the cetane numbers may be estimated indirectly from other properties. (a) The diesel index is derived mathematically from the fuel's API GRAVITY and its aniline point, which is the lowest temperature at which the fuel can be mixed with an equal volume of aniline. (b) The cetane index is derived from the fuel's API gravity and its volatility (i.e. ease with which it vaporizes). The cetane index is generally thought to give a better prediction of cetane number than the diesel index.

cloud point. At low temperatures, the paraffinic components present in diesel fuels start to crystallize and separate out as wax. The temperature at which this starts to occur is called the cloud point, which occurs at anything up to 10–20°C above the pour point (*see* QUALITY, CRUDE OIL), at which the fuel is no longer pumpable. Because many high-speed diesel engines have very fine filters, these are likely to become blocked by even small deposits of wax, and hence the cloud point is an important quality characteristic in diesel fuels for these engines. Lower-speed engines with larger filters are less affected by small wax deposits and can therefore tolerate fuels with less restrictive (i.e. higher) cloud points.

other properties. Sulphur content, corrosivity and carbon residue and ash content are other diesel oil quality specifications. The sulphur content can have a very wide variation, depending on the type of engine used (e.g., for automotive use, recommended sulphur content would be 0.4 percent or less, but larger, slower-speed engines can tolerate considerably higher levels). Corrosivity is tested for by the copper strip corrosion test, as for motor spirit (*see* QUALITY, MOTOR SPIRIT). Carbon residue and ash content are controlled to avoid deposits forming within the system.

quality, distillate and residual fuels. Despite the wide variations in the type of application of distillate and residual fuels, both are liquid fuels and share certain common features. Thus the same quality characteristics are important for both groups of fuels, although the levels at which they are specified may differ widely from one group to another. The distillate fuels are kerosine and gas oil (i.e. domestic heating oil). These are termed 'clean' fuels and present few problems in terms of storage and handling. Residual fuels are heavy, black, viscous oils which do not burn as cleanly as distillate fuels and frequently require heating in order to facilitate their handling and efficient combustion. The term 'fuel oil' is frequently, although not exclusively reserved for residual fuels, and the term is in any case an imprecise one, since residual fuels often contain a proportion of distillates in order to improve their quality (*see also* FINISHED PRODUCTS, FUEL OIL).

volatility. Before liquid fuels can be burned efficiently they need to be broken up into very fine particles and mixed with air. This can be achieved by vaporization or by atomization. Vaporization involves the fuel flowing into the base of a pot-type burner. It is thus heated by both the container and the flame, and rises in the form of a vapour to mix with the air drawn through perforations in the walls of the container. This method is used only for the clean distillate fuels, because residual fuels would leave too many deposits on the surface of the burner. Using atomization, the fuel is forced into the combustion chamber under pressure, which transforms it into a spray of fine droplets. Either distillate or residual fuels are technically suitable for the atomization process, but in view of the considerably greater cost of distillate fuels, residual fuels are more frequently chosen.

The maximum output of vaporizing-type burners is approximately 50 000 Btu/hr, which makes them particularly suitable for domestic central heating installations. Their cleanliness and ease of fuel handling also represent significant incentives in this context. The atomizing type of burner requires the fuel to be fed in at a fairly high rate, which results in a minimum output of approximately 50 000 Btu/hr. This type of burner is therefore used in larger central-heating installations and in many industrial applications.

viscosity. This is the most important quality characteristic of a liquid fuel for most purposes (*see* QUALITY, CRUDE OIL). In the case of liquid fuels it is important not just for ease of handling, but also for its relevance to the efficiency of the combustion process. If fuel oil is to be burnt in the most efficient way, it is important for it to be introduced into the handling and combustion system at the right viscosity. This can be achieved either by heating, which reduces viscosity and thus makes the oil flow more easily, or by specifying oil of a suitable viscosity such that it will not require prior heating. The alternative selected in any particular case depends largely on whether the price differential between fuels of high and low viscosity is sufficient to offset the capital cost of the investment in heating facilities.

It is essential to quote viscosity at a specific temperature. For fuel oils, the most commonly used form of measuring viscosity is in centistokes at 82.2°C (180°F) for heavy fuel oil and 37.8°C (100°F) for distillate fuels. This has now become the international unit of measurement of viscosity, replacing other previously common forms such as the Redwood No 1 Second (UK), the Saybolt Universal Second (USA) and the Engler Degree (continental Europe).

pour point. The pour point is defined as the temperature 5°F above that at which an oil just fails to flow. This property is, in principle, important for fuel oils if filter-blocking problems are to be avoided. In practice, however, it is not regarded as an entirely reliable indicator of fuel-handling characteristics and has tended to be replaced by more empirically based measures of minimum storage and handling temperatures.

sulphur content. The sulphur content of fuel oils is important for three reasons: (a) high-sulphur fuel oil can corrode equipment; (b) it can contaminate the finished product of a manufacturing process; or (c) it can lead to high levels of sulphur dioxide emission into the atmosphere. To a certain extent, the first two effects can be counteracted by improved technical design or the use of additives. The problem of conforming to legal limits of sulphur dioxide emission, however, can only be solved by using fuel oil with a medium or low sulphur content. At present, in most countries, this involves using fuel oil manufactured from medium- or low-sulphur crude oils, and this has led increasingly to the emergence of sulphur-based price differentials in both crude and fuel oil (*see* SULPHUR PREMIUM).

quality, fuel oil. *See* QUALITY, DISTILLATE AND RESIDUAL FUELS.

quality, gasoline. *See* QUALITY, MOTOR SPIRIT.

quality, gas oil. *See* QUALITY, DISTILLATE AND RESIDUAL FUELS.

quality, kerosine. *See* QUALITY, DISTILLATE AND RESIDUAL FUELS.

quality, lubricating greases. Lubricating greases are solid or semi-solid products, and their major quality requirements are therefore more akin to those of bitumen (*see* QUALITY, BITUMEN) than of liquid fuels, namely penetration and pour point, which measures the temperature at which it changes to a liquid and starts to flow.

quality, lubricating oils. Lubricating oils are a highly specialized group of products with a wide range of applications in industry and transportation (e.g., aviation, automotive and marine). Quality specifications differ according to the type of application, but some generalization is possible, and the main properties are outlined below, together with some indication of their relative importance in specific applications.

The basic function of a lubricating oil is to reduce friction and wear between metal surfaces that move with respect to each other. It does this by acting as a separating film between the two surfaces, and its most important quality characteristic therefore relates to its performance in this role.

viscosity. This is the single most important property of a lubricating oil. If the oil has too low a viscosity, it will be too thin to form a film between the surfaces. If, on the other hand, its viscosity is too high, too much heat will be generated in the system. Because viscosity decreases with an increase in temperature, an oil whose viscosity may be acceptable at normal temperature may, at higher temperatures, thin out too much to provide adequate lubrication. It is therefore important to establish the viscosity of a lubricating oil over a range of temperatures. The effect of a change in temperature on the viscosity of an oil is measured by the viscosity index (VI). This is an arbitrary scale based on the use of reference fuels to define the limits of the scale from 0 to 100; a high viscosity index indicates a low rate of change of viscosity with temperature. The use of additives known as viscosity index improvers makes it possible to achieve VI values in excess of 100, in which case they are designated on an extension scale (VIE).

VI is particularly important for applications where substantial changes in temperature are experienced, as for example in automatic transmission systems of cars, where lubricating oils must function adequately both for cold starting and at the higher temperatures of normal running (*see* SAE VISCOSITY).

cloud point. The cloud point is relevant where there is a risk of fine filters or narrow feed lines becoming blocked by small accumulations of wax (*see* QUALITY, DIESEL OIL).

pour point. This is particularly relevant in colder climates, where the ambient temperature may well be below the pour point of certain oils. In such cases (e.g., diesel oil) pour point depressants can be used, which inhibit the formation of wax crystals and hence maintain the oil in a liquid form at low temperatures (*see* QUALITY, DIESEL OIL).

quality, motor spirit. The most important characteristic of a motor spirit is that its combustion should be smooth and efficient. In efficient combustion, the flame from the sparking plug spreads evenly across the combustion space until all the motor spirit has burnt. As the flame spreads, the temperature of the whole spirit/air mixture rises. If combustion is not efficient, part of it will ignite spontaneously before the flame reaches it, causing detonation or 'knocking' (*see* KNOCK). Severe or prolonged knocking causes loss of power and can be harmful to the engine. It is therefore essential to ensure that motor spirits are not likely to cause knocking, and the preferred fuels are those with a high anti-knock rating. The tendency to knock differs widely from one type of hydrocarbon to another. As a broad generalization, the hydrocarbons with high anti-knock values, and hence efficient combustion, are the AROMATICS, the alkenes and the highly branched ISO-PARAFFINS. Most NORMAL-PARAFFINS have low anti-knock ratings. The other major property to influence motor spirit performance is the ease with which it vaporizes (i.e. its volatility).

anti-knock rating. A fuel's anti-knock rating is measured by comparing it with two reference fuels — isooctane and n-heptane — which are known to have very good and very bad anti-knock characteristics, respectively. Arbitrary values, or octane numbers, are ascribed to the reference fuels of 100 for iso-octane and 0 for n-heptane. The fuel being tested is then classified as having an octane rating equivalent to that of a blend of x percent isooctane and $100–x$ percent n-heptane. Thus a fuel having an octane number of 97 has the same anti-knock tendency as a fuel consisting of a blend of 97 percent isooctane and 3 percent n-heptane.

The testing is carried out in a standard engine known as a CFR engine (developed by the Cooperative Fuel Research Committee, USA). Two tests are normally carried out to simulate normal and more severe engine operating conditions, giving rise to two different octane ratings. (a) The research octane number (RON) gives a rating appropriate to the engine loading and temperature associated with most passenger cars and light-duty commercial vehicles. (b) The motor octane number (MON) relates to more severe operating conditions, namely high engine loading and temperature associated with high-speed, full-throttle operation. The actual performance of a motor spirit on the road frequently falls between these two figures, giving rise to a third rating, road octane number. In addition to the various MOTOR SPIRIT UPGRADING processes, anti-knock properties of motor spirit can be improved by the use of lead additives, principally TEL (tetraethyl lead) and TML (tetramethyl lead). The amount of lead that can be added to motor spirit is controlled by government legislation in most countries, as a result of debate about its possible toxic effects.

volatility. The principle of the motor spirit internal combustion engine depends on spirit vaporizing readily when mixed with air in the carburettor to form a combustible mixture. If it vaporizes too readily, however, it can form vapour in the fuel line before it reaches the carburettor. This affects the flow of fuel. The condition is known as vapour lock, and it can cause stalling of the engine. Volatility of motor spirits is measured and controlled by reference to its vapour pressure, which is the force that a vaporized liquid exerts on its container.The most common method of testing and measuring motor spirit vapour pressure is the REID VAPOUR PRESSURE test, and the results are normally expressed in units of pressure, such as pounds per square inch. However, the Reid method is not regarded as a completely reliable criterion of vapour-locking tendencies, and other methods are also used. The simplest of these specifies a minimum volume of vapour per unit volume of liquid at a stated temperature, because vapour pressure increases with increases in temperature.

gum content. In common with other products, motor spirit may deteriorate if stored for long periods. A general requirement for all products, therefore, is that they should be stable. Tests for stability involve testing for the by-products formed in the deterioration process, which is largely caused by oxidation as a result of exposure to the air. The main effect of the oxidation of motor spirit during storage is that a resinous material or gum is formed, which separates out from the liquid fuel as a solid. This can be detected in a gum content test. The formation of gum in stored motor spirit can be prevented by the use of ANTI-OXIDANT INHIBITORS.

corrosiveness. The refining process (*see* REFINING, SECONDARY PROCESSING) should have reduced the sulphur content of motor spirit to acceptable levels, but tests for corrosiveness are still carried out on finished motor spirit to detect the presence of sulphur compounds. The most commonly used are the DOCTOR TEST and the COPPER STRIP CORROSION TEST.

quality, residual fuel oil. *See* QUALITY, DISTILLATE AND RESIDUAL FUELS.

quality, waxes. Like most other special products with a diverse range of grades and end uses, waxes form a highly specialized and complex branch of petroleum products. The end uses range from the coating of foodstuffs such as cheese to the manufacture of polishes, as well as various uses in modelling and wax-casting. One major test that is common to all waxes and is used to establish a wax's quality and type is that of melting point. This defines the temperature at which a wax melts, or starts to melt as the point is often not a sharply defined one.

quality, wide-cut gasoline fuels. These fuels have volatility requirements between those of aviation spirit (*see* QUALITY, AVIATION SPIRIT) and aviation turbine kerosine (*see* QUALITY, AVIATION JET FUELS). Volatility is controlled, as with motor spirit and aviation spirit, by a combination of distillation tests and Reid vapour pressure (*see* QUALITY, MOTOR SPIRIT) within limits, which makes specification of a flash point redundant.

quality differential. An adjustment to the price of crude oil or products to take account of differences in quality. When the pricing system was determined by the oil companies (*see* CRUDE OIL PRICES) quality differentials were less important because each company tended to lift a wide range of crudes, and therefore anomalies were evened out in the process of composing the refining SLATE. However, as pricing came under the control of the producing countries in the early 1970s, because most producers sold a relatively narrow range of crudes the quality differential became important in determining a country's ability to sell its output. When supplies of crude oil were generally plentiful, it was extremely important that the quality differential was an accurate reflection of the PRODUCT WORTH. During this period, OPEC had a precise formula which specified differentials for gravity and sulphur differences from the MARKER CRUDE. In practice, however, the tendency was for crude sellers to set the differentials on the basis of what the market would bear. *See also* QUALITY, CRUDE OIL.

quality specification. *See* QUALITY.

Quoin Island. A small island at the entrance to the GULF which is used as a convenient starting point for fixing voyage costs. Thus voyage rates in the Gulf are quoted from the Gulf loading point to Quoin, and there is then a series of voyage rates from Quoin to the unloading point.

quota. (1) A specification of the maximum amount (usually expressed in physical units) of imports of a good allowed into a country. It is a more precise means of PROTECTION than the imposition of TARIFFS since the latter's effect is a function of PRICE ELASTICITY OF DEMAND. (2) The agreed quantity that a producer must sell when producers form a CARTEL. Invariably, the size of each quota is a matter for hard bargaining since it determines revenue share.

R

racking pipe. The stacking of drilling pipe on a DERRICK.

rack price. A price quoted by a refinery for truck loads of products.

raffinate. The refined product produced by solvent refining. Solvents are used to remove unwanted components either by dissolving or by precipitating processes.

random walk. Describing the time series path of a variable. It has been likened to a drunk walking down a street: one is reasonably certain that he will reach the end of the street, but there is no way of knowing in advance the precise route he will take and there is no way of explaining the route eventually taken. It is frequently used to describe the path of exchange rates and commodity prices. It is particularly relevant in markets where expectations play a key role in day-to-day trading.

ranking problem. A situation that occurs when evaluating a series of competing projects. The use of NET PRESENT VALUE and INTERNAL RATE OF RETURN as the criteria produce different results if there are significant differences in the initial capital outlays, lives or cash flow timing between projects. *See also* DISCOUNTED CASH FLOW.

rated capacity. When a plant such as a refinery or pipeline is built, its capacity is determined by the designers based upon a specified set of operating conditions which determine the engineering tolerances. This is the rated capacity at which the plant can safely operate and is essentially a prediction of how the plant will perform under specified conditions. Not until the plant is actually operational will the accuracy of its rated capacity be known.

rate of interest. The price of borrowed money. Although commonly used, there is no such thing as *the* rate of interest. Instead there exists a structure of interest rates reflecting four different influences — time preference, liquidity, risk and inflation. Money that is lent represents a sacrifice of current consumption. Since economists assume that people prefer to consume now rather than later, the longer the period over which the money is lent (i.e. the longer consumption is postponed) the greater is the reward they must receive (i.e. the higher the interest rate). The lender exchanges cash for an asset. The LIQUIDITY of the asset will vary, for example, from a bank deposit, which requires a few days' notice of withdrawal, to one that requires several months' notice. The more illiquid the asset, the greater must be the reward (i.e. the interest rate) to compensate for the lender's loss of liquidity. The lender of the money also faces the risk of default and the loss of his money. The greater the risk associated with the loan the greater must be the reward to induce the lender to accept that risk. There is also the fact that inflation will erode the PURCHASING POWER of the money lent. In theory, this should apply equally to all loans thereby not affecting the range of the interest rate structure. In practice, however, perceptions of future inflation can differ between groups and over time, and therefore can effect the range of interest rates, especially in an international context.

rate of penetration. The depth of DRILLING achieved in a specific time; it is normally measured in feet per hour. The rate of penetration is a function of the type of formation being drilled and the quality of the drilling BIT.

rate of return. A financial ratio that expresses the net profit after DEPRECIATION of a company or project as a percentage of the capital used. There are many variations depending upon the definitions of net profit (e.g., before or after tax) and definitions of capital used (e.g., equity capital, i.e. excluding loan capital, capital less working capital, etc.). Rate of return is used as a measure of the performance of a company or project, although in recent times much less so because the measure excludes the timing of capital outlays and earnings. In general, the INTERNAL RATE OF RETURN is used to account for the time value of money.

rate sensitivity. In many fields, depending upon the geology, the amount of oil eventually produced (i.e. RECOVERY FACTOR) can be influenced by the rate of production from the reservoir. If production is 'excessive' then the recovery factor falls. Such fields are described as rate-sensitive. *See also* CAPACITY.

rathole. *See* MOUSEHOLE.

ratio analysis. A process by which various ratios between firms are compared. It assumes this comparison will show up normality or abnormality, which will guide or determine subsequent management decisions. The ratios used may be: return on capital (profits/assets); rate of asset turnover (sales/assets); profit margin on sales (profits/sales), etc.

raw gas. The untreated gas produced from a gas-field. Normally some treatment is required before marketing.

real. NOMINAL values (i.e. values in current money) that have been adjusted to take account of changes in the PURCHASING POWER of money which arise from inflation (and possibly exchange rate fluctuations). Thus if the nominal price of oil rises by 10 percent in one year and inflation during that year is 10 percent, the real price of oil remains unchanged. If inflation were greater than 10 percent, the real price would have fallen. When considering any TIME SERIES of a variable measured in money values, it is the real value that should be used for most analytical purposes.

real asset. *See* ASSETS.

realized prices. *See* CRUDE OIL PRICES.

recession. A downturn in the level of economic activity which may or may not be part of the BUSINESS CYCLE. The use of the term implies that the downturn is of fairly short duration. A recession that persists is more accurately called a depression, although the dividing line is far from clear. During the 1950s and 1960s, when a strong link appeared to exist between economic activity and oil demand, a recession meant a reduction in oil demand. However, during the 1980s there is a growing view that this link has been broken which, if true, implies that recessions have limited relevance for oil demand.

recon crude (reconstituted oil). A crude input to a refinery which has been specifically blended for the refinery's configuration and market.

reconstituted crude. *See* RECON CRUDE.

recoverable reserves. The amount of oil or gas that will be produced from the oil-in-place. It is an imprecise concept because it makes no distinction between PROVEN RESERVES, PROBABLE RESERVES and POSSIBLE RESERVES.

recovery factor. The ratio of total production from a field to the oil-in-place. It can be measured *ex poste* or *ex ante*, in which case RECOVERABLE RESERVES are used instead of production. There are enormous variations between the recovery factors of different fields. The extent of the recovery factor is a function of the geology of the field, its RATE SENSITIVITY, the rate of development and production, and the extent to which recovery methods other than PRIMARY RECOVERY are used. *See also* ENHANCED OIL RECOVERY; SECONDARY RECOVERY.

recurring costs. *See* RUNNING COSTS.

recycling. (1) The lending of surplus money to deficit countries. Following the FIRST OIL SHOCK, the OPEC countries acquired large balance of payments surpluses matched by large balance of payments deficits which accrued to the oil-importing countries. It was feared that this imbalance would reduce world liquidity (i.e. the amount of money available to pay for world trade). Therefore it became a matter of some urgency that the OPEC surpluses should be recycled. A large part of the OPEC surpluses was automatically recycled to the USA and the EURODOLLAR MARKET by virtue of the fact that oil was priced in dollars and so the surpluses were initially dollars held in banks abroad. It was these banks which played a key role in the subsequent recycling by lending the dollar deposits to deficit countries. Other recycling measures involved the setting up of funds (by the IMF and OPEC) to provide balance of payments relief. In the event the recycling was successful, but it planted the seeds of the Third World debt problem which emerged in the 1980s. (2) The returning to the oil reservoir of oil products for which there is no market demand as a result of an unbalanced demand structure; this practice is unusual now an unusual one. (3) The recirculation of the parts of the feedstock that have remained unchanged during the refining process.

Red Line Agreement. An agreement whereby the partners in the Iraq Petroleum Company (IPC) agreed not to carry out oil operations independently in an area roughly matching the old Ottoman Empire. The intention was to prevent competition over concessions or purchase and refining of crude oil. The Agreement remained intact until 1948 when Jersey and Mobil withdrew to be able to join Socal and Texaco to create the Arabian American Oil Company (Aramco) to operate in Saudi Arabia. The title comes from a line drawn by Gulbenkian to show his understanding of the Ottoman Empire area.

Redwood No 1 Second. A unit for the measurement of VISCOSITY of liquid fuels and lubricants. It was widely used in areas of the world where British imperial units are employed. Values were determined using a Redwood viscometer. Other measures of viscosity were the SAYBOLT UNIVERSAL SECOND which was used in the USA and the ENGLER DEGREE which was used in Europe. All these units have now generally been replaced by the centistoke (cSt).

reef effect. A situation in which an offshore structure, usually as a result of increasing the sea temperature, provides a suitable environment for marine life.

reef trap. *See* GEOLOGY.

reference fuel. A standard fuel against which the QUALITY aspects of a specific fuel can be tested. *See also* QUALITY, DIESEL OIL; QUALITY, MOTOR SPIRIT.

refinery. A chemical plant where crude oil is refined (*see* REFINING).

refinery configuration. The general term for the types of units contained within a refinery. It is commonly represented as a flow diagram on which the various PRIMARY DISTILLATION UNITS and units of SECONDARY REFINERY CAPACITY are shown. During the 1970s refinery configurations tended to expand towards the secondary (upgrading) units to enable an increase in the proportion of LIGHT ENDS from a given barrel of crude.

refinery fuel. The fuel that is used to run a refinery. It can be oil products produced by the refinery itself or it may be 'imported' from outside in the form of gas, electricity or steam. If it is obtained from the refinery's own product production then between 3 and 10 percent of the crude run is used up as fuel, depending on the complexity of the REFINERY CONFIGURATION; the more complex the refinery the greater the need for fuel. *See also* REFINERY GAS.

refinery gas. The gaseous HYDROCARBONS produced as a by-product of REFINING. It is normally used as a REFINERY FUEL with any surplus being flared (*see* FLARING).

refinery location. Refineries may be located on the oil-field, at the market or at some intermediate point between the oil-field and the market. In general, the major sources of traded oil (e.g., OPEC countries) are geographically separate from the major markets (e.g., OECD).

Before World War II, the majority of refineries were located on oil-fields. This was for two reasons: (a) during DISTILLATION about 25–50 percent of the crude was lost which meant that a large proportion of the crude oil, if it were to be transported, would be lost (i.e. transported to no purpose); (b) the product markets had an unbalanced demand structure for products, consequently having shipped crude to the markets unwanted products would have to be reshipped to other markets.

After World War II, the location pattern changed with the majority of new capacity being built in the market regions. There were several reasons for this switch: (a) technical improvements reduced the loss of crude during refining to 4–10 percent; (b) the product demand structure altered to match the product proportions obtained from simple refining, especially in Europe; (c) there was a growing need for the 'market countries' to save foreign exchange, and less foreign exchange was required to import crude compared with products; (d) due to pipeline and tanker developments, the cost of transporting crude oil had fallen compared with that of transporting products (these developments were more a response to the change of location than a cause of the change); (e) the 'market countries' were concerned that their supplies would be maintained. While there was an excess capacity to pump crude, there was no excess refinery capacity, thus if oil-field supply was nationalized so too were the oil-field-located refineries, as happened in Iran in 1951; thus oil companies lost access to crude and refinery capacity. The crude loss could be replaced using other sources, but the refinery capacity was irreplaceable. Since the oil companies as disposers of crude had the right to choose where to refine, they chose increasingly to refine in market locations where the risk of nationalization was less. As a result of the relocation, by 1970 oil-field-based refineries in the Middle East and Venezuela accounted for less than 13 percent of the WOCANA total, whereas refineries in Europe and Japan accounted for nearly 63 percent.

In the early 1970s many producing countries began to believe that new factors had swung the economic locational advantage back in favour of the oil-fields. Land for refineries in market locations was becoming increasingly expensive. Also the capital costs of building refineries were rising rapidly because increasingly stringent pollution regulations required more equipment. In addition, the oil-producing countries had access to ASSOCIATED GAS at negligible cost with which to fuel the refineries. These countries also wished to develop DOWNSTREAM capabilities which would promote economic diversification and enhance their control over the oil market. However, these plans came at a time of excess refining capacity as a result of the massive refinery building programme in the early 1970s (between 1970 and 1975 world capacity increased by 34 percent) coupled with the fall in demand for oil following the FIRST OIL SHOCK. The consequence was that most plans were rejected and the market locations have remained dominant; the Middle East and Venezuelan percentage of WOCANA capacity in 1985 was similar to the 1970 value (*see above*).

refinery yields. The different proportion of products from the refining of a specific crude in specified refinery units. Thus the refinery yield is specific to the crude rather than to the refinery. Its most important function is in the calculation of PRODUCT WORTH.

refining. The treatment of crude oil in order to form FINISHED PRODUCTS. Refining is not a one-stage operation; it involves a number of manufacturing processes. Initially the crude is distilled to separate fractions with different boiling points. Most of the fractions thus separated then undergo one or more of the secondary processing stages, of which there are a wide range, but which fall into basically two types: (a) treatment or purification processes, which remove impurities (e.g., sulphur) to bring the product up to the required market standard or specification; (b) conversion processes, which involve changing the chemical composition of the fraction being processed, either to improve its quality (e.g., reforming), or to change it into another product (e.g., cracking). *See also* MOTOR SPIRIT UPGRADING; REFINING, PRIMARY PROCESSING; REFINING, SECONDARY PROCESSING.

refining, primary processing. The primary processing stages of refining involve washing the crude oil followed by distillation, which separates the crude oil into distillation cuts, or fractions. The distillation process is one of physical separation and involves no change in chemical composition.

desalting. This is a preliminary washing stage to remove any salt with which the crude oil may be contaminated.

distillation. This is the start of the refining process proper. The object of distillation is to separate the many compounds contained in crude oil into groups of similar compounds. The principle underlying this process is the fact that different liquids vaporize at different temperatures (called boiling points). If the liquid hydrocarbons that make up crude oil are subjected to heat, then as the resulting vapour cools different groups of compounds condense into liquids at different temperatures, which correspond to the different boiling points. As each lot of vapour condenses it can be collected in liquid form. In practice, this operation involves heating the crude oil and pumping it into a crude distillation unit (CDU), which is a column divided by a series of horizontal trays. The action of heat on the crude causes part of the oil to vaporize and rise to the top of the tower through holes in the trays. The tower is cooler at the top, which is furthest from the heat source, causing some of the vapour to condense. Liquids with a low boiling point can therefore be collected from the top of the column and those with successively higher boiling points at intermediate points lower down; compounds with the highest boiling points remain unchanged at the bottom forming a residue. The separation of low and high-boiling materials in this way is called fractional distillation (fractionation) and the distillation column in this context is called a fractionating column.

A group of compounds obtained within a certain range of boiling points is called a straight-run fraction or cut. A boiling range of temperatures is used in the crude distillation context because the distillation cuts do not comprise pure substances, but are mixtures of a number of hydrocarbons that have similar, but not identical, boiling points. The limits of the boiling ranges at which different groups of compounds are taken off are called cut points. In general certain cut points correspond to certain products, but it should be emphasized that significant variations are possible. A rough breakdown of straight-run components produced at different cut points is shown in Table R.1

The primary distillation phase is undertaken at atmospheric pressure, and the oil is not heated to temperatures greater than 350°C. This is because some high-boiling point compounds start to decompose when heated to their boiling point at atmospheric pressure. This basic distillation process is therefore sometimes referred to as distillation, and the residue, which boils at above 350°C, is known as the atmospheric residue.

Table R.1.

Boiling Range (°C)	Straight-run components
0	Gases
0–70	Light gasoline*
70–140	Benzine*
140–180	Naphtha*
180–250	Kerosene*
250–350	Gas oil
>350	Residue

If it is required to separate out cuts with boiling points above 350°C, a second distillation of the atmospheric residue is carried out under vacuum (i.e. vacuum distillation). The vacuum reduces the temperature required for these compounds to vaporize. The residue obtained from vacuum distillation has a cut point of 550°C and is termed vacuum residue. The main products of vacuum distillation are vacuum gas oil, which is used as cracker feedstock, lube oil feedstocks and bitumen.

A distinctive feature of oil refining at all its stages is flexibility. Crude distillation units are designed to cope with a variety of crudes and to tailor the yields, within limits, in response to market demand. The main way in which this is achieved is by different cut points. Because the range of straight-run components represents a continuous spectrum in terms of gradation of properties, any cut point represents a somewhat arbitrary division and therefore usually can undergo some upward or downward revision. For example, if a high kerosine yield is required, then a wide-cut kerosine can be obtained by encroaching on the heavier ends of naphtha and on the lighter ends of gas oil. By this means, straight-run yields on individual components can be varied by several percent.

refining, secondary processing. In secondary processing, units are designed so that feedstocks and operating conditions can be varied. Such a high degree of flexibility inevitably leads to considerable complexity in refining operations. The net result is that there is no unique manufacturing route to any single finished product. In addition, it must be remembered that most finished products are not single substances, but are blends of several compounds. The final stage of refining therefore, and an integral part of the overall process, involves the selective blending of basic components to give a wide variety of individual grades of the same product. This process is particularly sophisticated in the case of motor spirit, which may contain half a dozen components, each of which has followed a different manufacturing route. In view of this, the role of the individual secondary refining processes is best understood in relation to the contribution which each makes to the transition of straight-run components into final components for blending into finished products.

desulphurization. The sulphur which is present in varying degrees in all crude oils (*see* QUALITY, CRUDE OIL) is distributed throughout the range of STRAIGHT-RUN distillation cuts. As in the case of crude oil, the sulphur present is present in chemical compounds combined with other elements (e.g., carbon, hydrogen). Different sulphur-containing compounds have different properties and virtually all have a pungent smell, which is unacceptable for most end uses. In addition, many sulphur compounds are poisonous and/or corrosive. For most end uses, therefore, there is a need to eliminate the smell of the sulphur compounds, and in some cases to remove (or substantially reduce) the level of the sulphur compounds. Processes that improve the smell only are known as SWEETENING processes, whereas those that actually remove sulphur compounds are called desulphurization processes. There is a tendency for sulphur compounds to increase in complexity in the higher boiling ranges, and the type of process required therefore varies according to the fraction being treated.

(a) For LPG and light-end GASOLINE (up to about 80°C) the main type of sulphur compounds present are derivatives of hydrogen sulphide (H_2S) called mercaptans. In addition to having an unpleasant smell mercaptans are corrosive and must be removed. The process by which this is done is the simplest of all the desulphurization processes — the UOP Merox process — which involves washing the LPG and gasoline streams with a solution of caustic soda.

(b) The fractions boiling between 80 and 250°C (i.e. the remaining light distillate fractions of benzine, naphtha and kerosine) contain mercaptans of a slightly different type to those described above, which cannot be extracted by the Merox process. They can, however, be sweetened by means of oxidation, which converts them to a different type of sulphur compound — disulphides — which are non-corrosive and almost odourless. This can be done either by a modified version of the Merox process or by a similar process using copper chloride as a catalyst. Both these processes can be used to treat aviation kerosine, as well as the lighter fractions. The kerosine to be used for domestic central heating, however, frequently needs to undergo desulphurization rather than sweetening and therefore undergoes the same treatment as gas oil (*see below*).

(c) In the gas oil/diesel oil range (250–350°C), the sulphur concentration tends to be higher, and the sulphur compounds present are more complex than in the lighter fractions. This necessitates a different type of process, which involves firstly breaking down the sulphur compounds and then extracting the sulphur. The oil together with hydrogen is passed over a catalyst at high temperature and pressure, with the result that the sulphur compounds are broken down and the sulphur released is combined with hydrogen in the form of hydrogen sulphide, leaving the oil relatively free of sulphur. This process is called hydrofining or hydrotreating. Most straight-run middle distillate fractions (e.g., gas oil, diesel oil), as well as kerosine required for domestic heating produced from medium- and high-sulphur crudes require hydrofining to meet finished product sulphur specifications.

(d) A variation of the hydrofining process has been developed for the desulphurization of fuel oil residue called residue hydrodesulphurization (HDS), but it is largely unused outside Japan. There are two main reasons for this. (i) Although for many end uses fuel oil sulphur levels are controlled by anti-pollution legislation two important outlets — BUNKERS and certain industrial fuels (*see* FINISHED PRODUCTS, FUEL OIL) — remain for the use of high-sulphur fuel. High-sulphur fuel oil outlets can absorb high-sulphur residues from Middle East crudes, and it is possible to meet more restrictive fuel oil sulphur specifications by selective blending of residues from low-sulphur crude oils without recourse to residue HDS hydrodesulphurization. (ii) Because the sulphur levels in the residue are higher, and the sulphur is present in the form of more complex compounds than in the distillate fractions, residue HDS is technically more complex and consequently more expensive. In most countries, therefore, the process cannot be economically justified.

conversion processes. Distillation yields can be modified by up to several percent on each fraction (*see* REFINING, PRIMARY PROCESSING). Beyond certain limits, however, this approach ceases to be effective. For example, a typical distillation yield from Arabian light crude would give approximately 18–20 percent light distillates and around 40 percent fuel oil. This accorded fairly well with the demands in Western European in 1973, which was approximately 18 percent light distillates and 36 percent fuel oil. By 1982, however, the proportion of fuel oil required in Western Europe represented a much smaller share of the barrel (around 27 percent) and light distillates a higher share at about 22 percent of demand. The imbalance between straight-run fuel oil yield of 40 percent and market demand of less than 30 percent is beyond the limits of what can be achieved by manipulating cut points. To meet this sort of situation, it is necessary to use some form of conversion (or yield shift or upgrading) process in order to change the proportions of the various products that can be obtained from crude.

Cracking is the most common form of conversion. Its name is derived from the chemical process on which it is based, namely that of 'cracking' (or breaking up) of large molecules to form smaller ones. Since it is the heavier products that have the larger molecules, and since as a general rule the value of products increases the lighter they are, the process represents a means of converting a low-value feedstock (e.g., fuel oil) into a high-value product (e.g., motor spirit). The economic justification for the process thus depends on the relative value of the products involved.

(a) thermal cracking. This was the earliest form of cracking used. It involved heating the feedstock to very high temperatures, which caused the large molecules of the heavy feedstock to decompose into the smaller molecules of gas oil and motor spirit. The quality of the resulting motor spirit is not sufficiently high for modern specifications, however, and the process is now little used.

(b) catalytic cracking (cat cracking). This is the most common form of cracking currently employed. The use of a catalyst was found to improve the quality of the motor spirit, and there are now several forms of this type of process. The feedstock used can be fuel oil (ATMOSPHERIC RESIDUE) or straight-run gas oil or vacuum gas oil (gas oil produced from VACUUM DISTILLATION). The several different types of catalytic cracking most commonly used differ mainly in the type of catalyst used and the method in which it is handled, and many of the distinguishing names of various processes relate to these features (e.g., short contact time (SCT) catalytic cracking, FIXED-BED CATALYTIC CRACKING, etc.). The process most widely used currently is fluid catalytic cracking (FCC) in which the catalyst is powdered so that it can be moved through the unit as a fluid mixture.

(c) hydrocracking. A less common form of cracking is hydrocracking, which can be used to produce either motor spirit or gas oil from heavier fractions. As well as using the basic cracking process, it also involves the addition of hydrogen (i.e. hydrogenation), which means that, although the process offers considerable flexibility in terms of feedstock and yield, it requires a separate hydrogen plant, which adds considerably to the capital cost.

(d) visbreaking. This process, which is an adaptation of the thermal cracking process, breaks down the larger molecules of heavy, viscous (thick) residues to smaller ones which, while still being fuel oils, are lighter and less viscous. The alternative method of reducing fuel oil viscosity is to blend gas oil or kerosine with the viscous residue. Thus, although the visbreaking process does not itself produce light or middle distillates, it effectively increases the availability of these fractions for use as finished products by eliminating, or reducing, the need to use them as fuel oil dilutents.

In addition to the cracking process the following conversion processes may also be carried out.

(e) delayed coking. This a rather more specialized process that normally uses the residue from low-sulphur crudes to produce electrode-grade coke of the type used in the production of aluminium. Gas, motor spirit and gas oil are also produced.

(f) fluid coking. This is a development of the coking process in which coke production is a secondary objective, and hence the yield of the more valuable lighter products (e.g., motor spirit and gas oil) from the residue feedstock is maximized. *See also* MOTOR SPIRIT UPGRADING.

reflux. The distillate that is recycled into the top of a DISTILLATION COLUMN to control temperature. *See also* REFINING, PRIMARY PROCESSING.

reforming. The chemical process that rearranges the molecular structure of hydrocarbons. In refining it refers to the process in which STRAIGHT-RUN feedstocks are subjected to high temperatures and pressures often in the presence of a catalyst to change their structure in order to produce AROMATICS or to increase their OCTANE RATING for motor spirit blending. *See also* MOTOR SPIRIT UPGRADING.

regression analysis. The process of taking a dependent variable such as oil price and trying to establish an explanatory independent variable(s) by the use of statistical techniques. The technique may use only one independent variable (simple linear regression) or it may use several (multiple regression). Regression analysis is commonly used in ECONOMETRICS.

regressive taxes. A system of taxation, unlike one characterized by PROGRESSIVITY, that takes an increasing proportion of income or profits as income or profits fall thereby causing a disincentive effect. Royalties as a fixed barrelage tax are an example of such a tax: as oil price falls the burden of tax rises thereby threatening the economic visibility of smaller fields. *Compare* PROGRESSIVE TAX.

regular motor spirit (RMS). Motor spirit that has a RESEARCH OCTANE RATING around 90 or lower.

Reid vapour pressure. The commonly used method for measuring the VAPOUR PRESSURE of petroleum products. It is measured at 100°F and is reported in pounds per square inch. *See also* QUALITY, MOTOR SPIRIT.

relative price. The amount of one good (e.g., oil) compared with the amount of another good (e.g., coal) which the same amount of money can buy. It is relative prices which, in part, determine the choice of purchases and therefore the resource allocation in an economy. In terms of the oil and gas industry, relative prices are an important factor in FUEL SWITCHING.

relinquishment. The surrender by a licensee or CONCESSION holder of ACREAGE, if the company regards the acreage as not worth further exploration or development. Such relinquishment may be voluntary or it may be as part of a strict timetable of relinquishment contained in the licence/concession agreement. Early concession agreements contained no obligatory relinquishment, although some (e.g., the Aramco concession) allowed relinquishment at the company's discretion. Given the very large areas covered by these concessions (88 percent of Iran, Iraq, Kuwait and Saudi Arabia together) this lack of relinquishment was increasingly seen as a drawback. For example, in Iraq and Kuwait there was no territory available if the national oil companies of those countries wished to explore in their own territory. In 1948 Aramco agreed to a programme of relinquishment, but other companies were reluctant to follow. The watershed came in 1960 when Iraq invoked LAW 80, which sequestered all the unused acreage. Since then relinquishment has become more common, and obligatory relinquishment is now usual in most oil and gas agreements.

renewable resources. Resources that are either self-perpetuating, such as wind, sun or tide, or self-reproducing, such as animals and plants. The term is used to differentiate from EXHAUSTIBLE RESOURCES of which there is only a finite supply. Renewable resources can become exhaustible if the existing stocks are abused. Renewable energy sources are usually defined as solar, wind, sea (tidal, wave and ocean), hydroelectric, biomass and geothermal energy.

rent. *See* ECONOMIC RENT.

rental. Money paid for the hire of an ASSET. Economists tend to use the word to try to avoid confusion with ECONOMIC RENT.

rent bidding. An auction for acreage whereby RENTAL rates on the acreage are offered so as to make the bid appear as attractive as possible. *See also* AUCTION SYSTEM OF LICENSING.

replacement cost. *See* DEPRECIATION.

repressuring. The reinjection of gas into a RESERVOIR to maintain or increase pressure. *See also* PRODUCTION, SECONDARY RECOVERY MECHANISMS.

resale agent. *See* JOBBER.

research octane number (RON). The OCTANE NUMBER derived from tests on a standard internal combustion engine. Normally the RON is higher than the MOTOR OCTANE NUMBER, and it is the RON that is used for marketing purposes. *Compare* ROAD OCTANE NUMBER. *See also* QUALITY, MOTOR SPIRIT.

reserve currency. A currency which governments are willing to hold their gold and foreign exchange reserves and which is widely used to finance international trade. To qualify as a reserve currency it must be relatively stable in value, widely used, and efficiently and freely convertible.

reserves. The starting point by which to understand the concept of hydrocarbon reserves is OIL-IN-PLACE. This is the total amount of oil contained in a reservoir. It is generally agreed that volumetric methods for determining the amount of oil-in-place cannot be more accurate than ±10 percent and can often be outside the range of ±20 percent. This is even more true if there are political or economic motives for understating or overstating the reserves. This estimate of oil-in-place is simply a measure of the physical presence of oil.

The next stage is to make some estimate of how much of the oil present will eventually be produced (i.e. the recoverable reserves). Conventionally three categories have been identified. (a) Proven reserves are the part of the oil-in-place that will almost certainly be produced given existing technology and economics in the context of the knowledge about the specific field. (b) Probable reserves have a greater than 50 percent probability of production. (c) Possible reserves have a less than 50 percent probability of production. However, the key lies in the terms 'given existing technology and economics' and 'in the context of the knowledge about the specific field'. These elements change over time: technology advances; as production gets underway more is learnt about the field; the economics of production changes with (in addition to changing technology) changes in the prices of inputs and outputs.

Estimates of recoverable reserves are extremely uncertain figures and should be treated with the greatest of caution and scepticism. In an attempt to secure greater precision the US Geological Survey devised a set of definitions which divided the recoverable reserves into demonstrated (measured and indicated) and inferred. In 1977 the United Nations Centre for Natural Resources, Energy and Transport set up a working party to consider the adoption of the Geological Survey's definition. The conclusion was that such an approach required too great a level of detail.

reserves remaining. The RECOVERABLE RESERVES (based on proved and probable reserves) less past production.

reservoir. A stratum in which oil and gas is present, normally subject to a single pressure system. *See also* GEOLOGY.

reservoir pressure. The fluid pressure in an oil or gas RESERVOIR. Once the CAP ROCK has been breached and production has begun, the pressure naturally declines. As the pressure declines so too does the oil or gas flow. Theoretically once the internal pressure of the reservoir is equalized with the surface pressure no oil or gas would flow naturally. However, long before then, if the economics are justified, it would be expected that the pressure would be boosted or at least maintained by the use of secondary recovery methods or ENHANCED OIL RECOVERY. *See also* PRODUCTION, ENHANCED OIL RECOVERY; PRODUCTION, NATURAL DRIVE MECHANISMS; PRODUCTION, SECONDARY RECOVERY MECHANISMS.

reservoir rock. Geological term for the

rock in which oil is retained in an oil-bearing structure. In order to contain the oil or gas the rock must be porous (e.g., sandstone or limestone). For the rock to retain the oil or gas it must be overlaid with a non-porous rock. *See also* GEOLOGY.

reservoir simulation model. A mathematical model, usually prepared using a computer, that creates the physical characteristics of gas- or oil-field. It can therefore be used to assess RESERVES and possible production patterns.

residual fuel oil. A very heavy fuel oil produced by means of FRACTIONAL DISTILLATION rather than from the distilled fractions. *See also* FINISHED PRODUCTS, FUEL OIL; QUALITY, DISTILLATE AND RESIDUAL FUELS.

residue hydrodesulphurization (residue HDS). A process that removes sulphur compounds from fuel oil by treating it with hydrogen in the presence of a catalyst. The process is used mainly in Japan. *See also* REFINING, SECONDARY PROCESSING.

resources. Economists use this general term to describe all the factors of production (i.e. land, labour, capital and entrepreneurship) available to a community. They represent the inputs into the productive process. The main point about resources is that they are relatively scarce, and therefore the object is to ensure their most efficient use.

resource substitution. *See* CONSERVATION.

retained earnings. The part of corporate PROFITS that is retained by the corporation. These profits can then be used for new investment thus providing self-financing. For much of the history of the international oil industry, oil companies have been heavily reliant on such sources of finance. The companies have frequently cited such self-financing to justify the need for high levels of profit.

returns on assets. The net PROFIT expressed as a percentage of the total ASSETS owned by a company. There are numerous variations depending on the definitions of profits and assets. *See also* RATIO ANALYSIS.

returns to scale. The relationship between a change in output resulting from altering the scale of the operation by varying all the inputs in the same proportion. If output increases proportionally more than the increase in inputs the operation is characterized by increasing returns to scale. If the output increases proportionally less, this would be a case of decreasing returns to scale. If the output change is the same, this would be constant returns to scale. The nature of returns to scale is determined by the presence of ECONOMIES OF SCALE or diseconomies of scale. Over a very wide range of output, because oil and gas are capable of moving in continuous flow in three-dimensional space, both are associated with increasing returns to scale.

revaluation. An increase in the rate at which one currency is exchanged for another. This may occur as a result of government decree or as a result of market forces, depending on the exchange rate regime. Revaluation makes imports cheaper and exports more expensive. *Compare* DEVALUATION.

revenue. The payment received in return for supplying a quantity of a good or service. For an enterprise, total revenue (TR) is the total payment received for supplying a quantity Q of a good in a given time period. Average revenue (AR) is the revenue received per unit sold

$$AR = TR/Q$$

Marginal revenue (MR) is the change in total revenue resulting from selling an extra unit. For very small changes in output it can be specified in calculus notation as

$$MR = d(TR)/dQ$$

The relation between MR and AR is given by

$$MR = AR + Q(\text{slope of AR})$$

Consequently when the slope of the AR curve (the DEMAND CURVE) is zero, and the demand curve is horizontal, AR = MR (*see* Fig. R.1). This is the case for a firm in perfect competition (*see* COMPETITION). However, if the demand curve is negatively sloped, then MR lies below AR. When price changes, the price of all units demanded changes, not just the price of the last unit. This then affects total revenue. For a straight-line demand curve only, it can be shown that the MR curve will be twice as steep as the AR curve. Total revenue will be at its maximum value when its derivative, MR, is zero.

ricardian rent. *See* ECONOMIC RENT.

rig. Any movable structure used for drilling. The term is somewhat misleading since it can cover a small unit used for seismic work (*see* SEISMIC EXPLORATION) or an offshore PLATFORM.

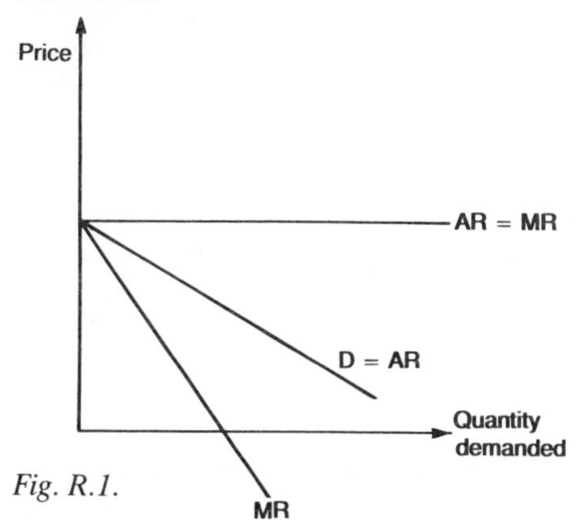

Fig. R.1.

rig floor. *See* DERRICK FLOOR.

rigging up. The loading of necessary equipment and supplies onto the RIG before drilling begins.

rights issue. The offering of new shares in a company to existing shareholders. The cost of raising capital in such a way is significantly lower than for a new issue of shares because there are no advertising or UNDERWRITING costs.

rig time. The length of time during which a RIG is actually used for drilling.

ring fence. A fiscal device whereby taxable profits from one activity cannot be reduced by losses from other activities within the same company.

In the case of the UK sector of the North Sea there exists a CORPORATION TAX (CT) ring fence around a company's oil extraction activities prohibiting the deduction of CT losses made outside the UK sector from CT profits in the North Sea.

riser (marine riser). A pipeline that connects the SUB-SEA WELLHEAD to a production PLATFORM.

risk. The term risk and uncertainty are frequently used as synonyms in everyday usage. However, in economics there is a significant distinction between the two terms. Both risk and uncertainty exist when a course of action can lead to different possible outcomes. If the probability of a specific outcome is known (or can be computed) then the situation is one of risk. If the probability cannot be known the situation is one of uncertainty. For example, throwing a dice involves risk since the probability of different numbers coming up can be computed. However, the probability of where the next war will break out cannot be computed and is therefore represented by uncertainty. The last example illustrates an interesting point since it might be argued that experts in such matters could qualitatively come up with probabilities thereby converting uncertainty into risk. However, in such a case a view would have to be taken on how objective a probability must be to create risk. At a practical level the distinction is important since it might be expected that risk can be insured against (because actuaries can compute probabilities) whereas uncertainty is uninsurable. Where probabilities cannot obviously be computed, decision makers can revert to the Monte Carlo method, which involves using repeated runs of a model to build up the probability values of the outcomes.

Risk and uncertainty are of crucial importance in the international oil and gas industry. Decisions in the industry depend upon taking a view of the future (e.g., prices, costs, physical availabilities, technologies, etc.). The future involves possible outcomes or events that are risky or uncertain. Because such projects have relatively long LEAD TIMES the further into the future the greater the uncertainty. In addition, the further away the future event, the more risk converts to uncertainty since in the far future, as many parameters may change, probabilities become meaningless.

RMS. *See* REGULAR MOTOR SPIRIT.

road octane number. A measure of the anti-knock quality of motor spirit derived from tests on a standard engine under motoring conditions. *Compare* MOTOR OCTANE NUMBER; RESEARCH OCTANE NUMBER. *See also* QUALITY, MOTOR SPIRIT.

road oil. An oil intended for cold application to road surfaces, as opposed to straight-run BITUMEN which requires heating. Road oil frequently consists of residues and light CUTBACKS.

rollover. A situation in which a loan repayment is held over for a specified period of time.

RON. *See* RESEARCH OCTANE NUMBER.

rotary drilling. A DRILLING process in which the entire DRILLING STRING is rotated as opposed to turbine drilling in which only the BIT rotates by means of a downhole multi-stage turbine powered by drilling mud. The string is rotated by the rotary table, which has a square hole that engages the string, and is driven by chains from the drawworks. The speed of rotation is measured by the rotary rpm.

rotary rpm. *See* ROTARY DRILLING.

rotary table. *See* ROTARY DRILLING.

Rotterdam market. The north-western European market in crude oil and products. It is based mainly on the ARA complex of refineries and storage. *See also* ROTTERDAM PRICES.

Rotterdam prices. Crude oil or product SPOT PRICES quoted in the ROTTERDAM MARKET. *See also* AFM PRICES.

roughneck. An assistant to the driller who works on the DERRICK FLOOR.

roundtrip (making a trip). The removal of the entire DRILLING STRING from the hole to replace the drilling bit. *See also* DRILLING.

roustabout. A general labourer working in a drilling/production operation.

Royal Dutch Petroleum Company. *See* SHELL.

royalty. A payment on the output made to the owner of the oil and gas — normally the government outside the USA. In most fiscal systems the royalty is a fixed amount either in value terms per physical unit of output (before profit sharing it was set for the Middle East at 4 gold shillings/ton) or in percentage terms based on the value of output (under the OPEC formula 12.5 percent). It is a form of REGRESSIVE TAX and may act as a disincentive to production. There is a case to allow for a remission of royalty on marginal fields, and some fiscal systems recognize this requirement in order to encourage the development of marginal fields and maintain production from a field that may be prematurely abandoned. This can be done either by allowing a purely discretionary system or by the use of sliding-scale royalties. During the 1960s, there was considerable debate over how royalties should be treated with respect to other forms of profits tax (*see* EXPENSING).

royalty bidding. When acreage is being auctioned companies may seek to obtain access by offering a greater than normal royalty. *See also* AUCTION SYSTEM OF LICENSING.

royalty crude oil. In some agreements the recipient of the ROYALTY (normally the government outside the USA) can opt to take its royalty entitlement in the form of oil. This is usually done for one of two reasons: (a) it provides the government with some objective measure of the worth of the crude oil; and/or (b) it may provide an opportunity for the national oil company to gain experience in the marketing of crude oil.

***R*-squared.** A statistical measure of the closeness of ‘fit’ between a dependent variable such as oil prices and an independent variable (or variables) such as supply/demand. It reflects how far changes in one variable match the changes in another variable. It is measured between zero and one. The closer the *R*-squared is to one, the better (i.e. closer) the fit. It is widely used in REGRESSION ANALYSIS.

rundown tanks. Tanks that receive the FRACTION or output direct from the refinery unit. They usually represent transit storage prior to further processing or blending.

running costs. (1) Recurring costs; expenditure on maintenance and operation over the life of a piece of capital equipment; a rather vaguely used term. (2) A synonym for OPERATING COSTS.

S

sacrificial anode. A piece of metal on an underwater structure designed to protect the immediate point of location from corrosion. The idea is that the seawater will attack the anode rather than the structure.

SAE viscosity. A system of classifying lubricants used in internal combustion engines or crankcases according to their VISCOSITY. In the system, which was derived by the Society of Automative Engineers (SAE) in the USA, the lower the number the thinner the oil (5W or 10W). A MULTIGRADE OIL has two numbers (20W/50) reflecting a change in viscosity according to temperature (i.e. SAE 20W at 0°F and SAE 50 at 210°F). *See also* QUALITY, LUBRICATING OILS; SAYBOLT UNIVERSAL SECOND.

salaried bulk agent. A US term for a gasoline (i.e. motor spirit) wholesaler who is an employee of the wholesale company.

sales gas. *See* MARKETABLE NATURAL GAS.

salt dome. A plug or dome of rock salt which has been formed under high pressure by being forced through overlying strata. Since salt is impervious to oil this can create a TRAP for oil or gas. Where the salt has not broken through the overlying rock strata, but has merely pushed the rock upwards, this is called a salt pillow. This can also form a trap. In a salt dome, if the salt is removed the resulting cavern may be used for the storage of crude oil. *See also* GEOLOGY.

salt gel. *See* ATTAPULGITE.

salt pillow. *See* SALT DOME.

salvage value. The value of a salvaged vessel that is agreed between the owner and the salvor prior to salvaging. It is the value on which the salvage agreement determining remuneration to the salvor is awarded.

saturated hydrocarbons. HYDROCARBONS with no unsaturated carbon–carbon bonds. *Compare* UNSATURATED HYDROCARBONS. *See also* ALKANES; CHEMISTRY.

Saybolt Universal Second. A measure of viscosity. It is measured using a Saybolt viscometer and is used to determined the SAE VISCOSITY. *Compare* ENGLER DEGREE; REDWOOD NO 1 SECOND. *See also* QUALITY, DISTILLATE AND RESIDUAL FUELS.

SBM. *See* SINGLE-BUOY MOORING.

SBP. *See* SPECIAL BOILING POINT SPIRITS.

SBR. *See* STYRENE–BUTADIENE RUBBER.

Scale No 1, 2, 3. Nominal freight scales; Scale No 1 was published by the London Tanker Brokers Panel from November 1952. Base rates were calculated such that a standard vessel would make the same return irrespective of route chosen. The chief innovation of the scale was its separate treatment of fixed differentials, especially canal dues. Many of the rates were identical to those of the US Maritime Commission rates applied during World War II to requisitioned ships. Scale No 1 was replaced by Scale No 2 in July 1954 and by Scale No 3 in 1958, each designed to reflect changing bunker and port charges.

scarcity. Although society's wants are unlimited, the resources to satisfy those wants are scarce. Scarcity therefore provides the basic problem of economics — which wants should be satisfied and which wants should be left unsatisfied (i.e. the basic economic problem is one of choice arising from scarcity). *See also* PRICE MECHANISM.

scenarios. Since the early 1970s, the perceived uncertainties of the future have increased and 'single number' forecasts (*see* FORECASTING) have gone out of favour. Instead, it is regarded as more appropriate to try to specify and explore the uncertainties of the future and to estimate their impact. In that way, a company contemplating a particular course of action can assess how robust its proposals are with respect to various future outcomes.

Sensitivity analysis can be used to explore the effects of future uncertainties, but inconsistencies may arise among the tests, and it may also be difficult to keep the number of tests (and the range of results) within sensible limits. Scenarios overcome some of these problems by presenting a small number of statements about the future, each of which is intended to be internally consistent.

For example, an international oil company might have one scenario which begins from a relatively optimistic view of world economic growth prospects, leading to a view about other relevant macroeconomic variables (e.g., interest rates) and to statements about conditions in the energy and oil industries (e.g., demand growth and relative fuel prices). Another scenario would start from a more pessimistic view of growth prospects and draw conclusions consistent with that view. Scenarios for individual countries, consistent with the world scenarios, would also be constructed.

It is sometimes argued that an even number of scenarios should be presented because, faced with an odd number, company management will choose the one in the middle and will therefore settle for something like single forecasting. Some scenario builders believe that two scenarios is the optimum number, although that depends on the nature of the market, the attitude of the company's management and the purpose of the scenario-building exercise.

SCFO. Steam-cracked fuel oil (*see* PFO).

SCNT. *See* SUEZ CANAL NET TON.

scrubber. *See* SCRUBBING.

scrubbing. The process of purifying a gas or liquid by washing it in a tower known as a scrubber. *See also* ACID GAS; AIR POLLUTION.

SCT cracking. Short-contact time cracking (*see* MOTOR SPIRIT UPGRADING).

SDR. *See* SPECIAL DRAWING RIGHT.

SDW. Summer DEADWEIGHT tonnage of oil tankers.

sealant polymer. A POLYMER that is added to the DRILLING FLUID in order to convert a porous formation into one that is impermeable.

sealed-bid auction (blind auction). An auction in which the bidders for the acreage submit confidential bids (i.e. sealed) by a specified time. A danger with such a procedure is that there is always a possibility of COLLUSION between the bidders. *See also* AUCTION SYSTEM OF BIDDING.

seasonal adjustment. *See* SEASONALITY.

seasonality. In the context of oil and gas, fluctuations in the demand resulting from changes in the ambient temperature associated with the seasons of the year. For example, in the winter it is cold, and hence more energy is generally consumed, in particular more of the oil products associated with space heating and the generation of electricity. Equally in summer, as the weather improves, the demand for motor spirit increases. As part of the analysis of TIME SERIES, it is possible to identify statistically seasonal variations over time and produce seasonal ratios. These reflect (normally on a quarterly basis) the increase or decrease around the norm expected as a result of seasonal factors. Seasonal ratios serve one of two functions: (a) they can be used in forecasting to adjust predictions of demand; (b) they can be used to remove the seasonal fluctuations from a time series. This seasonal adjustment enables the trend of consumption to be established. In the 1950s and 1960s as fuel oil replaced coal the seasonal pattern for oil became increasingly marked. However, after the FIRST OIL SHOCK, the switch away from oil (especially fuel oil) has significantly reduced seasonal fluctuations in demand.

seasonal ratio. *See* SEASONALITY.

seasonal specifications. Changes in quality specifications according to the time of year, generally reflecting seasonal changes in the ambient temperature.

secondary recovery. The PRODUCTION of oil using artificial means of maintaining RESERVOIR PRESSURE. The most common method is to reinject gas or water into the fields which effectively 'pumps up' the reservoir pressure. Such a process increases the costs of production. The extent of the cost increase is largely dependent on the availability of gas or water for reinjection. *See also* PRODUCTION, SECONDARY RECOVERY MECHANISMS.

secondary refinery capacity. The refinery upgrading plant which further processes the output of the PRIMARY DISTILLATION UNIT. It encompass cracking, reforming and isomerization (*see* MOTOR SPIRIT UPGRADING; REFINING, SECONDARY PROCESSING). The function of the processes is broadly to convert heavier products into lighter ones.

secondary refining processes. *See* REFINING, SECONDARY PROCESSING.

second best. A term used in WELFARE ECONOMICS. Under certain conditions (including perfect competition, *see* COMPETITION) a market economy is capable of reaching a 'first best' PARETO OPTIMAL resource allocation. However, if any of these conditions is infringed and the economy has one or more deviant sectors (e.g., that are imperfectly competitive or create EXTERNALITIES, or where there are PUBLIC GOODS) then a 'second best' Pareto optimal resource allocation is the best achievable. The theory of the second best shows that in the presence of extra constraints resulting from these deviations, the necessary conditions for a second best optimum may be different from the first best necessary conditions, even for the non-deviant sectors. Hence if any of the first best conditions cannot be satisfied it is not necessarily appropriate to fulfil all the rest; instead a new set of conditions is required. This has implications for government policy: for example in the area of control of public enterprises, where second best considerations may imply that such industries should not use unmodified MARGINAL COST PRICING.

Second Oil Shock. The explosion in crude oil prices between 1979 and 1981 associated with the Iranian revolution.

The Background

Before the Iranian crisis, which began in October 1978, the majority of oil traded (34 million barrels per day) was traded on the TERM MARKET, with only a small amount being traded on the SPOT MARKET. The majority of the crude oil traded on the term market was being sold on long-term contracts with the major oil companies acting as disposers. The key analytical point to note for subsequent events is that the price in the term market was an administered one. Thus, if there was a disequilibrium in the term market (arising, for example, from an excess of demand for crude relative to supply, or vice versa), this was not removed by price changes in the short term. Instead the imbalance between demand and supply spilled over into the spot market where the price was a market price in the sense that imbalances in supply and demand were rectified by price adjustments.

The Trigger

The crisis began with the striking of the Iranian oil workers, and reached its political peak with the subsequent revolution. Iran's exports dropped from 5.8 million barrels per day in September 1978 to close to zero by January 1979. Because of the uncertainty this generated, spot crude virtually disappeared from the markets. Some of the other OPEC members began to increase their oil supplies in an effort to make up the loss, with the result that the aggregate loss of crude was relatively small. The US Department of Energy estimated that the non-Communist world oil supply was only 2–3 million barrels per day below expected levels. However, as with the 1973 crisis (*see* FIRST OIL SHOCK), it was not the reality that was important but the perception of that reality.

At the time, the trade press consistently overstated the supply shortfall by about 1 million barrels per day. In addition, there was considerable uncertainty over future supplies. This was compounded by the actions of Saudi Arabia. In the later part of 1978 the Saudis increased production to 10.5 million barrels per day, but they began to experience severe technical problems in the Ghawar field and were forced to reduce output to 8.5 million barrels per day in January 1979. In February 1979 Saudi output increased to 9.5 million barrels per day, but in

March it fell once more by 1 million barrels per day following the Camp David discussions. All this added considerably to the uncertainty, which was acute given the low level of stocks prevalent at the time. Although in aggregate the supply loss was small, disaggregated the picture was rather different. Iran's exports had been fairly concentrated, with some countries being hit much harder than others. This added a certain competitive chaos to the market. Furthermore, when the Saudis began to increase production to offset the losses, they tended to deal with governments rather than with companies, thus compounding the companies' uncertainties over crude supplies. A final contribution to the uncertainty was the fear that as the price began to rise those countries who had previously expressed a dislike for the accumulation of foreign assets may simply have had in mind a revenue target. As the price rose they could reduce supply to meet that target, thereby forcing price even higher.

In the early stages of the crisis there was no physical shortage of crude because two months' supplies were already in transit (allowing three to four weeks for transportation and a similar time for refining and distribution). The initial impact fell upon the spot product market, and spot crude disappeared from the spot markets as holders of such crude awaited events. Consequently those refineries who felt threatened sought to build up their stocks. In February to March 1979 major refiners' stocks had fallen to 40 percent of the 1978 level, and they began recourse to the spot market. Inevitably spot prices began to rise. This was possible because the higher prices of spot products could be rolled into the average costs. This all added to a growing atmosphere of panic.

In the summer of 1978, the industrial countries' stocks of oil and products stood at about 90 days, which was only slightly lower than the level in the summer of 1977. However, the winter of 1978–79 was unusually cold, and this led to a heavier than usual stock drawdown to some 84 days. Thus, when the supply disruption occurred,the relatively low level of stocks aggravated the panic element in the market. The situation was compounded by the actions of governments. In February 1979, the USA was still buying for its strategic stockpile (*see* STRATEGIC RESERVES), and this continued until March. Japan too was trying to build up its stock levels during this period. In addition, the amount of stocks that governments allowed to be released was severely restricted because a large part of the national stocks were regarded as 'strategic' (estimates suggest up to 40–45 days of the supply); most governments refused to allow any drawdown on this strategic element.

The Market Responds

Eventually, the crude oil shortage appeared upon the market. Because the term price was an administered price, the excess demand spilled into the spot market where crude spot prices began to rise, with the rise being led by increasing product prices. As the spot crude prices began to rise because of a change in the marketing structure (*see below*) more crude became available on the spot market. Thus demand and supply in the spot market began to increase together, but demand was always in the lead because of the initial crude shortfall and the panic element in buying. While the shortfall was always relatively small, it was sufficient to keep on driving up the crude spot price. At the same time as demand in the spot market increased, demand also increased in the term market as companies and governments sought to replace their crude oil losses from Iran. The reason behind this was the effort to avoid the spot market, with the result that 'administered' prices began to attract premiums, which were officially allowed at the March meeting of OPEC. At no time in the crisis did product prices allow the recovery of these higher crude prices, but they could be justified by rolling them in with the existing 'normal' prices.

This process accelerated as governments began to compete among themselves. On 2 March the governing body of the International Energy Agency agreed that the countries would reduce their demand for crude oil by 5 percent during 1979, but little attention was paid to the decision. In April the US government suspended its oil import fee and encouraged US companies to seek imported oil, thus adding to the demand pressure.

Goverments Take Over Marketing

As the situation progressed, apart from the case of Saudi Arabia and one or two other oil producers, the administered price began to have little meaning. Suppliers were making TERM CONTRACTS on whatever prices the markets would bear, and because they were below the spot market, buyers at these higher prices on short to medium-term contracts were plentiful. However, as the spot crude oil price rose and oil producers were inundated with potential crude oil buyers, the producing countries began to lose their previously acquired fear of marketing crude on their own behalf. This was reinforced by the perception that many companies were lifting crude at the official price and then selling almost immediately on the spot market at a considerable profit. Thus the marketing function of the companies was eroded very rapidly. In the first quarter of 1979, six of the OPEC countries — Saudi Arabia, Kuwait, Iran, Libya, Venezuela and Nigeria — marketed only 9 percent of their production via government channels; by the end of 1980, this figure had risen to 27 percent. Effectively the VERTICALLY INTEGRATED structure of the oil industry had become seriously eroded.

The companies now bought their crude oil as any other buyers would. Also, during 1979–80, long-term contracts with the OPEC producers were largely discredited, and companies sought shorter-term arrangements. Given that the governments were now dealing with a relatively large number of small buyers, this added a new element of instability into the market. The years 1979 and 1980 saw a series of attempts by the producers to introduce new, more rigid contracts, but these were soon displaced in the post-crisis environment.

At the start of 1980 the netback worth (*see* NET PRODUCT WORTH) of a barrel of Arabian light was $18 per barrel, but by April this had reached $32. Inevitably, premiums on the government official selling price began to appear for crudes, since the governments were unable to increase the administered price unilaterally outside of the OPEC framework. As the spot price of crude and the netback value increased, pressure grew upon Saudi Arabia to change the administered price level to reflect the apparent market situation. During 1979–80 Saudi Arabia tried to resist this pressure and pushed up its production level in an attempt to hold the line. The pressure was too strong, and in March 1979 OPEC announced that the 10 percent increase decided in December 1978 for 1979 would be implemented immediately from 1 April, thereby raising the marker price to $14.546 per barrel. It was also agreed that premiums on official prices would be allowed 'at the discretion of the governments concerned'. This was effectively a licence for chaos, and individual countries pushed up their premiums to very high levels. In an effort to control this, OPEC announced in June 1979 that the marker would be increased to $18 per barrel, with the maximum for any premium being set at $2 per barrel and the maximum allowable differential (plus premium) from the marker at $3.50 per barrel. A six-month price freeze was also agreed, but in practice this was virtually ignored, most of the countries charging whatever they could get.

The Iran–Iraq War

The situation remained chaotic through to September 1980, in the sense that there was no meaningful 'official' price structure. In September the situation was aggravated by the outbreak of the Iran–Iraq war which implied that some 4 million barrels per day of exports would be lost from the market. However, the situation was very different from that immediately after the Iranian crisis (*see above*). Now stock levels were at all-time highs, and the lesson that cooperation was better than competition had been learnt well by the consuming countries. Although the spot crude price did initially rocket to $40 per barrel, this increase was short-lived, and by December 1980 the increased spot prices began to collapse. Nevertheless, the increase did pressurize Saudi Arabia again to increase the administered price.

The pressure was resisted and OPEC meetings in May and June of 1981 ended with no agreement on restoring any semblence of order to the price structure. Finally, after a series of unilateral price increases by Saudi Arabia, which were intended (but failed) to preempt the pressure and restore control, OPEC came up with a price agreement in October 1981. The marker was set at $34 per barrel with an upper differential for African crudes of $3–4 per barrel. This price arrangement was to be frozen until 1982 and at the same time Saudi Arabia announced a production cutback to 8.5 million barrels per day, or 1 million barrels per day below October levels. In the sense that this restored an effective OPEC pricing structure the meeting can be said to mark the end of the Second Oil Shock.

security. A paper that gives the title either to property or claims on income. Its most important characteristic is that it can be bought and sold.

sedimentary basin. An area in which layers of sediment have accumulated; commonly used to describe an area containing thicker and relatively undisturbed deposits of potential oil-bearing rock. *See also* GEOLOGY.

sedimentology. *See* GEOLOGY.

seepage. The leaking of oil and gas to the surface if the non-permeable CAP ROCK of a reservoir is eroded.

segregated ballast tanks. *See* TANKER TYPES.

seismic exploration. A method that is used to determine in detail the structure of rocks under a particular land area. The actual survey involves passing acoustic shock waves into the rock strata and interpreting the reflected signals. For example, the sound waves pass more quickly through hard rock compared with soft rock.

seismic survey. *See* SEISMIC EXPLORATION.

self-sufficiency. A situation in which a country's oil or gas production is as great or greater than its consumption. After the FIRST OIL SHOCK there was increased interest in oil self-sufficiency. This was partly in response to the deterioration in the balance of payments experienced by the oil importers. Of greater importance were the effects of the ARAB OIL EMBARGO, which for the first time caused a number of importing countries to consider the vulnerability of their energy supplies which previously had been taken for granted.

semi-submersible rig. A floating drilling platform. The buoyancy of the rig can be adjusted so the drilling platform can be lifted clear of the water for drilling operations. Normally such a rig is used only for exploration.

sensitivity analysis. Because the future is inherently uncertain, 'single number' or 'point' forecasts, which attach only one value to the forecast variable in each future time period, are misleading. One way of exploring the range of future uncertainty is to carry out sensitivity tests in which different values are given to key variables and the effect on the forecast variable is estimated. For example, if a forecaster were using a model in which residential gas demand depends on real personal disposable income, the price of residential gas relative to the prices of other fuels and the stock of gas-consuming appliances, he would vary his estimates of the future values of the three determining variables in order to estimate the sensitivity of gas demand to such variations. It is important to avoid inconsistencies among the chosen values of the determining variables when making such tests. *See also* FORECASTING; SCENARIOS.

service contract. A type of exploration/production contract. First introduced by Indonesia, it is a variant on the JOINT VENTURE arrangement. A foreign company undertakes the exploration at its sole risk, and if a COMMERCIAL FIELD is not discovered the foreign company bears the loss and withdraws. If a commercial discovery is made the company undertakes the development and production of the field as a contractor on behalf of the government, although the company is normally responsible for the necessary investment. Once production begins a proportion of the oil (often around 40 percent) is set aside to allow the company to recover its outlays on exploration, development and production. The remainder is split between the company and the government. Such agreements became increasingly popular during the 1970s, and it is easy to explain this popularity. From the government's viewpoint they achieve, assuming successful exploration, the development of a natural resource which they own and nominally control at no immediate cost to themselves. Without outside help such developments would not have occurred in the absence of indigenous technology. From the companies' viewpoint they gain access to crude oil with fairly minimal involvement by the government in the daily operations of the companies. In effect, the service contract has all the advantages of the old-style CONCESSION, but without the problems of diminution of sovereignty which were associated with the concession.

service lease. A variant on the SERVICE CONTRACT whereby a contractor bids to provide certain services to the owner of the acreage. *See* AUCTION SYSTEM OF LICENSING.

service station. A retail outlet for motor spirit.

seven sisters. A commonly used name for the MAJORS. Reputedly the expression was coined by Enrico Mattei, the creator of ENI, the Italian state oil company.

severance tax. Tax imposed on the output of a company producing natural resources such as oil and gas. Its purpose is to capture for the government some of the economic rent associated with the production.

shadow price. Prices which should represent the 'real' value of the good or resource. In theory, a market economy should produce an OPTIMAL allocation of resources. For this to be achieved it is necessary that every good or service or resource has a price and that that price is the competitive price which reflects the true MARGINAL value or OPPORTUNITY COST. If the market price of the good or resource is distorted by imperfections in the market or the existence of externalities, or if it is a public good which may have no price, then if governments are taking investment decisions (*see* COST–BENEFIT ANALYSIS) they should be using shadow prices. In many cases where the good or resources is tradeable so-called border prices are often used as shadow prices (i.e. the international price). For example, after the FIRST OIL SHOCK in many countries (most notably less-developed countries) the governments tried to protect consumers by holding down (usually by subsidies) the price of oil products. Clearly if the government was trying to decide on how to generate electricity it should not use the subsidized market price of oil but a shadow price to reflect the real foreign exchange cost. In practice the use of shadow pricing is often complicated by the problem of SECOND BEST.

shale oil. The oil distillates that are obtained from processing oil shales. Oil shales are clays containing organic matter of which KEROGEN is the main element. The shale is usually crushed and heated to about 600°C to produce a liquid similar to crude oil. The quality of the shale varies considerably. A yield of 20 gallons per ton is average, rising to 40 gallons per ton for the top-quality shale. The major problem concerning shale oil production is that it involves moving vast quantities of material which can be both damaging to the environment and expensive. Interest in shale oil production took off following the oil price rises of 1973 (*see* FIRST OIL SHOCK). Most of the projects, however, suffered from two problems: (a) scaling-up (i.e. moving from pilot plants to large plants) costs invariably overran; (b) extraction was very energy-intensive.

shale shaker. A machine that retains larger drill cuttings from the DRILLING FLUID before it is recirculated. *See also* DRILLING.

shared interest (carried interest). Describing a licence whereby the oil company explores at its own risk on behalf of the licensor. If a COMMERCIAL FIELD is discovered the licensor can then participate by paying some share of costs incurred. If no oil or gas is found the loss falls entirely on the oil company. *See also* AUCTION SYSTEM OF LICENSING.

Shell. The generic name for a group of companies which has two parent companies. One is British — The Shell Transport and Trading Company — and the other is Dutch — The Royal Dutch Petroleum Company. The group came into being in 1907 largely in an effort to restrict the activities of the Standard Oil Company in the Far East. Generally Shell is regarded as the second largest of the oil companies after EXXON.

Sherman Anti-Trust Act. *See* ANTI-TRUST LEGISLATION.

shoe. The heaviest of all the drill collar sections of pipe, above the bit. *See also* DRILLING.

shooting. The practice of firing an explosive charge down a producing well in order to fracture the producing formation thereby improving the flow of oil. *See also* PERFORATING GUN; PRODUCTION.

short contact time cracking. *See* MOTOR SPIRIT UPGRADING.

short term (short run). A period of no definite length where at least one factor of production is fixed. *Compare* LONG TERM.

short-term capacity. *See* CAPACITY.

shot. An explosion used in order to carry out a seismic survey (*see* SEISMIC EXPLORATION).

show. Indications of the presence of oil or gas or water in the formations being drilled in an exploratory well.

shut-in well. A well that either has not been used for production or from which production has ceased.

shuttle tanker. A tanker that makes regular round trips normally between a producing field and an onshore terminal or refinery. In the Gulf War between Iran and Iraq the Iranians opened a shuttle route from their Kharg Island terminal to Sirri and later Larak so that the tankers exporting the crude faced less risk of air attack.

sidetrack hole (dog-leg). A hole drilled to bypass the main hole when this has become unworkable due to lost equipment, etc. *See also* DRILLING.

sidewell cores. Sections of rock or sediment obtained by shooting cylinders of steel into the wall of the well. The sections are then used for geological evaluation.

signature bonus. A payment to the owner of the acreage (outside of the USA this is normally the government) made when the licence permitting exploration is signed. It is most common in a situation where there is demand for acreage from a number of companies and therefore is offered as an inducement to secure access to the acreage. *See also* BONUS PAYMENTS.

simple linear regression. *See* REGRESSION ANALYSIS.

simplex method. *See* LINEAR PROGRAMMING.

single-buoy mooring (SBM, single-point mooring). A system that enables a vessel to moor outside a harbour and discharge its cargo either to smaller vessels or directly to the shore by pipeline. This arrangement is advantageous to the owner in conditions where deep-water loading or unloading terminals are not available.

single flash. The sudden release of gases/vapours from oil rather than their removal in a number of stages.

single-point mooring. *See* SINGLE-BUOY MOORING.

sinking fund. The setting aside of money at regular intervals in a fund in order to repay debt at the end of a period or to provide sufficient DEPRECIATION to replace some asset. The sum of the regular payments is less than the total amount to be repaid or depreciated because each payment is invested to provide additional proceeds for the fund. *See* ANNUAL CAPITAL CHARGE.

skidding the rig. The moving of a RIG from one location to another with minimal dismantling of the rig equipment.

skimmer. A device that, rather like a vacuum cleaner, removes oil from the surface of the sea following accidental spillage.

slate. *See* CRUDE SLATE.

slips. Wedges that are used on the drilling table to hold the free end of the drill pipe in place.

slop tank. *See* LOAD ON TOP SYSTEM.

slow steaming. A means of saving fuel costs. For a given vessel, fuel consumption varies approximately with the cube of the vessel's speed. When fuel costs are high considerable savings may be effected by reducing operating speed below the design speed. Tankers built before 1973 tended to have design speeds of approximately 15.75 knots. In mid-1985 average travelling speeds have been estimated to be around 10.0 knots only. This yields a total loss of oil-carrying capacity of approximately 25 million tons of oil for the world fleet.

sludge. Sludge can be of three types. (a) Acid sludge, also known as acid tar, is a by-product of the treatment of oils by sulphuric acid (*see* REFINING). (b) Engine sludge is the insoluble degradation product of lubricating oil formed in internal combustion engines. (c) Tank sludge is the material (including water) that settles at the bottom of oil/product storage tanks. Tank sludge is also known as BS&W.

slurry. A mixture of crushed coal normally in water, although experiments involving methanol are becoming increasingly common. Originally slurry was the by-product of a coal-washing plant, but interest in its fuel potential grew after the oil price rises of 1973–74 (*see* CRUDE OIL PRICES). In particular it held out the possibility of creating the CRUDE OIL CHARACTERISTIC of being liquid which goes a long way to explain the favourable economics associated with oil use. *See also* COAL LIQUEFACTION.

slurry oil. *See* DECANT OIL.

slush pumps. *See* MUD PUMPS.

smoke point. The height to which the flame in a kerosine lamp can be raised before smoke results. This is carried out in the smoke point test and is used to provide a measure of the quality of kerosine. *See also* QUALITY, AVIATION JET FUELS.

smoke volatility index (SVI). The SMOKE POINT of aviation fuels in the case of wide-cut gasoline related to volatility. *See also* QUALITY, AVIATION JET FUELS.

Snake, The. The scheme devised in 1971 by the European Community in order to restrict fluctuation in exchange rates. In 1979, the EMS superseded The Snake.

snubbing. The running of pipe or casing into the well while the surface equipment is under pressure from the well.

Socal (Standard Oil Company of California). A US MAJOR which came out of the 1911 dissolution of the STANDARD OIL Trust. More recently, the company has become known as the Chevron Corporation. In 1984 it took over GULF; this was the largest corporate acquisition in US history. Outside the USA, Socal has been closely associated with TEXACO through their joint operation Caltex.

social costs. *See* COST–BENEFIT ANALYSIS.

Socony Mobil. *See* MOBIL.

Socony–Vacuum. *See* MOBIL.

softening point. The temperature at which bitumen changes its consistency. *See also* QUALITY, BITUMEN.

soft loan. A loan made at zero interest or at interest rates below the going commercial rate. A soft loan is frequently used as a means of providing aid to developing countries.

solar oil. An early name for GAS OIL, so called because it was used to produce illuminating oil by direct cracking. *See also* FINISHED PRODUCTS, GAS OIL.

sole-risk exploration. A situation in which a government explores for oil in its own territory without the involvement of an oil company.

soluble cutting oil. A blend of mineral oil and emulsifiers used as an aqueous cutting fluid for metals, for example, on a lathe.

solution gas. *See* NATURAL GAS.

solution gas drive. A type of natural recovery mechanism enabling oil to be produced without artificial assistance. *See also* PRODUCTION, NATURAL RECOVERY MECHANISMS.

solvent extraction. The process of removing undesirable constituents in products by dissolving them in a solvent. *See also* RAFFINATE.

solvent refining. *See* RAFFINATE.

solvents. Hydrocarbon solvents are derived directly from petroleum fractions boiling below 250°C and rely for their solvent power on their chemical composition. They are classed as aromatic, naphthenic or paraffinic. Petrochemical solvents, however, are synthesized from petroleum feedstocks by chemical processing. The main categories of such solvents are alcohols, ketones, esters, ethers and chlorinated solvents. *See also* FINISHED PRODUCTS, SOLVENTS.

Sonatrach. The Algerian national oil company which was created in 1983.

source rock. A geological term for the rock in which oil was originally formed, as distinct from the rock within which it is found in the reservoir. *See also* GEOLOGY.

sour crude. A crude oil that has a high sulphur content as a result of the presence of hydrogen sulphide and MERCAPTANS. Since sulphur in products is generally undesirable because of pollution emissions when burnt, sour crude commands a lower value than SWEET CRUDE, which has low sulphur content. This is because either the sulphur must be removed (adding to refining costs) or a high-sulphur product (and hence a lower-priced product) results. Sour crudes also have a tendency to create corrosion problems in handling equipment.

sour gas. NATURAL GAS or ASSOCIATED GAS that contains undesirable sulphur compounds such as hydrogen sulphide and methylmercaptan.

spacing. The density of producing wells on a FIELD. Unregulated, the density will be determined by a combination of technical factors and economic requirements. However, if spacing is fixed by legislation, it can be used, as it was in the US PRORATIONING system, to restrict rates of production.

spar. A floating storage tank used to store oil from the production platform prior to tanker loading.

SPD. Supplementary Petroleum Duty (*see* PETROLEUM REVENUE TAX).

special boiling point spirits. Solvents manufactured within a very narrow boiling range. The best known is white spirit (turps substitute). *See* FINISHED PRODUCTS, SOLVENTS.

special drawing right (SDR). A unit of account. Its value is linked to a basket of currencies (since January 1981: US dollar, Deutsche mark, French franc, pound sterling and the yen) weighted to reflect their relative importance in international trade and finance. It was first introduced by the International Monetary Fund in 1970 as an additional source of international LIQUIDITY; later it became the Fund's own unit of account and main reserve asset of the international monetary system. From time to time it has been suggested that SDRs should be used in the pricing of crude oil. In June 1975 OPEC decided to implement the use of SDRs, but no further action was taken.

special products. Petroleum products that are used in non-energy applications. They include such products as BITUMEN, WAX, SOLVENTS, etc.

specific gravity. *See* QUALITY, CRUDE OIL.

specific tariff. *See* TARIFF.

specific tax. *See* AD VALOREM TAX.

spiked crude. A crude oil to which another product has been added.

spindle oil. Any low-viscosity mineral lubricating oil. The name was derived from such an oil's use in the textile industry.

spinning. In the winter of 1984–85 the official price of North Sea oil paid by the British National Oil Corporation (BNOC) to the oil companies for PARTICIPATION OIL was greater than the SPOT PRICE of the oil. Some oil companies sold their oil to BNOC who could only sell the oil at a loss on the SPOT MARKET where it was bought back by the oil companies. This became known as spinning. BNOC made considerable losses during this period.

spinning line. A wire rope or chain wrapped round a section of drilling pipe in order to screw it into another section. *See also* DRILLING.

sponge coke (green coke). A product of petroleum feedstocks high in ASPHALTENES. It is a low-grade coke frequently used mixed with coal coke.

spot charter. The hiring of a tanker for a single specified voyage. *See also* SPOT FREIGHT RATES.

spot freight rates. The spot tanker freight market has increased in importance as owners have reduced the sizes of their owned or time-chartered fleet. Spot rates are fixed for voyage trips and are expressed in terms of a standard WORLDSCALE (WS) rate. When rates are falling it is advantageous for the charterer to operate in the spot market, but not for the shipowner; the converse is true when rates are rising. The mid-year average spot rates in WS and US$ per ton of oil from the Gulf to north Europe are summarized in Table S.1. Source: John I. *Jacob's World Tanker Fleet Review*.

Table S.1.

	1980	1981	1982	1983	1984	1985
WS	38.8	30.2	20.7	20.8	42.2	20.1
$/ton	8.4	8.43	6.0	5.65	11.36	5.59

spot market. A market in which an item is traded for immediate delivery as opposed to a FUTURES MARKET. However, in the oil industry the term is not used in this way.

In terms of petroleum a spot transaction is a one-off sale/purchase of a stock (barge or tanker) of crude or products. This distinguishes it from a TERM CONTRACT in which a flow of oil (i.e. so many barrels per day per time period) is bought/sold. Before the 1980s the spot market in which spot transactions were carried out was relatively small in volume terms and tended to be dominated by products. In effect it represented the marginal market, and its prime function was to provide a balancing service between supply and demand. As such trading tended to be erratic, and the spot price of crudes and products, which was a free market price determined as a result of a bargain, showed considerable fluctuations. Although spot markets tend to be identified by a specific geographic location (e.g., Rotterdam) they are in fact regional markets rather than a specific institutional location. The 'Rotterdam' spot market is not in Rotterdam, it simply refers to transactions carried out in terms of delivery in north-western Europe based loosely on the ARA refinery and storage complex at the head of the Rhine. To be a 'Rotterdam trader' all that is required is to have access to a telex, be it in Brazil, Bangladesh or Borneo. This raises a key problem. The spot market consists of the transactions carried out during a day relating to delivery in a specific region. There is no requirement that the transaction details — price and quantity — be recorded in any central place. Hence no one knows either the spot volumes or the spot prices. There exists a series of specialist agencies, the most famous being PLATT'S, whose function is to 'trawl' for information and come up with a best estimate of prices. This means that sometimes the spot price can be very misleading if the volume traded is small and involves DISTRESSED CARGOES.

During the SECOND OIL SHOCK the spot market played a crucial role. Following the loss of Iranian output spot crude virtually disappeared as both refiners and traders held on to their access. Faced with low stocks and a cold winter it was spot products which responded first as refiners/distributors tried to build up stocks in anticipation of a future shortage. These rising spot product prices fed into rising spot crude prices which in turn fed into rising GOVERNMENT OFFICIAL SALES PRICES. Refiners were willing to pay the higher spot crude prices because spot crude supplied only a proportion of the total CRUDE SLATE, and therefore the higher cost could be rolled into the lower-cost crude supplied at official prices.

Following the general discrediting of long-term contracts as a result of the Second Oil Shock, the size of the spot market (especially in crude) increased significantly. Oil companies secured increasing amounts on the spot market, although precise data are unobtainable due to the absence of compulsory reporting of deals. At the same time the number of market traders also increased significantly.

This expansion of the market helped to some extent to reduce the market's volatility. On the other hand, the development of future markets in both crude and products which inevitably had links with the spot market tended to add to the volatility. From time to time calls are heard for a greater control over the spot markets together with the need for greater transparency of transactions. To date nothing has resulted from these demands. *See also* AFM PRICES.

SPR (strategic petroleum reserve). Although the USA had always held STRATEGIC RESERVES, they came to prominence in the Carter National Energy Plan of April 1977. As part of this programme it was intended to stockpile the equivalent of 10 months of imported crude oil in underground caverns. In July 1978 the programme was accelerated (with stocks at 32.5 million barrels) and four Gulf Coast salt dome sites were acquired with a capacity of 275 million barrels. During the SECOND OIL SHOCK the reserves were not used — apparently because the oil was pumped into the sites, but there was no mechanism for retrieval. The reserve still exists, and more than once it has been suggested that the reserve could be used to help stabilize the market as a BUFFER STOCK.

spudding in. The start of DRILLING operations.

spur line. A small-diameter feeder pipe linking a producing well to a larger pipeline or storage facility.

squeeze. The injection of cement under pressure between the CASING and the well bore. *See also* DRILLING.

SRB. STRAIGHT-RUN benzine.

SRG. STRAIGHT-RUN gasoline.

SRN. STRAIGHT-RUN naptha.

SRP. *See* SUGGESTED RETAIL PRICE.

stabbing board. A retractable platform on the DERRICK which enables the two sections of drill pipe to be joined to be aligned correctly by the DERRICK MAN.

stability. Petroleum products stored for any length of time may deteriorate for example as a result of oxidation processes. Stability of quality is an important characteristic of oil products, and there are various tests to establish stability. *See also* QUALITY, MOTOR SPIRIT.

stability test. *See* QUALITY, MOTOR SPIRIT.

stabilized gasoline. GASOLINE in which the VAPOUR PRESSURE has been reduced by fractionation, thereby removing propane and some butanes to a specified maximum.

staged approval procedure. A procedure in the UK sector of the North Sea. It requires the operator of a field to present a staged plan of development, production and abandonment for government approval. *See also* ANNEX A; ANNEX B; DISCRETIONARY SYSTEM OF LICENSING.

Standard Oil. Between 1873 and 1882 J.D. Rockfeller and his group — Standard Oil Company established in 1870 — formed a series of alliances with other oil men to create the Standard Oil Trust. This effectively controlled the majority of the refining and pipeline capacity of the USA. Control was by setting up 'Standard' companies in each state which were bound together by means of a trust agreement. Because of legal problems the trust method was changed in 1892 to a holding company arrangement. By 1904 it refined and marketed over 85 percent of the throughput of US refineries. In 1911 the Supreme Court ruled that 33 of its subsidiaries must become legally separate companies. *See also* ANTI-TRUST LEGISLATION.

Standard Oil Company of California. *See* SOCAL.

Standard Oil Company of New Jersey *See* EXXON.

Standard Oil Company of New York (Socony). *See* MOBIL.

Standard Oil Trust. *See* STANDARD OIL.

standard tankship. A concept that is useful for freight rate fixing and for forecasting tanker demand. The standard tankship used by WORLDSCALE, for example, is one of 19 500 dwt with a speed of 14 knots. For forecasting purposes, the World War II tankers, the T-2 — a vessel of 16 600 dwt with a speed of between 14 and 14.6 knots — is often used. Given information on seaborne trade between individual exporting and importing regions, it is possible to express the requirements for tankers in terms of T-2. This can then be matched with a T-2 measure of commercial tanker availability which for mid-1985 has been estimated at around 16 820 T-2 equivalents.

star rating system. A system of identifying the OCTANE RATING of motor spirit which was introduced into the UK in 1967. Each star category represented a minimum RESEARCH OCTANCE NUMBER (two star = 90; three star = 94; four star = 97; five star = 100). The most widely used is four star.

statement of value. A return used to calculate UK ROYALTY payments. It is required two months after the end of each chargeable period.

state oil companies. Most countries now have a company or companies owned by the state (partially or wholly) concerned with aspects of petroleum. Such companies range from those of the producing countries (e.g., ARAMCO, KOC, NIOC, INOC, etc.) to those of the consuming countries (e.g., ENI, ERAP, OMV, Statoil, etc.). Their background and functions vary between countries, but in most cases their creation was a reflection of the view that somehow oil was too important to leave to the oil companies.

Statoil. The Norwegian state oil company.

steam-cracked fuel oils. *See* PFO.

steam cracker gasoline. *See* PYROLYSIS GASOLINE.

steam cracking. *See* REFINING, SECONDARY PROCESSING.

steam cylinder oil. A lubricating oil that is used in steam engine cylinders. It is normally a viscous oil of high FLASH POINT.

steam drive (steam flooding, steam injection). A technique of enhanced oil recovery. Steam is injected into the reservoir. This heats the oil, thus increasing its fluidity and the pressure. *See also* PRODUCTION, ENHANCED OIL RECOVERY.

steam flooding. *See* STEAM DRIVE.

steam injection. *See* STEAM DRIVE.

steam reforming. A process whereby hydrocarbon gas, steam and a catalyst produce carbon monoxide, hydrogen and low-molecular-weight hydrocarbons. Steam reforming is used to convert naphtha into MANUFACTURED GAS.

steel jacket. The steel support system used to build a steel offshore platform.

stem testing. *See* DRILLING.

step-out well (outstep well). A well drilled away from a discovery well. It is used for appraisal.

sterilization. A BALANCE OF PAYMENTS surplus, or deficit, if left will increase, or decrease, the domestic money supply with implications for inflation. Sterilization effectively offsets the balance of payments effect on the money supply. For example, when an oil producer sells a barrel of oil then in the monetary system the increase in foreign assets (i.e. an asset) is offset by an increase in government deposits (i.e. a liability). However, if the government spends the revenue, the fall in government deposits may be offset given fixed liabilities by an increase in money supply. Hence for the large oil producers, monetary and fiscal policy are closely interlinked. There are many different policy measures for sterilizing the effect of trade imbalances.

stinger. The boom (i.e. crane) used to lower a pipeline on to the seabed.

stocks. Stores of goods, including intermediate inputs. Such stores are held for a variety of motives. Several examples will serve to illustrate. (a) Deliveries occur at discrete intervals, whereas production or consumption is a continuous process. Thus stocks provide a continuous feed, or if shipments are discrete they allow continuous production. (b) There may be cost advantages in producing at a constant rate of output, whereas sales/shipments may fluctuate. Thus stocks take excess production to supply excess demand. (c) If forecasting requirements are difficult for any reason, stocks provide a means to offset any errors in the forecast. In effect they provide insurance.

Apart from the physical cost of storage and any possible loss from deterioration or spoilage stocks represent money which could have been used elsewhere. Thus assuming the REAL value of the stock remains unchanged the cost of holding the stock, aside from storage costs/losses, is given by the OPPORTUNITY COST OF CAPITAL. The x dollars tied up in a barrel of crude oil in a tank could have (at the least) been earning interest. If the price of the good changes so too does the value of the stock. Thus a company, for no obvious reason, can show fluctuations in 'profitability'. Different accounting methods such as LIFO and FIFO can affect such fluctuations.

Stock levels give the absolute amount of stocks. Stockbuild refers to an increase in stocks, stockdraft or stockdraw refers to a fall in stocks.

Because gas is high-volume and low-value, and because of the costs of high-pressure or cryogenic storage, gas stocks tend to be minimal. Oil stocks are large and during the 1970s began to play an increasingly important role in the oil market.

Oil stocks are notoriously difficult to measure. In the OECD/EEC, stocks include crude oil, major products and partially processed products held in storage tanks (refinery, bulk and pipeline), in barges, coastal tankers, tankers in port and inland ship bunkers, in storage tank bottoms and working stocks, and by large consumers. The last in terms of definition and reporting varies significantly between countries. Therefore these exclude considerable elements of real stocks — in transit (pipelines, rail and road tankers, tankers at sea plus their bunkers), held by other consumers either directly (including the military) or indirectly in service stations and retail units. These are excluded purely for reasons of problems of measurement, but are significant.

The size of stocks is usually expressed in terms of days of supply in relation to consumption or imports. For example, the International Energy Agency set a minimum target of 90 days consumption for its member countries. However, there is a danger in taking such numbers too literally. There exist two constraints which would prevent the use of the stocks to cover 90 days' consumption. (a) An element of stocks (some have suggested up to half) would be regarded as 'strategic' and would not be allowed for 'normal use', but would be held back for only the gravest emergency such as war. This is despite the normal exclusion of military stocks. (b) A certain minimum operating level of stocks is required to fill, for example, pipelines. Hence, the number of days consumption represented by stocks is almost certainly overstated, although it should be remembered that due to the exclusions mentioned above official stocks are understated.

Apart from the obvious strategic importance, stocks are important in another sense. Crude oil (and products) are produced. Because the producers are identifiable, reasonably good estimates of global supply can be obtained, although there are difficulties when strategic or production quota considerations lead to misleading disclosures. This supply, however, can either be consumed or added to stocks. Because there are many consumers and many exclusions from measuring stocks how much is actually burnt is very difficult to measure with any degree of accuracy. An apparent increase in consumption following a price cut, for example, may simply be due to increased storage. Thus stocks add considerably to market uncertainty because of their potential distortion of demand information.

storage tank bottoms. *See* BOTTOMS.

straight-chain paraffins. *See* NORMAL-PARAFFINS.

straight-line method. A method by which an item of capital investment is depreciated. The value is recovered by setting aside an equal sum for each year of the item's life. This is slightly unrealistic since the value of the capital item is likely to fall exponentially since mechanical wear is exponential, as is technological obsolescence. *See also* DEPRECIATION.

straight-run. Describing products that are produced directly from crude oil via distillation without any CRACKING or REFORMING. *See also* REFINING, PRIMARY PROCESSING; REFINING, SECONDARY PROCESSING.

strategic petroleum reserves. *See* SPR.

strategic reserves. The formalization of the holding of STOCKS developed after the FIRST OIL SHOCK either by individual governments (notably the USA) or via the INTERNATIONAL ENERGY AGENCY. 'Strategic' refers to their ability to influence/control the international oil market, usually in cases of emergency disruption. *See also* BUFFER STOCKS.

stratigraphic trap. One of the types of rock formation in which oil or gas are found. *See also* GEOLOGY.

stratigraphy. *See* GEOLOGY.

stream day. The number of days per year that a plant actually operates allowing for routine maintenance. In a non-leap year a plant that required 20 days' maintenance would have 345 stream days. Stream days are a more realistic measure of capacity than calendar days. *See also* LOST TIME.

strike price. *See* OPTION.

string. *See* DRILLING STRING.

stripped gas. *See* WET GAS.

stripper well. A well in the last stages of its life from which small quantities of oil (usually 20 barrels per day or less) are pumped. It is very high-cost production, although part of the high cost is the fuel for the pump (e.g., NODDING DONKEY). Thus lower oil prices, although making stripper wells less attractive, do also reduce their production costs. Stripper wells are common in the USA, and in the early 1980s accounted for up to one-fifth of US production.

structural trap. An oil-bearing rock formation. *See also* GEOLOGY.

structure. (1) A large arching or curving of rocks in a sedimentary series where oil and gas may be located. (2) *See* MARKET STRUCTURE.

styrene. A widely used ETHENE-based MONOMER. It is one of the basic building blocks from which petrochemicals are formed (e.g., STYRENE–BUTADIENE RUBBER).

styrene–butadiene rubber (SBR). A synthetic rubber; a polymer of STYRENE and butadiene.

sub-sea completion. The installation of a wellhead on the seabed. *See also* SUB-SEA WELLHEAD.

sub-sea wellhead. A WELLHEAD on the seabed that is remotely controlled. There is considerable interest in such sub-sea technology, which promises field development and production at much lower costs than conventional PLATFORMS.

subsidy. A payment usually from government to producers of a good. A subsidy will, other things being equal, increase supply, thereby reducing market price. This reduces the price to the consumer, effectively increasing the consumer's real income. A subsidy also increases the income of the producer.

subsistence fuel. *See* TRADITIONAL ENERGY.

substitute. A good whose demand changes in response to the price change in another good. In one sense, because all goods and services ultimately produce satisfaction, all goods and services are substitutes. However, the term is more commonly used with respect to close substitutes such as coal and fuel oil for power generation. The extent of the closeness of substitute availability influences a good's responsiveness to price changes — its ELASTICITY. For example, a sharp increase in the price of fuel oil eventually means less is used due to increasing use of coal. A similar price increase in jet fuel may also reduce its use, but due to using more efficient aircraft since there are no substitutes. *See also* DEMAND.

substitution effect. If the price of a good falls, other things being equal, more satisfaction is obtained per unit of currency spent on the good. This makes the good more attractive relative to other goods, and thus more of the good is bought. The substitution effect is used to make a distinction from the income effect which arises because the fall in the price of the good has effectively increased REAL income. *See also* DEMAND.

sub-surface pumping. *See* PRODUCTION, SECONDARY RECOVERY MECHANISMS.

Suez Canal. A canal that connects the Mediterranean Sea and the Red Sea. The Suez Canal derives its economic significance from the distance saving it provides to shipping travelling between Europe and Asia. For example, it is 6500 miles between Kuwait and London via Suez compared with 11 300 miles via the Cape of Good Hope, a distance saving of 42 percent. Smaller savings are also available for goods travelling between the Middle East and the east coast of the USA.

The Suez Canal is 168 kilometres long, running from Port Said in the north and to Suez in the south. It was opened in 1869 following 10 years' construction by Ferdinand de Lesseps and was substantially improved in 1876. It remained open until the Arab–Israeli (Six-Day) War of June 1967. Reopened in June 1975, it has since 1979 been available to all ships including those of Israel.

The major restriction on use of the canal is the draft limitation. At present ships of 53 feet maximum draft (e.g., loaded tankers of 150 000 dwt and 240 000 tons ballast) can use the canal. Ambitious plans to increase the maximum draft to 63 feet have been shelved, but widening work is being continued which will allow large vessels to utilize the canal.

In 1984, total tanker traffic amounted to 131.3 million dwt, consisting of 49.6 million tons northbound (laden) and 68.6 million tons southbound (mainly in ballast).

Suez Canal net ton. (SCNT). The gross registered tonnage of a vessel less certain deductions which vary from ship to ship. It is possible to make a very rough conversion between deadweight tonnage (dwt) and SCNT on the basis of 1 SCNT equals 2 dwt.

Suez Canal tolls. *See* CANAL TOLLS.

Suez Premium. A WINDFALL PROFIT gained by owners of Mediterranean-sourced crude oil. In June 1967 following the Six-Day War between the Arabs and Israelis, the SUEZ CANAL was closed. This meant the cost of shipping crude oil from the GULF to Europe increased substantially, and the cost of crude CIF in Europe increased accordingly. The increased value of crude oil in Europe also applied to Mediterranean-sourced crude oil, which did not face increased transport costs.

suggested retail price (SRP). The price recommended by the manufacturer to the seller. It has little standing in the UK and cannot be enforced unless resale price maintenance has been agreed for the good.

sulphur. A yellow, solid, non-metallic element. The free element does not occur in crude, but sulphur-containing compounds may be present.

sulphur content. The extent to which sulphur compounds occur in crude oil and its products. Burning fuels that have a high sulphur content (i.e. sour) causes AIR POLLUTION due to sulphur dioxide being formed. Therefore the sulphur present must be removed by HYDROFINING. *See also* QUALITY, CRUDE OIL; QUALITY, DIESEL OIL; QUALITY, DISTILLATE AND RESIDUAL FUELS.

sulphur premium. Crude oils that are sour (i.e. those containing high concentrations of hydrogen sulphide or other sulphur compounds) have a lower sulphur premium than SWEET CRUDES. This is because the crude oil must be treated to remove the sulphur (*see* REFINING, SECONDARY PROCESSING), which is expensive, or its products have more restricted uses and are therefore worth less. A sweet crude attracts a premium reflecting its higher value compared with a sour crude. For example, in 1973, OPEC adopted a formula based on approximately 7–8 cents per barrel for each 0.1 percent of sulphur below Arabian light, which has a sulphur content of about 1.7 percent. *See also* QUALITY, CRUDE OIL; QUALITY, DIESEL OIL; QUALITY, DISTILLATE AND RESIDUAL FUELS.

Sumed. A large-diameter (42 inches) pipeline from Port Suez to Alexandria built to complement the SUEZ CANAL in 1976. Pipeline tariffs were originally designed to encourage operators not able to use the Suez Canal and to offload oil at the Suez end of the pipeline and resume shipment in the Mediterranean Sea so avoiding the long-distance Cape of Good Hope route.

sunk costs (fixed costs). Costs which once spent cannot be recovered, hence as long as the producer is covering his variable costs he will continue to produce. In the oil and gas industry, because of the capital intensity a high proportion of costs are sunk costs. *See also* COSTS.

super-normal profit. Any profit above NORMAL PROFIT. In theory super-normal profit can be removed without affecting a supplier's willingness to supply since he must only receive normal profit to induce him to remain in business.

super-tanker. The largest tankers in the fleet. Since the introduction of the ULCC and VLCC it is no longer used by the industry, although the term is often used in the media.

Supplementary Petroleum Duty. *See* PETROLEUM REVENUE TAX.

supply. The quantity of a good or service that producers are willing and are able to supply at a given price. An individual producer's supply for a particular good per period of time is a function of the price of that good and its relative profitability. A supply curve plots the quantity supplied against the good's own price, holding all the other variables constant, and is normally upward-sloping (*see* Fig. S.1). A change in the quantity supplied of in good A implies a movement along the supply curve. However, if any of the other variables changes, the supply curve will shift. For example, if the producer's costs rise, thereby reducing profits, the supply curve would shift to the left, indicating at each price less willingness to supply.

The responsiveness of producer supply is of particular interest. The measure economists use is called elasticity. Price elasticity of supply measures the responsiveness of the quantity supplied to small changes in the good's own price. It is usually positive because most supply curves are positively sloped. In practice, actual measurement is extremely difficult. The reason for this lies in a general problem associated with PARTIAL EQUILIBRIUM ANALYSIS. If the quantity supplied is a function of price, then other determinants must be held constant if the relationship between price and quantity supplied (the price elasticity) is to be established. This is why only small changes can be considered. The implication is that if a large change in price is experienced other variables will also change, making it next to impossible to disentangle the effects of the different variables.

Often a distinction is made between short-term supply, when production capacity is fixed, and long-term supply, when capacity can be varied. In oil and gas the distinction is important in terms of ability to supply. In the short run the ability to produce oil and gas is fixed by the number and capacity of the fields together with the handling capacity. The actual numbers for short-term capacity are always uncertain. Unused capacity requires maintenance; DESIGN CAPACITY can be exceeded if risks are taken and fields can be overpumped if RATE SENSITIVITY is ignored. In the long run the ability to supply is determined by the extent to which new discoveries are developed and existing fields are depleted. In viewing long-term supply, much is often made of the fact that the oil and gas in place is finite thus 'supply' will eventually run out. However, such thinking about supply is misleading since when, and if, scarcity emerges price will rise choking off demand thereby 'conserving' the remaining supply.

Supply is determined by a producer's willingness to supply as well as his ability. Economists start from a simplifying assumption that more profit is preferred to less and that production will continue as long as profits are made. In the case of oil and gas such a view has to be adjusted because they are EXHAUSTIBLE RESOURCES, hence more profits may be made by producing later rather than sooner. There is another qualification relating to the economics of the profit maximization approach. It assumes the UTILITY (or satisfaction) from the extra revenue is positive. If, however, the producer is a TARGET REVENUE EARNER then extra revenue may have a low utility thereby reducing willingness to supply. Once political elements enter the supply picture then the 'willingness' issue becomes even more complex.

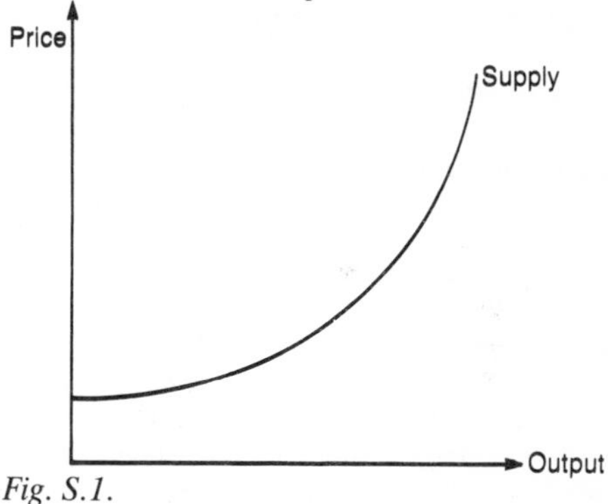

Fig. S.1.

supply curve. A graphical representation of the SUPPLY SCHEDULE (*see* Fig. S.1) with the price of the good on the vertical axis and the quantity supplied on the horizontal axis (*see* SUPPLY). A change in quantity supplied means a movement along the curve with other non-price supply determinants (e.g., costs, profitability of other goods, number of producers, etc.) held constant. A change in supply means the whole supply curve shifts in response to a change in a non-price variable.

supply function. A mathematical equation representing SUPPLY. Supply is on the left-hand side and supply determinants (e.g., costs, etc.) are on the right-hand side. Econometric estimation techniques (*see* ECONO-

METRICS) enable the mathematical symbols to be replaced by estimated numbers. If in log form the coefficient attached to each variable gives the ELASTICITY of that variable. Such functions are an important means of forecasting supply by simply inserting forecasts of the determining variable into the estimated equation and performing the calculations.

supply price. The lowest price at which a supplier will supply. The concept attracted considerable attention during the oil price collapse of 1986 since the supply price effectively represents, in economic theory at least, the FLOOR PRICE. Economic theory sets the supply price at MARGINAL cost, but only under conditions of perfect competition.

supply schedule. Data on price and quantity in tabular form from which a SUPPLY CURVE can be constructed.

surface tension. An open surface of a liquid is under a state of tension which is caused by the forces of attraction which exists between the molecules of the liquid. This tension can cause portions of the surface to break up or separate.

surface string. *See* DRILLING.

surrender provisions. *See* RELINQUISHMENT.

SVI. *See* SMOKE VOLATILITY INDEX.

swabbing. The inadvertent sucking-in of oil or gas into a well during DRILLING operations.

sweet crude. Crude oil that has a low sulphur content as opposed to SOUR CRUDE, which has a high content. The presence of sulphur compounds is undesirable, and these compounds must be treated to prevent air pollution due to sulphur dioxide. Sweet crude attracts a premium (*see* SULPHUR PREMIUM). *See also* QUALITY, CRUDE OIL.

sweetening process. A general term for any process which changes a product from sour to sweet (i.e. reduces the sulphurous smell, although not necessarily reducing the sulphur content). *See also* REFINING, SECONDARY PROCESSING.

sweet gas. Gas that has a low sulphur content. *See also* SWEET CRUDE.

swing producer. A variant of behaviour in an oligopolistic market structure (*see* OLIGOPOLY) usually referred to as the dominant producer model. In effect, the major producer sets the price and allows other suppliers to supply what they will at that price; the balance of demand is then met by the major producer. The production of the major producer therefore swings to match supply and demand in the market in order to protect price from potential oversupply (or indeed undersupply).

Many would argue that Saudi Arabia adopted such a role in the mid-1970s. After the SECOND OIL SHOCK there is little dispute that Saudi Arabia did adopt such a role in the market. This was particularly true when OPEC introduced production quotas when Saudi production levels were used to absorb both cheating on quota and error in setting total OPEC output.

If demand is declining the swing producer sees its production, and hence market share, eroding which imposes a growing price. The swing producer sacrifices, whereas the other market producers gain from the protected price. It was such pressures that forced Saudi Arabia to drop the swing role in September 1985.

swivel. The rotating link between the rotary base and the DRILLING STRING. *See also* DRILLING.

synthesis gas. A hydrogen-rich gas that is used in the synthesis of ammonia, hydrocarbons and methanol.

synthetic natural gas (SNG). A gas that is manufactured from coal or oil. It has the same chemical and burning characteristics as natural gas. *See also* NATURAL GAS.

systems analysis. A variant of LINEAR PROGRAMMING.

T

T-2. *See* STANDARD TANKSHIP.

take. *See* GOVERNMENT TAKE.

takeover. *See* MERGER.

tanker categories. A commonly used classification of tankers, based on AFRA, together with estimates of tonnages in each category. As of mid-1985 this is shown in Table T.1. *See also* TANKER TYPES.

Table T.1.

	Range (dwt)	Million dwt
Specialist vessels	<16–500	3.6
General-purpose	16 500–24 999	7.9
Medium-range	25 000–44 999	24.9
Large-range	45 000–159 999	81.6
VLCC	160 000–319 999	101.7
ULCC	>320 000	25.7

tanker freight rates. Most of the crude oil consumed in refineries throughout the world is imported using sea-going oil tankers. Thus the cost of freight is an essential, although by no means constant element in delivered oil costs. Oil transport costs help to determine the relative locational advantage of the major oil exporters and are explicitly included in the computation of price differentials by the OPEC producers. They are also included in internal profitability calculations by the integrated oil companies (*see* INTEGRATION). Generalizations regarding rates are difficult, however, due to the fact that freight services are offered by the shipping industry for which oil, although important, is only one among the many goods requiring transportation services.

Oil freight rates vary with distance to be covered, with the duration of the period for which transport services are required and with the availability of alternative routes, including canals and pipelines, to the cargo destination. These variables make it difficult to assess the general level of and changes in rates pertaining at any one time. Comparisons over time are further hindered by changes in the location and importance of the major producing and consuming centres and by changes in the technology of shipping including vessel scale.

Attempts to standardize rates began during World War II when some governments requisitioned oil company tanker fleets in return for a rental or TIME CHARTER payment. The need to establish some method of charging for the use of surplus capacity led to the establishment of schedules of freight rates which would generate the same net return per day for a standard-sized vessel irrespective of voyage length. In 1946 the UK Ministry of Transport (MOT) and the US Maritime Commission (USMC) were the first rate schedules of this type. They were very aggregative in their port groupings and wrongly included the constant element of canal dues in the scheduled rates.

In the post-war period, the MOT and USMC were used widely by tanker operators as a basis for charter negotiations, with agreed rates being expressed as a percentage plus or minus the scheduled rate. Responsibility for the schedules passed eventually into the hands of the International Tanker Nominal Freight Scale Association which in 1962 produced the first comprehensive port-by-port schedule (*see* INTASCALE). This schedule, which took into account changes in ship sizes, port charges and bunker costs, remained in use with revisions until 1969. It was, however, based on sterling and suffered due to the currency fluctuations of the late

1960s which culminated in the devaluation of sterling in November 1967. Intascale was replaced in 1969 by Worldscale, a schedule agreed between London and New York brokers and published jointly by the Worldscale Association (London) Ltd of London and the Worldscale Association (NYC) Inc. of New York. Initially it was a dual-currency scale, but in 1971 it began quoting exclusively in dollars. Worldscale standardized the treatment of port time and canal time, utilized a fixed daily hire rate and provided for an annual (since 1980 biannual) revision of rates to take account of changes in bunker fuel and other costs. Unlike Intascale actual rates are quoted as indices for which the Worldscale rate is taken as base 100. Thus, for example, if the Worldscale rate for the route from the Gulf to north-western Europe via Cape of Good Hope is given as $15.70 per ton, an actual market rate of $4.10 per ton would be quoted as WS 26 (approximately true for February 1987). This facilitates comparisons between routes by suppliers.

Tanker freight rates have been characterized by long periods of low levels interspersed by a small number of sharp peaks. In the 36 years since 1950 peak values have occurred in only six years — 1951, 1956, 1967, 1970, 1973 and 1979. Four of these have been associated with significant military events — the Korean War (1951), Suez (1956), the Arab–Israeli War (1967), the Yom Kippur War (1973) — and in all of these crises severe restrictions were imposed on the supply or routing of the world's tanker fleets. It may be seen that the present period of depressed freights, which has lasted over eight years, although longer than the average trough period length of 5.6 years is within the historic range (3–11 years). The association of freight rates with crisis events makes it extremely difficult to estimate rate functions for forecasting and investment purposes.

The economic analysis of tanker freight rates has focused on two main issues: (a) short-run rates and the role of expectations; (b) the explanation of long-run developments in rates for forecasting and planning purposes. The reason why rates rise excessively for short periods but remain low for much of the time requires some understanding of the term structure of rates and the part played by expectations in the market. When rates are low there is no need for operators to charter more than for immediate requirements. As rates rise, there is an incentive to secure services over a longer period if future rises are foreseen. A fear of being caught short may motivate charterers to purchase TIME CHARTERS. The effect of this is to reduce the future supply of spot vessels which may put further pressure on rates. Koopmans in 1936 argued that risks of possible oil sales loss and loss of prestige were factors that might lead charterers to retain ships on time charter when not strictly needed. Furthermore, in a rising freight market, those oil companies who own tankers are able to inflict damage on rivals by withholding tonnage from the market. In Koopmans' view the oil companies only exercise their market power in anticipation of shortage, but behave competitively at other times. Rate collapses occur when the companies release surplus vessels onto the spot market following a revision of market expectations.

In the Zannetis model of 1964, both sellers and buyers of tanker services are seen as having elastic price expectations. Thus a rate increase (decrease) is taken to indicate the likelihood of future rate increases (decreases), and prices are bid upwards (downwards). The wide fluctuations, according to this model, are due to asymmetry in expectations between buyers and sellers, but within certain limits. The lower limit is reached when sellers are forced to withdraw at very low prices. At high prices, either buyers withdraw or new supplies come forth and expectations are reversed. Evidence for this theory is somewhat mixed. Zannetis's own empirical results, although consistent with elastic price expectations, were unable to predict turning points. Glen, Owen and van der Meer in 1981 tested the impact of spot on time charter rates for 3000 fixtures over the period 1970–77. The results suggested a geometric learning process rather than unit elastic price expectations. Further work, however, needs to be done to assess the links between expectations and short-term rates.

It is more difficult to determine what freight rates will be in the long run. It is necessary to make a number of observations. (a) The market for tanker services is a competitive (or at least a contestable) one where independent shipowners outnumber the oil companies and where there are few entry barriers. (b) Costs such as lay-up costs are incurred when vessels are withdrawn from the market. (c) There are dynamic links between current freight rate levels and future shipping capacity through (often subsidized) investment, and also links between oil freight and dry freight markets. These observations imply that tanker rates in the long run are likely to be related to costs, but are also likely to be cyclical in nature depending on the match between future supply and demand for transport capacity. Among the costs of transport can be distinguished costs connected with individual voyages, mainly bunker fuel costs, operating costs consisting of crew, insurance, maintenance and marketing costs, which cannot be varied easily in the short term, and capital costs connected with the vessel's purchase. Such costs, for the currently most efficient vessel type, constitute, with suitable discounting, the long-run marginal cost of oil transport provided that future demand levels are anticipated correctly. In practice the unpredictable nature of demand produces deviations from this steady path. In the short run, the combination of a highly priced inelastic demand for oil freight services together with a steeply rising, short-run, marginal cost schedule up to capacity utilization can generate significant movements around the long-run rate path. The market response to a given rise in freight rates might be as follows. Initially, higher freight rates lead owners to order new vessels given the ready availability of subsidized finance for shipbuilding. The average time taken to build a vessel obviously varies with the shipyard and vessel type, but may be taken to range between one and two years. At the same time higher rates raise the profitability of existing tanker operations, and scrapping decisions are deferred. As a result an expansion of capacity occurs within a relatively short period. Once capacity has increased the initial rates increase is reversed due to short-run demand inelasticity. However, the durability of tankers hinders supply adjustment, and a period of high rates is followed by protracted over capacity and rates falling to below voyage cost levels. This is the conclusion of most work in the Koopmans/Zannetos tradition. The main policy measures which have been suggested to reduce the variability of rates consist of schemes to increase the cost of holding idle capacity. These include the removal of investment subsidies together with inducements to accelerate scrapping of redundant vessels. However, such measures often have only a limited attractiveness to countries keen to promote shipbuilding activities for employment and development purposes. *See also* TANKER RATE SCHEDULES.

Tanker Owners' Voluntary Agreement on Liability for Oil Production. *See* TOVALOP.

tanker rate schedules. Standard rate schedules were first introduced by the UK and US governments during World War II to establish the remuneration for requisitioned vessels. Rates were calculated such that whatever the route the government as owner would receive the same net return per day. This principle of equating return on a standard vessel over all routes still governs the tanker industry's rate schedules. In 1939, the UK government set up the MOT (Ministry of Transport) rate based on a 12 000 dwt vessel whereas the USA established its US Maritime Commission (USMC) rate. The MOT schedule was replaced in November 1952 by Scale No 1 which in turn was superseded by Scale No 2 in July 1954 and by Scale No 3 in December 1958. Meanwhile the US schedule was replaced by the American Tanker Rate Schedule (ATRS) in 1956. By 1961 the need was seen in London for a fresh schedule, and this resulted in the production of INTASCALE in May 1962 followed in September 1969 by WORLDSCALE, an agreed schedule between US and UK tanker owners and operators. Conversion between schedules is difficult since different-sized vessels with different speeds and fuel consumptions were used. *See also* TANKER FREIGHT RATES.

tanker types. Crude oil tankers range in size from 10 000 to 500 000 dwt. Vessels smaller than 10 000 dwt tend to be used for specialist purposes (e.g., the carriage of bitumen). Tankers in which the cargo may be held in many separate tanks are known as parcel carriers. These are used to transport a wide variety of petroleum products. Products carriers are vessels that are capable of carrying clean petroleum products. Certain petrochemicals (e.g., benzene and xylenes) may also be carried on such ships. When used in this way they are called chemical tankers. Increasingly stringent safety measures are required of tanker owners so that now many tankers are fitted with segregated ballast tanks (SBT), crude oil washing systems (COW) or inert gas systems (IGS) to lessen the risk of explosion or pollution. *See also* TANKER CATEGORIES.

tank farm. A collection of tanks for the storage of crude oil or products.

tank sludge. *See* SLUDGE.

Tapline (Trans-Arabian Pipeline). A 1000-mile crude oil pipeline between Ras Tanura in Saudi Arabia and Sidon in Lebanon that passes through Jordan and Syria. It was built in 1950 and remained open until 1975. Now it is only used for occasional deliveries to Jordan. Its theoretical capacity is 0.5 million barrels per day. During its operational life, it suffered from disputes over the transit fees to be paid.

tare weight. The weight of an empty container (e.g., a road or rail product tanker).

target revenue earner. A supplier who is seeking a specific revenue level and once that level is achieved will supply no more. If price were to rise he would then actually reduce the quantity supplied, thereby causing the SUPPLY CURVE to bend backwards (*see* Fig. T.1). Some have argued such behaviour has characterized some oil-producing countries, especially after the sharp increase in price of 1973 (*see* CRUDE OIL PRICES). Thus an inability to use more revenue at home coupled with an aversion to the accumulation of assets abroad encouraged them to reduce supply, thereby aggravating the perception of oil scarcity.

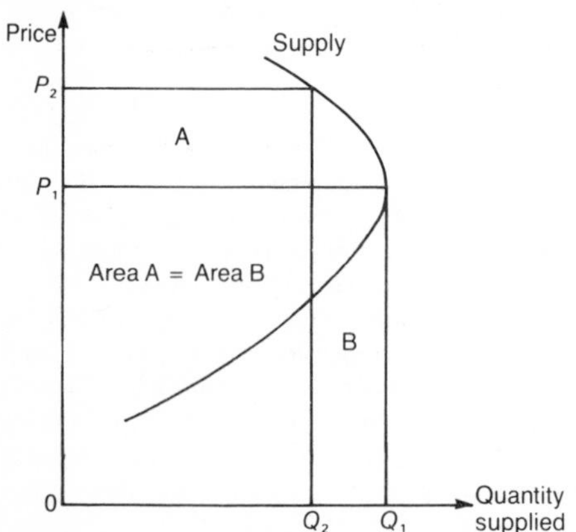

Fig. T.1.

tariff. A tax imposed upon an imported good or service. It can be imposed on a unit basis (a specific tariff) or as a percentage of value (an ad valorem tariff). The term tends to be used when the intention is to reduce the volume of imports and is often linked to the idea of QUOTAS. Taxes used to raise revenue are often referred to as customs duties, although the distinction is very casual. Since the oil price collapsed in 1986, there has been talk, especially in the USA, of imposing a tariff on imported crude oil and products in order to maintain a high domestic price which would encourage further exploration and development as well as further conservation.

tars. (1) Very heavy crude oils (about 10°API gravity). (2) Heavy liquid residues from REFINING. (3) A black viscous liquid resulting from the distillation of solid material (e.g., coal or wood).

tar sands. Sedimentary deposits of sand that contain up to 12 percent BITUMEN. The most prominent deposits are in Canada (Athabasca) and Venezuela. In theory it is possible to remove the tar by application of steam to provide heat and pressure. The resulting tar tends to be low in hydrogen compared with crude oil and thus generally requires HYDROGENATION and SWEETENING PROCESSES to produce a synthetic crude oil. During the perceived energy crises of the 1970s interest was shown in developing the tar sands. However, the economics of such projects has not been attractive because of the very high processing costs.

tax. A compulsory payment to the government. Taxes can be direct, normally on wealth or income, which means the person or institution paying the tax cannot shift the burden of the tax (i.e. the tax incidence). Indirect taxes are surcharges on price which although initially paid by producers or sellers can be passed on to the final consumer. Taxes can also be classified as progressive where the rate of tax increases with ability to pay (*see* PROGRESSIVITY) or REGRESSIVE where the rate is fixed. Governments use taxes to raise revenue, to redistribute income and/or to discourage use. Tax is a critically important variable in many aspects of the international oil and gas industries.

tax allowance. Items that a company may

deduct from its income thereby reducing its TAX BASE and hence its tax liability. Such allowances vary considerably between countries. In international oil, for example, many companies could offset taxes paid to producer host governments against their tax bill to their parent governments. Another important tax allowance was the US DEPLETION ALLOWANCE whereby a company can offset against taxable income a sum to reflect the DRAWDOWN of its oil reserves. This was equivalent to a company's DEPRECIATION allowance, intending to reflect the fact that oil is an EXHAUSTIBLE RESOURCE.

taxation neutrality. A tax system that is intended to have no distorting impact on economic behaviour is a neutral tax. For instance, a tax that falls only on SUPERNORMAL PROFITS and does not affect economic decision-making is said to be a neutral tax. However, a tax on crude oil production, for example, may lead to a reduction in output which would not be neutral in the sense that it would carry implications for resource allocation.

tax base. The size of the 'pot' to be taxed. In direct taxation it is flows of income or stocks of wealth. In indirect taxation it is the value of sales of a product reflecting both price and volume. Obviously for revenue raising, the larger the base the more attractive is the income/wealth/good or service.

tax credits. A refund of part of the amount of tax paid. It is an alternative to TAX ALLOWANCE.

tax haven. A country with low or negligible company tax. Tax havens are usually very small countries, the absence of tax being aimed at encouraging companies to locate 'offices' in that country. The company may do so for purposes of tax avoidance, which is legal, or tax evasion, which is illegal.

tax incidence. Describes who actually pays the tax as distinct from on whom it is levied.

tax-paid cost (TPC). The cost of a barrel of crude oil to the lifting company when the company has an equity stake in the producing company (i.e. the cost of EQUITY CRUDE). It includes the costs of physically producing the crude oil and loading it (for use or export) together with all payments due to the government in the form of ROYALTIES and taxation. The difference between the market value of the crude and the tax-paid cost is the equity margin and should represent an element of preferential access for the parent company (i.e. it should normally be positive).

tax rate. The rate at which a TAX BASE is taxed. It is normally expressed as a percentage.

tax reference price. When a good (e.g., crude oil) has no market price because, for example, it is moving through the vertically integrated channels (*see* VERTICAL INTEGRATION) of a large multinational corporation there is no mechanism to identify the TAX BASE. To be able to compute a tax base therefore requires a tax reference price. Such a price was the POSTED PRICE introduced with fifty/fifty profit sharing in the late 1940s and early 1950s. More recently a tax reference price is still used to compute a company's tax liability under many tax regimes, but is normally (via some formula) based upon 'market' price.

tax revenue. The TAX BASE multiplied by the TAX RATE. If the collection cost of the tax is subtracted this gives the tax yield.

tax yield. *See* TAX REVENUE.

TBA. (1) *See* TERTIARY BUTYL ALCOHOL. (2) Tyres, batteries and accessories sold by service stations.

TBP curve (true boiling point curve). *See* QUALITY, CRUDE OIL.

TBP distillation (true boiling point distillation). *See* QUALITY, CRUDE OIL.

TCE. Tonnes of coal equivalent (*see* TOE).

technical progress. The increase in technical knowledge as related to production techniques. It is assumed to increase with time. In neoclassical economics, technical progress was assumed to be disembodied. Modern economists tend to view technical progress as being embodied in the company's PRODUCTION FUNCTION, either directly or as an inherent characteristic of the factors of production.

The implication of technical progress on the company's level of output can be looked at from two angles. Either the company can maintain its current level of production, but as a direct consequence of technical progress it will require a lower level of inputs to do this or, by maintaining current input levels, the company will be able to increase production. This is really two ways of looking at the same idea: what essentially happens is that technical progress improves the marginal product of the factors of production or inputs. These improvements fall into two categories which can be illustrated more clearly with the aid of diagrams. For the sake of simplicity it is assumed that there are only two factors of production — capital (K) and labour (L).

In Fig. T.2 and Fig. T.3, Q is the company's initial production curve representing a given level of output for all possible combinations of capital and labour for a given technology. Technical progress has the effect of shifting the production curve from Q to Q'. The two curves can be thought of as contours on the company's production map with Q' reflecting a higher level of output than Q. In Fig. T.2 the shift is a parallel one where the relative marginal products of labour and capital do not change. In economics this is referred to as Hicks neutral technical progress. In Fig. T.3 the shift is a non-parallel one, and technical progress has improved the marginal product of one input more than the other one. This case is referred to as Hicks non-neutral technical progress.

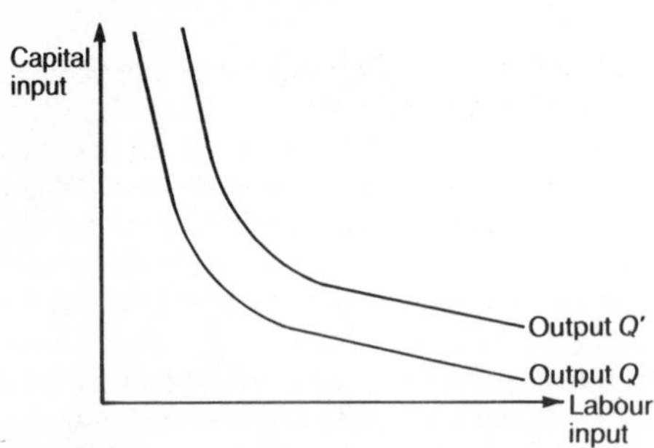

Fig. T.2

Technical progress is an important variable in terms of the oil and gas industry. On the demand side it influences the thermal efficiency of energy-using equipment. On the supply side it influences the rate of new discoveries and developments and the size of the RECOVERY FACTOR. Because of the time lags involved in the widespread adoption of the fruits of technical progress, observers have a noticeable tendency to neglect its impact.

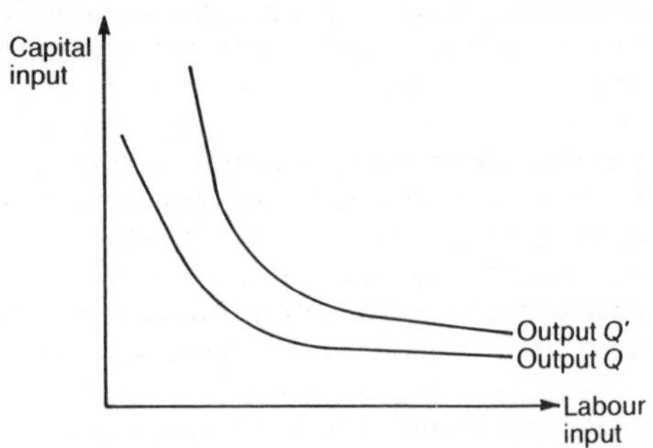

Fig. T.3

Teheran Agreement. A five-year agreement signed in February 1971 between the Gulf oil producers and the oil companies that is similar to the TRIPOLI AGREEMENT. The main terms of the Teheran Agreement were a general increase of 35 cents per barrel in Gulf posted prices for 40°API crudes, with heavier crudes attracting up to a further 5 cents per barrel. These prices were to be escalated at 2.5 percent per year (for inflation) plus 5 cents for refined product price increases. Income tax was set at 55 percent.

TEL. Tetraethyl lead (*see* LEAD ALKYLS).

temperature effect. Unusually cold days compared with the seasonal average lead to higher than usual energy consumption. When a TIME SERIES of energy demand is being considered, it is advisable to adjust the trend to take account of the variations in temperature. This is a variant of removing SEASONALITY from the data with the qualification that no set pattern of temperature deviations can be expected.

term contract. A purchasing contract for crude oil or gas that buys a flow of oil over a specific time period. An example is a contract to sell 10 000 barrels per day for one year. Before the SECOND OIL SHOCK, the vast majority of crude oil and products was bought and sold under term contracts usually covering quite long periods of time. After the Second Oil Shock, the SPOT CONTRACT became much more important and most term contracts became relatively short-term, although how much of the traded oil and products move on spot contract is extremely uncertain with estimates ranging from 30 to 50 percent.

terminal. An onshore unit that receives oil and/or gas by pipeline or tanker for storage and further transportation.

terminal platform. An offshore unit that collects oil and/or gas and then pumps it ashore.

terminal price. The FOB price quoted at a TERMINAL.

term limit pricing (TLP). An agreement in which the price of the transaction is specified for a specific period of time.

term market. The ARM'S LENGTH market in which a flow of oil is bought for a specified time period. It is used to distinguish the market from the SPOT MARKET.

term price. The price associated with the buying/selling of crude oil, products or gas by TERM CONTRACT. *Compare* SPOT PRICES.

terms of trade. The ratio of an index of export prices to an index of import prices. There are a variety of methods of measurement. An improvement in the terms of trade means export prices rise faster than import prices, which is equivalent to an increase in purchasing power of the country (i.e. per unit of exports it can buy more units of imports). A deterioration in the terms of trade is the reverse (i.e. to obtain the same import units, more must be exported). The two oil shocks of the 1970s (*see* FIRST OIL SHOCK; SECOND OIL SHOCK) saw a spectacular improvement in the OPEC members' terms of trade. It is a contentious area, but it is widely thought that THIRD WORLD countries as exporters of PRIMARY PRODUCTS face a deteriorating long-term trend in their terms of trade in relation to the industrialized countries.

territorial waters. The area offshore of a country that is regarded, or claimed, as part of that country's territory. The significance in this context is that oil and gas in the territorial waters belong to the country of ownership. Hence territorial claims can represent very serious issues. A commonly used solution to disputes over territorial water is the use of the MEDIAN LINE as, for example, between UK and Norway.

tertiary butyl alcohol (TBA). An additive to high-octane motor spirit. It can also be used as an anti-icing additive.

tertiary recovery. *See* PRODUCTION, ENHANCED OIL RECOVERY.

test drilling. The procedure that is used to provide empirical confirmation of the extent of the find determined initially using various seismic techniques (*see* SEISMIC EXPLORATION). It is a stage in the development of the field, although it may occur prior to the declaration of COMMERCIAL DISCOVERY.

test market. *See* FORECASTING.

tetraethyl lead. *See* LEAD ALKYLS.

tetramethyl lead. *See* LEAD ALKYLS.

Texaco. One of the seven MAJORS; a US company which was one of the few to survive the Spindletop find in EAST TEXAS. Since 1936 Texaco has been closely linked with SOCAL outside of the USA in their jointly owned company Caltex.

Texas Railroad Commission. The most important of the US State Commissions. It was responsible for setting crude oil production levels to prevent over-production due to the LAW OF CAPTURE.

theory of games (game theory). A theory developed during the World War II by the mathematician John von Neumann and the economist Oskar Morgenstern. The theory can only be applied to a specific kind of game situation with certain characteristics. There must be a finite number of players in competition with each other, each player being faced with a finite number of alternative strategies from which he may choose. It is assumed that each player will act in a rational manner and will attempt to resolve the conflict of interests in his favour. For each alternative strategy there is a specified result; these results can then be expressed as a payoff matrix. The fundamental idea is that a player will aim to minimize his maximum loss or conversely to maximize his minimum gain. The theory is most commonly applied to the very restricted case of a two-player ZERO SUM GAME where the sum total of all gains and losses is zero.

Game theory has been recently applied to the field of energy economics in an attempt to model, for example, the decisions regarding strategic oil reserves or the development of alternative energy sources. The oil reserves problem can be viewed in the context of a game between major oil-consuming countries and OPEC, and is a good example of a non-zero sum game. In this case one player's gain will be expected to differ from the other player's loss. The alternative source development problem has been applied as a three-player game between OPEC, the USA or other user countries and an entrepreneur.

thermal cracking. A refinery process whereby a heavy residue or distillate is heated under pressure in order to break down (i.e. crack) the heavy molecules into lighter molecules thereby producing products at the lighter end of the barrel. *See also* REFINING, SECONDARY PROCESSING.

thermal efficiency. The amount of useful heat derived from a fuel. It is normally expressed as a proportion of the total CALORIFIC VALUE of the fuel. *See also* APPENDIX.

thermal pollution. The release of heat into the environment that has an adverse effect on the environment. The most obvious example is the discharge of cooling water from power stations.

thermal recovery. A type of enhanced oil recovery that involves the ignition and burning of the oil within the reservoir. The burning generates gases which increases pressure and also heats the oil, encouraging it to flow more easily. *See also* PRODUCTION, ENHANCED OIL RECOVERY.

thermal reforming. A refinery process developed around 1930 used to raise the OCTANE RATING of straight-run naphthas. *See also* MOTOR SPIRIT UPGRADING.

thermal tar. A heavy aromatic oil derived from THERMAL CRACKING. It is used as a feedstock to produce PETROLEUM COKE.

Third Oil Shock. (1) The $5 per barrel cut in the crude price which came out of the LONDON AGREEMENT of March 1983. (2) The oil price collapse of 1986 which followed the December 1985 declaration by OPEC that it would seek a 'fair share' of the international oil market. Since the fall in price far exceeded the 1983 price cut — SPOT PRICES fell from $27–28 per barrel to less than $10 — it is likely that the term will be associated normally with 1986.

third-party sales. ARM'S LENGTH sales of crude oil, products or gas between independent companies as opposed to sales between affiliates of the same company.

Third World. The developing countries. The term generally finds favour among the countries themselves which are often sensitive to some of the other generic expressions used (such as less-developed countries) because of their perjorative implications.

throughput. The amount of raw material (e.g., crude oil) that is processed in a plant (e.g., a refinery).

thruster propeller. A small propeller on a vessel or floating platform that is used to maintain a given position.

tied oil. In the UK oil on which no tax has been paid, but which can be delivered to a bonded user or distributor.

time charter. A form of tanker CHARTER in which the hirer has the use of the vessel for a specified period of time (normally at least two years). The shipowner provides the crew and provisions unless it is a BAREBOAT CHARTER in which case these are supplied by the hirer.

time credits. The period of time between delivery of crude oil (and products) and payment for that delivery. The longer this period the greater the 'discount' off the sales price. For example, at a general interest rate of 10 percent with 90-day credit, oil priced officially at (say) $30 per barrel carries a discount of 75 cents. This is because the $30 could earn 75 cents during the three months before payment is due. When the oil market is tight, as for example during the SECOND OIL SHOCK, time credits will reduce to cash on delivery or, in extreme cases, prepayment.

time lags. The amount of time that elapses between an event and its consequences; it is the reverse concept to a LEAD TIME. In the international oil and gas industry time lags are of crucial importance in analyzing developments, yet they are often ignored and frequently unknown with any degree of precision. One example is 'production' in Saudi Arabia which is effectively measured for purposes of world trade when the barrel is loaded at Ras Tanura. However, it takes at least six to eight weeks before that barrel is translated into physical products at a refinery in north-western Europe. Another example is a sharp increase in oil prices which reduces the quantity of oil used. This effect is composed of behavioural changes that are likely to occur quickly. However, it also involves changes to the oil-using equipment which takes a longer time (i.e. long lags) since the new fuel-efficient equipment must be designed and produced, and then the stock of equipment must be turned over. Thus for cars, the time lag for the full effect to be felt may be in excess of 15 years.

time preference. A positive time preference refers to the economist's assumption that people prefer to consume now rather than in the future. Thus saving implies giving up present consumption which requires compensation in the form of interest. An alternative consequence of this time value of money is when a sum of money is expected in the future. If time preference is positive then the expected future sum of money should be more than its present value to compensate for the delay in receiving the money.

time series. Time series data represent values of a variable over a period of time, an example being motor spirit consumption in the UK between 1950 and 1980. Cross-section data represent values of a variable at a given point of time in different locations, an example being motor spirit consumption in the UK, France, Germany, Italy, USA and Japan in 1980. A combination of time series and cross-section data is also possible. Time series data can be broken down into three elements: (a) SEASONALITY (e.g., in summer more motor spirit may be consumed as driving for pleasure outings increases); (b) TREND (e.g., over time as a country's per capita income rises motor spirit consumption is also likely to rise); (c) random fluctuations (e.g., a sharp increase in price may induce an immediate behavioural change that causes motor spirit consumption to fall).

time value of money. *See* TIME PREFERENCE.

TLP. *See* TERM LIMIT PRICING.

TML. Tetramethyl lead (*see* LEAD ALKYLS).

TNC. Transnational corporation (*see* MULTINATIONAL CORPORATION.

TOE. Tonnes of oil equivalent. When aggregating different types of energy, it is necessary to measure the different fuels using a common unit. An obvious such unit would be the CALORIFIC VALUE of the fuels. However, a commonly used method is to make such a measurement not in JOULES or BTU, but in a specific fuel. A typical tonne of coal has 29 gigajoules of energy, any amount of fuel having 29 gigajoules of energy is thus expressed as 1 tonne of coal equivalent. Variability between different crudes (or different types of coal), however, makes such average figures of limited value. The reason for the use of TOE rather than heat units is that within the oil and gas industries people are more familiar in thinking in terms of a weight (or volume) equivalent. It is possible to convert TOE into million barrels per day equivalent. *See also* APPENDIX.

toluene. An important AROMATIC. It is used as a petrochemical feedstock. *See also* CHEMISTRY; PETROCHEMICALS.

tongs. Hydraulically operated grabs which hold sections of drilling pipe while the DRILLING STEM is being screwed together or unscrewed. *See also* DRILLING.

tonnage (ships). By international agreement merchant ships, including tankers, are measured in tons of 100 cubic feet to provide the gross tonnage. When the non-carrying space (i.e. engine room, bunker space, etc.) is subtracted this gives the net register tonnage. However, most often the ship size is given in deadweight tonnage (dwt), which is the total carrying capacity (in tons of 2240 pounds) when loaded to the maximum allowed in the summer season. The effective carrying capacity is the deadweight tonnage less bunkers, stores and water. For tankers, as a very rough rule of thumb, deadweight tonnage is 2.5 times the net register tonnage, whereas gross tonnage is 0.66 of deadweight tonnage.

tonne. A metric unit of weight. Although the barrel was the most commonly used measure in the oil industry, ton(ne) was also used. Unfortunately it was never clear if it was a short ton (i.e. 2000 pounds), which is commonly used in the USA, or a long/imperial ton (2240 pounds), which is commonly used in the UK and areas that used imperial measurements. During the 1970s the industry has moved increasingly to the use of metric tonnes (i.e. 0.984 of a long ton or 2205 pounds). *See also* APPENDIX.

tool pusher. The supervisor in charge of a drilling rig.

topped crude. A crude oil from which the lighter (i.e. more volatile) ends of the barrel have been distilled away.

topping plant. A simple distillation unit that is used to remove the more volatile fractions of a crude oil to produce TOPPED CRUDE.

Total. *See* CFP.

total costs. *See* COSTS.

tour. A drilling shift.

TOVALOP (Tanker Owners' Voluntary Agreement on Liability for Oil Pollution). An agreement that was first signed in 1969; two years later most tanker owners had joined. The agreement bound them to provide compensation for pollution at a rate of $100 per ton with a maximum limit of $10 million per accident/tanker.

town gas. *See* MANUFACTURED GAS.

TPC. *See* TAX-PAID COST.

trade investments. Shares held by one company in another. *See also* ASSETS.

trade press. Publications that regularly cover aspects of the oil and gas industries. Their frequency varies from daily through weekly to monthly. Among the longest established and most influential are *Middle East Economic Survey* (*MEES*), *Oil and Gas Journal*, *Petroleum Intelligence Weekly* (*PIW*) and Platt's *Oilgram*.

traditional energy. Fuel that either moves in small local markets, or fuel that is not bought or sold (i.e. subsistence fuel). Traditional energy can also defined as energy that is collected and used by the consumer; it involves no input of foreign exchange to produce it. In many developing countries traditional energy represents a large proportion of the primary energy input into the economy. Examples of traditional energy include wood and wood-based fuels, BIOGAS, animal and vegetable residues, and human and animal power. Traditional energy is often mistakenly called NON-COMMERCIAL ENERGY (NCE); part of traditional energy is commercially transacted, especially wood and charcoal. Clearly NCE is a subset of traditional energy. Often the animal and vegetable residues are called 'wastes'. This is also an error since it implies no alternative use. However, such residues do have an OPPORTUNITY COST such that if burnt as fuel they cannot be used as fertilizers.

The distinction between commercial and traditional energy is of importance in developing countries because of the possibilities of substitution between the energy forms coupled with the high percentage of traditional energy in total energy use. Thus, for example, policy measures used by countries in the 1970s to reduce the oil import bill often either failed or produced unexpected side effects because the substitution linkages were neglected. In addition, commercial and traditional energies, although linked, may require different policy objectives and different instruments to achieve those objectives. Prior to the oil price rises of the 1970s (*see* FIRST OIL SHOCK) traditional energy was ignored (or at least neglected). Among other effects, its omission gave rise to the myth that developing countries had less ENERGY INTENSITY per unit of output than industrial countries. In fact, as shown by numerous studies, because of low CONVERSION EFFICIENCIES the reverse is true. *Compare* COMMERCIAL ENERGY.

traditional methods of investment appraisal. The methods of investment appraisal that were widely used before the development of DISCOUNTED CASH FLOW techniques. The most obvious ones are RATE OF RETURN and PAYBACK.

tramp. A vessel owned by an independent (non-oil company) operator that trades in the short-term (spot) or medium-term (TIME CHARTER) markets. Tramps are more commonly known as 'private owner' vessels.

tranche. An amount of money taken at a point in time as a loan under the terms of a loan agreement.

transaction costs. The costs associated with any transaction (buying/selling or contractual agreement). They may cover the time costs of bargaining, the costs of specialist advice, legal or otherwise, the costs of gathering information, etc. It is argued by some that the existence of such costs encourages the development of VERTICAL INTEGRATION on the grounds that inter-affiliate dealings involve lower transaction costs. A long-term crude oil supply contract between affiliates involves less administrative costs than buying the equivalent volume on the spot market tanker by tanker.

transfer earnings. *See* ECONOMIC RENT.

transfer payments. Payments to people for which no service or good is forthcoming. Examples of such payments include unemployment pay, sick pay and supplementary income for the poor.

transfer price. The price at which one company affiliate transfers its goods or services to another company affiliate in a vertically integrated company (*see* VERTICAL INTEGRATION). The manipulation of these prices clearly influences the accounting profits of the affiliates which means that MULTINATIONAL corporations can choose in which countries to maximize their accounting profits in order to minimize the amount of tax paid. Thus an affiliate located in a country with a relatively favourable tax regime would be expected to make more profit than an affiliate located in a country with a harsh tax regime. Because the international oil industry has always been highly vertically integrated transfer pricing has been important.

When host governments began to tax oil companies operating in their territories in the late 1940s and early 1950s, because little crude was sold in the open market — it moved through the affiliates — there was no crude 'price' to determine profits. To overcome this problem the concept of POSTED PRICE was used, which was intended to approximate to a market price. Following the break-up of the vertically integrated structure after the SECOND OIL SHOCK, transfer prices lost some of their importance and in many cases, for the first time, REFINERY MARGINS or profits became real figures. This led, within the companies, to greater emphasis on 'centres of profitability' rather than group profitability.

transhipment. The transfer of a cargo from one ship to another. In the oil industry it most often refers to the process whereby a large tanker discharges its cargo to smaller tankers which then deliver the oil to ports whose size limits the arrival of the larger tanker. Normally the large tanker unloads to a TERMINAL.

transit lines. The relatively large oil or gas pipelines that frequently cross international boundaries.

transnational. *See* MULTINATIONAL.

transnational corporation. *See* MULTINATIONAL.

transport capacity. *See* CAPACITY.

trap. A geological structure that holds oil and gas. *See also* GEOLOGY.

travelling block. A part of the hoisting equipment used to hook onto the drill pipes when these have to be raised or lowered. *See also* DRILLING.

treatment. All refining processes aimed at removing small proportions of undesirable constituents such as sulphur, etc. *See also* REFINING, SECONDARY PROCESSING.

trend. Changes in the value of a variable — a TIME SERIES — allowing for seasonality and random variation over time. It is frequently measured statistically by a line of best fit derived by means of simple linear regression. In graphical form the variable (e.g., oil demand) is measured on the vertical axis and time is measured on the horizontal axis. Trends are frequently used as the basis for forecasting the future path of the variable.

tributary refineries. Refineries that are able to supply the main regional trading centres such as Rotterdam or Singapore.

Tripoli Agreement. A five-year agreement on oil prices signed between the Mediterranean oil producers and the oil companies in April 1971. A similar agreement (*see* TEHERAN AGREEMENT) was drawn up between the Gulf oil producers and the oil companies. During 1970, Libya had successfully negotiated an increase in the POSTED PRICE (effectively the tax reference price) of oil with its oil companies. This was the first time that the oil companies had allowed any role for governments in the pricing decision. The Tripoli Agreement led to an OPEC resolution in December 1970 that a uniform general increase in price would be negotiated with the oil companies. The original idea was that these negotiations would take place on a regional basis. Immediately the companies opposed the idea of regional negotiations, fearing that such deliberations would allow the governments to divide and rule. This is what the Libyan government had done in the 1970 negotiations, leading to the creation of the LIBYAN PRODUCERS AGREEMENT. The companies, with the necessary anti-trust waiver from the US government) set up the LONDON POLICY GROUP to try to force joint global negotiations. In the event, at the last minute the US government shifted its position to favour regional negotiations, which duly took place. The US government's decision remains very controversial, many seeing the policy reversal as a stab in the back for the oil companies, and the cause of higher oil prices. It has been argued by some that the US government thought that higher international oil prices would favour US economic interests. Others have argued that it was based upon the perception that such global negotiations would be impossible. More research is needed to resolve the debate.

The Tripoli terms were similar to those of the TEHERAN AGREEMENT, but effectively gave a larger price increase to acknowledge the short-haul advantage of Mediterranean crudes as reflected in the SUEZ PREMIUM. Both the Tripoli Agreement and the Teheran Agreement were overtaken by events with renegotiations for exchange rate fluctuations (*see* GENEVA I; GENEVA II) and eventually the FIRST OIL SHOCK.

true boiling point (TBP). *See* QUALITY, CRUDE OIL.

turbine drilling. A drilling method in which the DRILLING STRING remains stationary while the drilling BIT, which revolves, is powered by the flow of the drilling mud. *Compare* ROTARY DRILLING. *See also* DRILLING.

turbine oil. A well-refined petroleum distillate used to lubricate steam turbines.

Turkish Petroleum Company. *See* IPC.

turnkey contract. A contract in which the contractor undertakes to build a plant (e.g., a refinery or a petrochemical plant). The contractor also undertakes to start up the plant and begin its operation. Once the plant is operating satisfactorily, it is handed over to the customer who takes over responsibility.

turnover. The total sales revenue of a company.

turps substitute. White spirit (*see* FINISHED PRODUCTS, SOLVENTS.

two-stroke fuel. A blend of motor spirit and a lubricating oil (5–15 percent) used in two-stroke engines.

U

UKCS. The UK continental shelf.

UKOOA (United Kingdom Offshore Operators Association). An industry lobby organization comprising the oil and gas companies that are licence operators in the UK sector of the North Sea.

ULCC (ultra-large crude carriers). The designation for tankers in excess of 320 000 dwt introduced in 1979. Such tankers have limited routes because of their very large size. This frequently involves them in TRANSHIPMENT exercises. *See also* TANKER TYPES.

ullage. The space in a tank not filled by oil or products. It is therefore a measure of unused capacity and is normally measured as the gap between the level of liquid and the top of the tank.

ultimate recoverable reserves. The expected production of oil or gas from a RESERVOIR. Normally the figure is more strongly influenced by geographical/engineering constraints rather than the economic context. There is no guarantee the expectations inherent in the figure will actually be realized.

ultra-large crude carrier. *See* ULCC.

uncertainty. *See* RISK.

underlifter. *See* OFFTAKE NOMINATIONS.

underwriter. (1) Someone who for a fee agrees to bear some or part of a specific risk. (2) In the context of new share issues someone who agrees to buy any shares not bought by the public.

underwriting. The business of insuring against RISK. *See also* UNDERWRITER.

unfinished oils. In the USA semi-finished products which require further processing or blending.

United Kingdom Offshore Operators Association. *See* UKOOA.

unitization. Where an oil- or gas-field lies under more than one block and adjoining blocks are held by different licensees the field is developed as a single unit so as to avoid wasteful duplication of facilities or damaging overproduction. The various licensees negotiate a new unit operating agreement covering the whole field. Where an international boundary is involved, negotiations may occur at government level.

unleaded motor spirit. Lead-free motor spirit. *See also* LEAD ALKYLS.

unsaturated hydrocarbons. HYDROCARBONS that contain less than the maximum potential number of hydrogen atoms per carbon atom. *See also* CHEMISTRY.

UOP Merox process. The simplest of desulphurization processes used for liquid petroleum gas and light gasoline. It involves washing the product streams with a solution of caustic soda. *See also* REFINING, SECONDARY PROCESSING.

updated book value. *See* BOOK VALUE.

upgrading. The alteration of the REFINERY CONFIGURATION to allow further processing of the heavier ends of the barrel. This involves the building of CRACKING or REFORMING units and is essentially an expansion of SECONDARY REFINERY CAPACITY. In the mid-1970s the heavier ends of the barrel became more difficult to sell and there was a significant upgrading of much of the world's refinery capacity. *See also* MOTOR SPIRIT UPGRADING; REFINING, SECONDARY PROCESSING.

uplift. (1) The total amount of fuel supplied to an aircraft at a single refuelling. (2) The amount of oil that can be recovered from a field on an annual basis before it is necessary to pay taxes.

upper-tier crude. In the US system of pricing regulations as of February 1976, new crude oil and STRIPPER WELL crude. After September 1976 it included only new crude.

upstream. In the context of the VERTICAL INTEGRATION of a company, describing activities towards the crude-producing end of the chain as opposed to expansion towards the final consumer, which is known as downstream. Frequently upstream is used as a synomyn for crude oil production and downstream for crude oil refining and product distribution.

USAC. The US Atlantic coast.

USBM. The US Bureau of Mines.

user cost. An economic concept commonly found in depletion theory as applied to an EXHAUSTIBLE RESOURCE such as oil or gas. If a barrel of oil is produced today it cannot be produced tomorrow. Thus the user cost attempts to reflect the OPPORTUNITY COST of producing today. The cost to future generations of the use a non-renewable resource by the present generation results from the assumed higher cost of the eventual replacement for these resources. In the case of oil, user cost can be estimated by subtracting the marginal cost of extracting conventional oil from the marginal cost of the technology which is assumed eventually to replace conventional oil extraction and discounting back to the present. *See also* DEPLETION.

US fuel oil grades. There are six standard grades of fuel oil identified in the USA, ranging from No 1 to No 6. No 1 and No 2 are the distillates, No 3 is no longer used and No 4, No 5 and No 6 are the heavier grades. *See also* FINISHED GOODS, FUEL OIL.

US Gulf Plus pricing system. *See* GULF PLUS.

utility. The satisfaction a consumer obtains from consuming goods and services. Although some attempts have been made to define the measurement in terms of units of satisfaction, it is more easily understood as being measured in money terms (i.e. how much money the consumer is willing to pay to consume the good or service). This is then used as a measure of its value (satisfaction giving) to the consumer.

V

vacuum distillation. A DISTILLATION process that is carried out under reduced pressure in order to separate out more components than can be derived from ATMOSPHERIC DISTILLATION alone, using ATMOSPHERIC RESIDUE as feedstock. *See also* REFINING, PRIMARY PROCESSING.

vacuum gas oil. One of the principal products of VACUUM DISTILLATION. It is used as CRACKER FEEDSTOCK. *See also* REFINING, PRIMARY PROCESSING.

Vacuum Oil Company. *See* MOBIL.

vacuum residue. One of the main products of VACUUM DISTILLATION. It is used in BITUMEN production.

value added. The total output of a plant or company less its intermediate inputs. Thus it measures the value 'added' by the firm as a result of further processing. On an aggregate basis it can be used to measure gross national product (*see* GROSS DOMESTIC PRODUCT).

vanadium. A metal which is frequently used as a catalyst that may be present as a contaminant in crude oil. *See also* QUALITY, CRUDE OIL.

vaporizing oil. *See* POWER KEROSINE.

vapour lock. *See* QUALITY, MOTOR SPIRIT.

vapour pressure. The pressure exerted by the vapour escaping from a liquid. Vapour pressure is effectively a measure of how easily the liquid boils. It is measured by the REID VAPOUR PRESSURE. *See also* QUALITY, AVIATION SPIRIT; QUALITY, MOTOR SPIRIT.

variance. A statistical measure that estimates how far a set of numbers are dispersed around their arithmetic mean.

VAT (value-added tax). A general tax applied at each point of exchange of a good or service as the good is processed to a final product. It is a form of INDIRECT TAXATION that can be applied to oil products or gas.

very large crude carrier. *See* VLCC.

vertical integration. The result when a firm extends its activities from one stage in an industry to another stage in the same industry: for example, a crude-producing company may move into refining or a refinery company may move into pipelines or tankers. As a company moves towards the crude-producing end of the business this is moving UPSTREAM whereas moving towards the final consumer is a move DOWNSTREAM. Vertical integration has been a crucial element in the international oil industry since the very beginning, and many oil companies, especially the MAJORS have shown very high levels of integration although often the various activities from production to transport to refining to distribution have been unbalanced. Companies who had access to more crude than refinery capacity are termed CRUDE LONG COMPANIES whereas those who had more refinery capacity than access to crude were termed CRUDE SHORT COMPANIES.

There are three reasons that may explain a tendency towards vertical integration. The first reason is transaction costs. This argues that it is too difficult to write a secure contract for the supply of an input in a situation where the costs of a failure to obtain inputs — the OUTAGE COSTS — are very high. Therefore it is better to gain secure access to the input by the development or acquisition of an AFFILIATE to produce the input. The second reason concerns information costs. It is argued that a vertically integrated firm has a significant advantage in that it will know faster and more cheaply when and where investment must be made. The third reason is that vertical integration allows restriction of competition because it allows different prices to be charged to different customers. This last argument is controversial both on theoretical grounds and on empirical grounds, and it is not clear to what extent vertical integration has restricted competition.

Since the start of the oil industry in 1860, oil companies have developed a vertically integrated structure epitomized in the development of STANDARD OIL. For the MAJORS, it reached its peak in the 10 years following World War II. The existence of vertical integration has generated problems over transfer pricing (*see* TRANSFER PRICE) which in turn has led to disputes over taxation. The vertically integrated structure of the industry remained effectively intact until the SECOND OIL SHOCK despite expectations that the takeover of the operating companies by the governments which began in the early 1970s (*see* NATIONALIZATION; PARTICIPATION) would cause a break up of the structure. However, the governments' decision to allow the oil companies to continue marketing the crude oil on their behalf retained a form of vertical integration. When the governments took over the crude marketing during the Second Oil Shock, this effectively destroyed to a great extent the vertically integrated structure, although companies still retained some of their OWNED CRUDE. *See also* ARM'S LENGTH.

VI. *See* VISCOSITY INDEX.

virgin naphtha. A straight-run distillate produce from crude oil that falls within the BOILING RANGE of NAPHTHA.

visbreaking. *See* VISCOSITY BREAKING.

viscosity. The property of a liquid that determines its flow. Normally the higher the temperature of the fluid the more easily it flows (i.e. the lower its viscosity). Viscosity is measured in CENTISTOKES. This property is extremely important for oil and its products since it affects crude oil recovery rates and the cost of moving crude oil and products at the heavier end of the barrel. *See also* QUALITY, CRUDE OIL; QUALITY, DISTILLATE AND RESIDUAL FUELS; QUALITY, LUBRICATING OILS.

viscosity breaking (visbreaking). A refinery upgrading process. It involves lowering (i.e. breaking) of the viscosity of RESIDUE by CRACKING at fairly low temperatures. The product is distilled to separate gas, motor spirit and light distillate from a fuel oil with a significantly lower viscosity than that of the residue feedstock. *See also* REFINING, SECONDARY PROCESSING.

viscosity index (VI). An arbitrary number that is used to characterize the rate at which the viscosity of a lubricating oil changes as temperature changes. A high index number indicates relatively small changes in viscosity as temperature changes; the lower the index number the greater the change in viscosity. *See also* QUALITY, LUBRICATING OILS.

viscosity index improver. A compound that is added to lubricating oils in order to reduce the rate of change of VISCOSITY with temperature. *See also* QUALITY, CRUDE OIL.

visible trade. In the calculation of the BALANCE OF PAYMENTS the export and import of goods (e.g., oil or gas). This distinguishes trade in goods from the trade in services which are part of INVISIBLE TRADE.

VLCC (very large crude carrier). A crude oil tanker of between 150 000 dwt and 319 999 dwt classified as long range 3 for AFRA purposes. The existence of very great economies of scale in tanker building and operation coupled with rapid growth in the demand for oil transportation led in the 1960s and early 1970s to a rapid rise in the size of tankers and especially to the development of VLCCs. Since 1973, however, the VLCC has become less attractive to owners as large-sized cargoes are more difficult to obtain, and bunker fuel costs have risen. Many are now laid up or used for oil storage. *See also* TANKER TYPES.

volatility. The ease with which a product begins to vaporize. A liquid with a high VAPOUR PRESSURE (i.e. one that boils at a low temperature) is described as one with high volatility. *See also* QUALITY, AVIATION SPIRIT; QUALITY, MOTOR SPIRIT.

voyage charter. The hiring of a tanker for a single voyage.

W

W. *See* WATT.

waiting on cement (WOC). The period when drilling ceases to allow casing cement to set.

wash oils. Petroleum fractions that are used to absorb the heavier (more easily liquefiable) components in a gaseous mixture.

water coning. A situation when a well starts to produce water rather than oil. It arises because the OIL/WATER CONTACT tends to rise more rapidly in the vicinity of the well. *See also* PRODUCTION, NATURAL DRIVE MECHANISMS.

water drive. A type of natural recovery mechanism (*see* NATURAL DRIVE), enabling oil to be produced without artificial assistance. *See also* PRODUCTION, NATURAL DRIVE MECHANISMS.

water flooding. *See* WATER INJECTION.

water gas. A manufactured gas comprising carbon monoxide and hydrogen. The gas, which is produced by blowing steam through a bed of burning coke, has a relatively low CALORIFIC VALUE — about one-third of that of natural gas.

water hammer. The energy created by the sudden halting of a fluid in motion.

water injection (water flooding). A type of secondary recovery mechanism that maintains reservoir pressure in order to permit further recovery of oil. *See also* PRODUCTION, SECONDARY RECOVERY MECHANISMS.

watt (W). A unit of power equal to one JOULE per second. *See* APPENDIX.

wax content. Wax is a solid (at ambient temperatures) hydrocarbon found in some crude oils, particularly if the crudes are paraffinic (*see* PARAFFINIC CRUDE). If the wax content of a crude is high it can result in handling problems thereby reducing the value of the crude. *See also* QUALITY, CRUDE OIL.

waxes. Fractions that are extracted from heavy GAS OIL by means of VACUUM DISTILLATION. *See also* FINISHED PRODUCTS, WAXES; QUALITY, WAXES.

weathering. The loss of LIGHT FRACTIONS due to evaporation when oil is exposed to the atmosphere.

weather window. That part of the year when the weather can normally be expected to be sufficiently favourable for offshore work (e.g., pipeline laying, installation of a platform) to be carried out.

weighed average cost of capital. *See* OPPORTUNITY COST OF CAPITAL.

welfare economics. Economics that is concerned with the evaluation of alternative economic states or situations and the formulation of criteria for economic policy-making. It tries to set up criteria for judging whether economic welfare in one economic situation is higher or lower than in another. It deals with such questions as: what would be an ideal resource allocation?; what forms of social and economic organization can lead to such an allocation?; how can any economic change from one situation to another be evaluated?; if an ethical or social 'welfare function' is needed in order to help answer these questions, what sort of properties might be appropriate for it? Much of welfare economics is based on the concept of the

PARETO OPTIMAL, which offers a definition of efficiency — an allocation of resources is efficient if it would be impossible to make anybody better off without simultaneously making somebody else worse off — and conditions for achieving efficient allocations of resources. Distributional considerations are also important because in finding optimal social welfare it is necessary to choose between different efficient allocations of resources, each associated with a different distribution of economic welfare across the population. In this approach, a society's economic welfare is assumed to be dependent only on the happiness, satisfaction or welfare of the individuals in the society, as they perceive it for themselves. Thus each person is assumed to be the best judge of their own satisfaction. A Bergson social welfare function takes values which depend on all the variables that affect social welfare in different states of the world. It can be used to order different situations. However, an Arrow social welfare function is a collective choice rule or constitution that specifies how to pass from individual rankings of states of the world to society's ranking of these states, subject to a set of mild-looking conditions which turn out to be stronger than they seem; a result in economics known as Arrow's impossibility theorem.

Welfare economics offers some theoretical bases for economic policy-making, as well as some empirical techniques (e.g., COST–BENEFIT ANALYSIS) for evaluating policy choices. It has never been without controversy, not least because it cannot avoid dealing with issues where ethical values and ideological views are important. *See also* COMPENSATION PRINCIPLE; COMPETITION; PUBLIC GOOD; SECOND BEST.

well bore. The hole in the rock made by the drilling BIT. *See also* DRILLING.

wellhead. An installation at the top of a well that houses valves and other pressure-control equipment.

well log. A record that is kept for each well during drilling. It gives details of technical operations, production potential, etc. *See also* DRILLING.

wet barrel. The various terminologies for barrels — paper, prompt, wet — are derived from the development of a futures market (*see* FORWARD MARKET) in both crude oil and products. A paper barrel is crude oil or products with no specified delivery date: for example, an April paper barrel refers to a barrel for delivery sometime in April, although the seller must give 15 days' notice of the delivery date to the buyer. The delivery date at the seller's option can have a two days' range. Once a delivery date has been assigned the paper barrel becomes a wet barrel. If the delivery date (i.e. the conversion from paper to wet) is specified within a few days of the transaction then the barrel becomes a prompt barrel. The range between the prices provides a reflection of the state of the market. If stocks are low and supply constrained the prompt price would be higher than the wet price which in turn would be higher than the paper price. A glutted market with high stock and oversupply would exhibit the reverse pecking order.

wet gas. Natural gas that contains large amounts of easily liquefiable components (i.e. condensate). Often very wet gas is associated with the vicinity of an oil reservoir. When the condensate is removed the gas (sometimes called stripped gas) effectively becomes dry gas (i.e. mainly METHANE).

wet tree. *See* SUB-SEA WELLHEAD.

whipstock. A wedge set at the bottom of a hole that is used to cause the drilling string to change direction. It is used in DIRECTIONAL DRILLING.

white noise. A TIME SERIES of a variable that has virtually no discernable structure or pattern. In other words, the figures are purely random. A white noise series of observations is of no value in forecasting.

white oils. (1) The products of drastic refining of light lubricating oils that are colourless. These products are used for medicinal purposes and in the manufacture of toiletries. (2) Light crude oils that contain a high proportion of light distillates.

white oil ships. Ships used to carry light-coloured products up to gas oil.

white products. *See* BLACK PRODUCTS.

white spirit (turps substitute). A refined

distillate that comes between gasoline and kerosine. It is used mainly as a paint thinner. In the USA it is called petroleum spirit. *See also* FINISHED PRODUCTS, SOLVENTS.

wide-cut gasoline. *See* FINISHED PRODUCTS, AVIATION FUELS.

wildcat drilling. *See* WILDCAT WELL.

wildcat well. An exploratory well that is used to establish whether oil or gas is present in an unproven area. *See also* DRILLING.

wild gasoline. A very light petroleum spirit which contains components that are normally gaseous at atmospheric temperatures.

willingness to pay. A measure of the consumer's expected satisfaction from the opportunity to consume an extra unit of a commodity. It indicates the person's marginal valuation of the extra unit and can be represented by a point on the consumer's DEMAND CURVE. The area under the demand curve, yields an estimate of total willingness to pay for a given total quantity, whereas the area under the demand curve between two prices indicates the benefits expected from a price change (e.g., lower gas prices). Where a uniform price is charged, a consumer with a downward-sloping demand curve will be willing to pay more for some of the units than the uniform price which actually has to be paid for all units consumed. This difference, a triangular area under the demand curve is the consumer's surplus. Estimates of willingness to pay are used in COST-BENEFIT ANALYSIS to provide measures of benefits to society expected from public investment projects (e.g., a new dam or power station). The most appropriate estimates use 'compensated' demand curves, which take account only of the SUBSTITUTION EFFECT of a price change.

windfall profits. A large increase in profits resulting from an unexpected increase in price. Since the gainer has apparently done nothing to deserve such extra benefits, windfall profits are favourite targets for government taxation. In both the oil shocks of the 1970s (*see* FIRST OIL SHOCK; SECOND OIL SHOCK), for example, such profits accrued to the holders of crude oil STOCKS, which almost overnight increased in value. Similarly, since PRICE DEREGULATION in the USA was expected to lead to such profits, part of the deregulation package included such a tax on the oil companies.

WOC. *See* WAITING ON CEMENT.

WOCA (world outside communist areas). The world excluding the communist bloc. It is frequently used by those who find the term 'free world' somewhat misleading.

WOCANA (world outside communist areas and North America). Commonly used to represent the region in which the international oil industry operates.

working capital. That part of current assets that is financed from long-term funds. In effect, it represents the difference between current assets and current liabilities.

workover. Any work on a well (such as routine maintenance) performed after completion.

work programmes. *See* AUCTION SYSTEM OF LICENSING.

Worldscale (Worldwide Tanker Nominal Freight Scale). Worldscale superseded both INTASCALE and the US ATRS in 1969 as a schedule of nominal tanker freight rates. In November 1971, it was decided to quote rates entirely in US dollar terms. The scales are revised regularly — initially once per year, but more recently twice a year — in the light of changes in bunker fuel prices and port charges. Actual contract rates are negotiated in terms of an index with the flat rate calculated for each route being set at WS = 100, and these are quoted in the shipping press. To find the actual freight cost per ton of oil transported from Ras Tanura to Rotterdam via Cape of Good Hope route, for example, multiply the flat rate for this route ($27.8 in 1985) by the quoted market rate of 20.1 percent (June 1985) to get a rate of $5.59 per ton. *See also* AFRA.

World Tanker Nominal Freight Scale. *See* WORLDSCALE.

Worms system. *See* LNG/LPG carriers.

X Y Z

X efficiency. A term used by economists to indicate the general managerial and technical efficiency of a company. A company operating at maximum X efficiency is effectively minimizing costs. X inefficiency is therefore when a company, for whatever reason, is producing at a higher cost than may be necessary. It is suggested that in times of boom, companies accept X inefficiency in order to maximize TURNOVER. Such an argument has been used with respect to the relatively high cost of developing North Sea oil capacity. The very high oil price during and immediately after the SECOND OIL SHOCK encouraged oil companies to develop the capacity quickly almost irrespective of cost. A lower price could encourage cheaper methods of development. It is for this reason that the lower oil prices in 1986 (and following) might have less of an inhibiting effect on further developments in the higher cost areas such as the North Sea than many believe.

xylene. A important AROMATIC. It is used as a petrochemical feedstock. *See also* CHEMICALS; PETROCHEMICALS.

yield. (1) The proportion of products obtained from REFINING. (2) The income from an ASSET measured as a proportion of its current value. It is most commonly associated with financial assets. Thus, for example, the dividend yield is the current dividend expressed as a percentage of the market price of a security. There are various definitions of yield: nominal yield is the nominal (coupon) annual interest divided by the current market price; redemption yield, or yield to maturity (usually calculated for fixed-interest securities) is the rate at which we need to discount cash flows over the remaining life of the security, including the redemption value, in order to equate their sum to the purchasing price.

yield analysis. A measure of the value of crude oil by the value of its YIELD. It is essentially netback (*see* NET PRODUCT WORTH) pricing.

yield shift process. The alteration of the REFINING CONFIGURATION of a plant or region to produce different proportions of products. *See also* REFINING.

zero sum game. A situation, usually a bargaining process, in which a gain by one side leads to a loss by the other. Put simply, the size of the 'cake' to be divided is fixed. Therefore the gains and losses of those sharing the cake will add up to zero. *See also* THEORY OF GAMES.

Chronology

1859
Oil production in the USA commences following the well drilled by Edwin Drake at Titusville, Pennsylvania.
Building of the Suez Canal commences.

1861
American Civil War starts.

1862
Oil production commences in Canada.

1864
Schleswig–Holstein war starts.

1865
American Civil War ends.

1866
Austro–Prussian War.

1867
USA purchases Alaska from the Russians for $7.2 million.

1868
Meiji restoration begins in Japan.

1869
Suez Canal formally opened.

1870
John D. Rockefeller and associates incorporate the Standard Oil Company in Ohio.
Franco–Prussian War starts.

1871
Franco–Prussian War ends.
Trade Unions legalized in Great Britain.

1872
First oil concession granted in Persia.

1875
Great Britain purchases Khedive's shares in Suez Canal.

1876
Bell invents the telephone.

1878
Edison and Swan produce the first successful incandescent electric light bulb.
First national oil company (Compañía Nacional Minera Petrolia del Tachira) formed in Venezuela.

1882
Triple Alliance (Germany, Austria, Italy) formed.
Cairo occupied by British troops.
Standard Oil Trust created.

1884
France establishes complete protectorate in Indo–China.
Berlin Conference defines rights of European powers in Africa.
Parson invents his turbine.

1885
Crude oil production in Indonesia commences from Toenggal Well in Northern Sumatra.

1886
Burma Oil, oldest British international oil company, established as a joint stock company to conduct oil operations in India and Burma.
Daimler produces his first motor car.
Canadian Pacific Railway completed.
Gold discovered in the Transvaal.

1888
Neutrality of Suez Canal declared.

Oil concencession granted by Ottoman Empire to a group controlled by Deutsche Bank.

1890
Sherman Anti-Trust Law passed in the USA.
Royal Dutch Petroleum Company formed.

1891
The United States of Brazil formed.

1893
Oil production commences in Peru.

1895
Kiel Canal opened.
Marconi sends message over one mile by wireless.
Roentgen discovers X-rays.

1896
Gold discovered in the Klondike.

1897
Shell Transport and Trading Company formed.

1898
Radium discovered by Marie and Pierre Curie.

1899
Boer War starts.

1901
Trans-Siberian Railway opened.
Lucas Gusher discovered at Spindletop, Texas — the 'first great well drilled by an engineer'.
D'Arcy awarded oil concession by Persian government.

1902
Boer War ends.

1903
First controlled flight in heavier-than-air machine.
First Exploration Company takes control of the drilling programme in Persia.

1904
Turkish Sultan Abdul Hamid II transfers Mesopotamian oil lands from state ownership to his own privy purse.
Russo–Japanese War starts.
Norway separates itself from Sweden.

1905
Concessions Syndicate Ltd formed taking over the First Exploration Company's assets in Persia; its director is D'Arcy.
First Morocco crisis.
San Francisco destroyed by earthquake and fire.
Simpleon Tunnel opened for rail traffic.
Vitamins discovered by F.G. Hopkins.

1907
Anglo–Russian Convention recognizes Persia's independence, but divides Persia between British and Russian spheres of influence.
Gulf Oil Corporation formed to acquire the stock of two other US companies.
Royal Dutch and Shell Transport merge to create Royal Dutch Shell.

1908
Crude oil production commences in Trinidad.
Oil found in Masjid-i-Sulaiman, Persia.

1909
Phenol–formaldehyde resins discovered.
Anglo–Persian Oil Company formed.
Bleriot flies across the English Channel.

1910
Young Turk Revolution in the Ottoman Empire; Sultan Abdul Hamid overthrown.

1911
Organization of African and Eastern Concessions Ltd formed to get access to Mesopotamian oil.
Anti-Trust Decree declared by US Supreme Court breaks up the Standard Oil Trust.
Tripoli taken by Italy from Turkey.
US gasoline sales exceed kerosine sales.

1912
African and Eastern Concessions Ltd changes name to Turkish Petroleum Company.
Cushing oil-field discovered in Oklahoma.
Henry Ford uses assembly line techniques to build the Model T.

1913

Anglo–Persian Oil Company commission pipeline to Abadan Das and build a small refinery at Abadan on the Shatt-al-Arab.

Crude oil production commences in Brunei.

Depletion allowances introduced in the USA.

1914

British Foreign Office sponsor an agreement with Turkish Petroleum Company fusing Deutsche Bank, Anglo–Saxon Petroleum Company Ltd and Anglo–Persian Oil Company.

World War I starts.

Panama Canal opened.

Cellulose acetate used industrially.

British government invests £2 million in Anglo–Persian Oil Company and enters into long-term contract for supply of oil to Royal Navy.

1917

Crude oil production commences in Ecuador and Venezuela.

Bolshevik Revolution.

Balfour Declaration.

1918

Great Britain gains effective control of 75 percent of Turkish Petroleum Company following expropriation of the Deutsche Bank's 25 percent holding, which was placed in the hands of a British trustee.

World War I ends.

1919

First drafts of Anglo–Persian agreement make attempts to stabilize Persia's and Great Britain's 'alliance' and keep Russia out of Persia.

Gold parities for currencies abandoned.

Einstein's theory of relativity confirmed experimentally.

J. Alcock and A.W. Brown fly across the Atlantic Ocean.

Treaty of peace signed with Germany at Versailles.

1920

Petrofina Oil Company established in Belgium.

San Remo Conference discusses Middle East oil and politics.

National revolution in Iraq.

First meeting of the League of Nations.

Prohibition (of alcohol) in USA starts.

1921

Ratification of 1919 agreement in Iran after Reza Khan coup d'etat.

Oil production commences in Colombia.

Anti-knock properties of tetraethyl lead discovered.

First Iraqi government formed.

1922

Exploration incentives given in Venezuela by a new petroleum law.

1924

Lenin dies.

Compagnie Française des Petroles established, concluding an agreement with the French government which placed under its control the French international petroleum interests, including its interests in Turkish Petroleum Company.

Urea–formaldehyde resins discovered.

1925

Iraq Petroleum Company confirms claims to Mesopotamian oil with new concession approved by Iraq's King Faisal.

Sheikh of Bahrain grants a two-year exploration licence to the Eastern and General syndicate.

Great Britain creates Oil Fuel Board to keep the question of oil and tanker requirements under constant review.

Great Britain returns sterling to pre-World War I parity.

1926

Azienda Generale Italiana Petroli (Agip) created in Italy to look for oil.

General Strike in Great Britain

1927

Second largest oil-field in Persia discovered at Haft Kel.

Shell obtains 23.75 percent interest in Turkish Petroleum Company.

Kirkuk oil-field discovered in Iraq.

Lindbergh flies solo across Atlantic Ocean.

Kingsford–Smith flies across Pacific Ocean.

1928

Standard Oil Company of New Jersey, Anglo–Persian and Shell mastermind the Achnacarry Agreement.

Gach Saran oil-field discovered in Persia.

1929

Black Tuesday on Wall Street; US stock market collapses.

1930

Smoot–Hawley Act raises tariffs which invites foreign retaliations and triggers the Great Depression.

Socal obtains concession in Bahrain.

1931

France refuses to hold sterling as monetary reserves; Great Britain forced to break sterling–gold price link.

1932

King Abdul Aziz Ibn Saud assumes the title of King of Saudi Arabia.

Iraq grants concession to the British Oil Development Company Ltd later taken over by the Iraq Petroleum Company.

Iraq becomes independent.

Persian government cancels concessions to Anglo–Persian Oil Company.

British government requests the League of Nations to include the Persian cancellation on their agenda.

Crude oil production commences in Bahrain.

Franklin D. Roosevelt elected President of the USA (the 'New Deal').

1933

Hitler appointed Chancellor of Germany.

Roosevelt refuses to negotiate currency stabilization or international tariff reduction at London Economic Conference.

New concession agreement signed between Persia and the Anglo–Persian Oil Company.

US Congress approves leaving the gold standard.

Socal signs a concession agreement in Saudi Arabia, and assigns it to its wholly owned subsidiary California Arabian Standard Oil Company.

Prohibition in the USA ends.

1934

Ruler of Kuwait grants a concession to Kuwait Oil Company, approved by the British government, covering the whole of Kuwait.

1935

Acrylic resins marketed.

Persia changes its name to Iran.

Connally Hot Oil Act enables the Texas Railroad Commission to proration production.

1936

Crude oil production commences in Saudi Arabia.

Spanish Civil War starts.

Socal and Texaco begin cooperation when Socal agree to buy one-half interest in Texaco's marketing facilities east of Suez.

1937

Polyvinyl resins marketed.

First commercial cat cracker commissioned.

New economic downturn begins in USA.

1938

Californian Arabian Standard Oil discovers oil in Saudi Arabia and signs a supplemental agreement extending the concession's area and life.

Mexico nationalizes foreign oil companies.

1939

World War II commences.

Crude oil production commences in China.

The Agha-Jari oil-field discovered in Iran.

BP discovers UK onshore oil at Eakring, Nottingham.

1940

British government help compensate Saudi Arabia for loss in oil royalty from Californian Arabian Standard Oil. Later the US government also provides aid.

Nylon produced.

1941

USA introduces controls on petroleum exports.

British and Soviet forces occupy Iran.

Japanese attack Pearl Harbour.

1943

Foundation of United Nations discussed at Moscow Conference.

1944

Californian Arabian Standard Oil changes name to Arabian–American Oil Company (Aramco).

Crude oil production commences in Algeria.

United Nations comes into existance, with 50 nations joining.

Allies invade Europe.

1945

May. World War II ends in Europe.

June. United Nations charter signed in San Francisco.

July. Aramco organizes the Trans-Arabian Pipeline Company (Tapline) to construct a pipeline from the Mediterranean Sea to the Gulf.

August. Atomic bomb exploded at Hiroshima.

World War II ends in Asia after Japan surrenders.

December. International Monetary Fund and the International Bank for Reconstruction and Development (World Bank) formed.

1946

Crude oil production commences in Kuwait.

League of Nations formally wound up.

1947

January. British coal mines are nationalized.

June. General Marshall proposes US aid to stimulate recovery in Europe ($17 million April 1948–June 1952).

July. Federal District Court of Southern New York, in response to a request from Aramco, quashed the subpoena obtained by James Moffat against Aramco earlier in the year when the court had ordered Aramco to produce the original concession of 1933.

Saudi Arabian government grants pipeline construction and operation to the Trans-Arabian Pipeline Company.

October. General Assembly of the United Nations opens in New York.

1948

Exxon and Mobil become partners in Aramco.

Arab–Israel War threatens Tapline.

British Railways nationalized.

Ghandi assassinated.

April. British electricity industry nationalized.

First European aid shipments sail from USA under the 'Marshall aid' programme.

Organization for European Economic Co-operation (OEEC) established.

May. State of Israel declared.

July. Berlin airlift carried out.

1949

Iran's first seven-year plan adopted.

January. Syria and Lebanon reach agreement to share pipeline royalties on Tapline.

February. Pacific Western Oil Company sign concession agreement with Saudi Arabia for the Neutral Zone.

Crude oil production commences in Qatar.

April. USSR completes control of Eastern Europe.

North Atlantic Treaty Organization formed.

May. UK gas industry nationalized.

July. Anglo–Iranian Oil Company and Iran sign agreement supplemental to 1933 concession.

September. UK government devalues sterling from $4.03 to $2.80.

Mao Tse-Tung proclaims People's Republic of China.

October. Creation of Benelux, economic union of Belgium, Luxembourg and The Netherlands.

1950

Tapline is completed.

World crude oil production attains 10 million barrels per day.

May. Petrol rationing ends in UK.

June. Korean War starts as North Korean troops advance into South Korea.

December. Saudi Arabia and Aramco sign an agreement inaugurating the 50/50 sharing agreement in the Middle East.

1951

Iran nationalizes its oil industry; state-owned National Iranian Oil Company formed.

April. Treaty setting up the European Coal and Steel Community (forerunner of the European Economy Community) signed in Paris.

December. Libya becomes an independent state.

Kuwait announces Decree No 5, including a tax rate of up to 50 percent for the Kuwait Oil Company.

1952

Jet airliner (UK, de Havilland Comet) enters services.

February. Iraq Petroleum Company signs agreement with Iraq; government adopting 50/50 sharing.

July. Military coup d'etat in Egypt.

October. First UK atomic weapon exploded.

November. US hydrogen bomb exploded.

General Eisenhower wins US presidential election.

1953

Abu Dhabi's first oil-field (Bab) established.

Ente Nazionale Idrocarburi (ENI) formed.

Major anti-trust suit is brought against five US major oil companies. The case lasted until 1968.

March. Stalin dies.

July. Korean War ends.

British Steel Industry denationalized.

December. Agreement signed for laying first transatlantic telephone cable.

1954

War of National Independence of Algeria starts.

Agreement between Iran and consortium of major oil companies that succeeded the Anglo–Iranian Oil Company.

Crude oil production commences in Neutral Zone.

February. First breeder pile in operation in UK at Harwell.

April. UK withdraws from Suez Canal Zone.

May. Dien Bien Phu falls to Viet-Minh.

June. First electric power station using atomic energy begins working in USSR.

October. UK, France, USA and USSR agree to end the occupation of Germany.

Exports of Iranian oil to world markets restored. The Iranian Consortium of eight oil companies starts operation.

1955

Libya enacts its first petroleum law.

February. Turkey and Iraq sign Baghdad Pact.

Plans announced to build 12 atomic power stations in UK during next 12 years.

March. Iraq Petroleum Company and Iraq sign agreement lowering discount to 2 percent and 2 shillings per ton for Iraq Petroleum Company.

June. USA and UK agree to cooperate on atomic energy.

August. International conference on peaceful uses of atomic energy held.

November. Iraq joins Baghdad Pact.

Iraq Petroleum Company and Syria agree to supplement the Iran 1931 agreement on pipelines.

1956

Oil discovered in Nigeria.

Middle East Emergency Committee created.

Tunisia becomes independent from France, Morocco from Spain and France, and Sudan from UK and Egypt.

May. First atomic power station in UK starts up at Calder Hall.

June. Egypt states that it will not renew Suez Canal Company's concession after the 1968 expiry date.

UK troops leave Suez Canal base.

July. President Nasser nationalizes Suez Canal.

September. Submarine telephone cable linking UK and USA comes into operation.

October. Suez War starts.

UK, refused entry to the EEC, proposes to set up a free trade area.

Israel attacks Egypt.

November. OEEC Petroleum Emergency Group (OPEG) formed for Europeans.

December. Petrol rationing in UK.

United Nations troops start clearing Suez Canal.

1957

January. OPEG makes first allocation of emergency oil to OEEC member states.

Eisenhower doctrine for Middle East announced.

Eisenhower asks US Congress for power to use US forces in Middle East in the event of USSR aggression.

Trans-Iranian oil pipeline from Adaban to Teheran completed.

March. Small ships able to use the Suez Canal again.

Treaty of Rome creates EEC.

July. Electricity Bill enacted in UK, appointing a new Central Electricity Generating Board and Electricity Authority.

Iran's Petroleum Act opens the possibility of joint ventures.

October. First earth satellite launched by USSR.

November. Bahrain created an independent Arab state under UK protection.

1958

Crude oil production commences in Alaska, Israel, Angola, Gabon and Nigeria.

January. First US earth satellite (Explorer I) lauched.

March. Intergovernmental Maritime Consultative Organization (IMCO) formed for cooperation among governments on shipping matters.

May. Military and colonist insurrection in Algeria.

Nuclear reactor at Dounreay, UK, starts up.

June. Indonesia bans Shell operations.

UK Clean Air Act bans emission of dark smoke into the atmosphere.

July. Iraq monarchy overthrown during Iraq Revolution; Colonel Qassim takes over power.

December. Last of the four nuclear reactors at Calder Hall, UK brought into operation.

1959

Castro takes over in Cuba.

Oil discovered in Libya.

Groningen gas-field discovered in The Netherlands.

January. Tariff reduction of 10 percent and quota enlargements in the EEC.

March. USSR and Iraq sign economic and technical aid agreement.

USA replaces voluntary restrictions with mandatory oil import quotas.

April. Venezuela creats Coordination Commission for Conservation of Hydrocarbons to act as watchdog over oil sales and production.

June. Iraq withdraws from Sterling Area.

August. Announcement of plan for oil pipeline network between USSR and East European countries.

November. Dounreay fast breeder reactor comes into operation.

December. Twelve-power treaty on Antarctica signed in Washington.

Opening of 400-mile Sahara pipeline.

1960

Kuwait National Petroleum Company (KNPC) formed.

OPEC formed.

Crude oil production commences in Nigeria.

January. Army rising in Algeria.

March. Last steam locomotive of British Railways named.

May. Army seizes control in Turkey.

August. Gabon becomes independent from France.

October. Nigeria becomes independent from UK.

1961

Kuwait becomes (*de jure*) independent of UK; Iraq attempts a takeover.

Crude oil production commences in Libya.

Tidal power used to produce electricity at La Rance, France.

April. Major Yuri Gagarin makes first successful flight into space.

May. Alan Shepherd makes first US flight into space.

June. Iraqi government announces that Kuwait is part of Iraq.

July. UK forces land in Kuwait following appeal from ruler.

September. Kuwait becomes a member of League of Arab States.

December. Gas-field discovered at Stochtern, The Netherlands.

Law 80 issued by Iraqi government effectively expropriating most of the Iraq Petroleum Company concession areas. It leaves the company with only its producing fields.

1962

Death of Enrico Mattei, creator of ENI, in a plane crash.

Exploration licences awarded for Danish sector of North Sea.

Crude oil production commences in Abu Dhabi.

March. War in Algeria ends; Algeria gains independence.

Oil companies in Iraq agree to comply with Law 80.

April. International agreement by 40 countries to stop oil pollution of seas and beaches.

July. First experimental communications satellite (Telstar) launched.

August. Maiden voyage of the first nuclear-powered merchant ship (*Savannah*).

September. Imamate overthrown in Yemen; a republic proclamined.

November. General Petroleum and Mineral Organization (Petromin) created in Saudi Arabia.

1963

Dubai Petroelum Company founded by Continental Oil Company of the USA.

Denmark grants the first North Sea concessions.

Pipeline between USSR and East European countries opened.

January. UK refused entry to EEC.

February. Iraqi government overthrown; General Kassin executed.

March. Beeching Report on British Railways published.

April. Saudi Arabia and Aramco sign new marketing allowance agreement.

Iraq and United Arab Republic form union.

May. Kuwait joins the United Nations.

August. Partial nuclear test ban treaty signed in Moscow by America, USSR and UK.

OPEC member countries appoint Secretary-General Rouhani to negotiate with the oil companies on the members' behalf regarding royalty expensing.

September. United Nations calls for ban on arms and oil supply to South Africa.

New state of Malaysia formed.

October. Nigeria becomes a republic.

November. President Kennedy assassinated in Dallas, Texas.

1964

Oil companies, Middle East governments and OPEC begin negotiations over new royalty rate payments.

Iraq National Oil Company (INOC) formed.

Algerian liquefied natural gas delivered to the UK.

November. UK grants licences for a full-scale search for oil and gas on the continental shelf.

King Saud of Saudi Arabi deposed; his brother Faisal acceeds.

Lyndon Johnson wins US presidential election.

1965

Dr al Bazzaz become Iraq's first civilian prime minister since 1958; the appointment lasts less than one year.

February. First US retaliatory raids against North Vietnam.

Syrian government and USSR sign an agreement for drilling in North Syria.

March. EEC propose, as from 1 July 1967, all members' import duties and levies paid into a budget.

Syria nationalizes the country's nine oil products distribution companies.

April. Saudi government and AUXIRAP sign an agreement granting Red Sea exploration and production rights.

June. President Ben Bella of Algeria deposed by Revolutionary Council under Colonel Boumedienne.

September. West Sole gas-field discovered in North Sea.

Abortive coup d'etat in Indonesia led by the Communists.

Libya establishes the Petroleum Prices Negotiating Committee to negotiate prices and discounts with the oil companies.

November. Unilateral Declaration of Independence in Rhodesia; United Nations Security Council urged oil embargo.

December. Oil rig *Sea Gem* collapsed in North Sea.

Standard Oil Company of New Jersey buys distribution network of Agip — the marketing subsidiary of Ente Nazionale Idrocarburi — in the UK.

First Norwegian oil concession granted in the North Sea.

1966

Algerian goverment's own company (Sonatrach) builds pipeline from Hassi Messaoud to Arzew.

'Production sharing contract' in operation through Pertamina in Indonesia.

Socony Mobil changes its name to Mobil Oil thus dropping the last hint of its Standard Oil origins.

Discovery of first oil-field, Fateh, in Dubai.
Leman, Indefatigable and Hewett natural gas-field discovered in the UK sector of the North Sea.
March. Labour wins UK General Election.
Executive power of Indonesian government given to General Suharto.
August. Indonesian–Malaysian peace agreement signed.
Agreement between National Iranian Oil Company (NIOC) and Enterprise de Recherches et d'Activités Petrolieres (ERAP) for an oil joint venture.

1967
Venezuelan government replaces concessions to oil companies by service contracts.
First state-owned oil company in Venezuela — Corporación Venezolaña de Petroleo (CVP).
Tax discount allowances on Mediterranean crude exports eliminated.
Crude oil production commences in Oman.
February. President Sukarno of Indonesia surrenders rule to General Suharto.
March. BP produces gas from West Sole.
Torrey Cannon wrecked off Cornwall; 117 000 tons of crude oil spilt.
April. Syria and Israel clash.
May. Egypt puts armed forces on a state of emergency.
President Nasser closes Gulf of Aqaba to Israeli shipping.
June. Arab–Israeli War starts.
Arab embargo on oil exports to USA and The Netherlands.
July. Nigerian Civil War stops oil exports.
United Nations Security Council agrees to place observers on Suez Canal.
UK announces withdrawal from East of Suez by mid-1970s.
September. Publication of Law 123 leads to complete reconstruction of the Iraq National Oil Company.
November. Devaluation of sterling to $2.40 to the pound.
Contract lasting 20 years between INOC and ERAP announced in Iraq.

1968
National Oil Company formed in Libya.
Prudhoe Bay oil-fields discovered in Alaska.
January. OAPEC formed by Saudi Arabia, Kuwait and Libya.
June. Cod gas-field discovered in the North Sea.
Algeria becomes a member of OPEC.
July. EEC abolishes last internal industrial tariffs; common external tariff comes into effect.
Iraq's Arif overthrown by Ba'th (Arab socialist movement).
Novemeber. Richard Nixon wins US presidential election.

1969
Offshore Californian oil leak from an oil-field causes extensive environmental damage and effectively halts US offshore exploration for several years.
April. De Gaulle resigns as President of France.
May. Tapline, from Jordan to Syria, shut down for 112 days due to sabotage.
September. Libyan Arab Republic formed when the king was deposed.
Colonel Ghaddafi becomes Chairman of the Libyan Revolutionary Command Council.
November. Start of negotiations where Algerians demanded revision of ASCOOP fiscal terms to include 55 percent tax based on posted prices.
December. Montrose oil-field discovered in UK sector of North Sea.
Ekofisk oil-field discovered in Norwegian sector of North Sea.

1970
European Conservation Year.
January. Nigerian Civil War ends.
Colonel Ghaddafi advises 21 oil companies to accept Libya's demands for an increase of at least 10 cents in postings.
February. Opening of Eliat to Ashkelon oil pipeline to by-pass the Suez Canal.
May. Tapline 'accidently' ruptured near Deraa in Syria by a bulldozer; repair initially barred by Syria.
Iraq, Libya and Algeria form Mediterranean oil exporters' common front to negotiate with oil companies.
Libyan government introduces series of allowable production restrictions.
June. Conservatives gain overall majority in UK General Election.

July. Algeria raises tax reference price for French oil companies by 77 cents a barrel to $2.85 per barrel.

August. Deputy Libyan Prime Minister Jallud calls for final offers from Exxon and Occidental for price increases.

September. Death of President Nasser.

Occidental President, Armand Hammer, signs agreement with Libya for 30 percent increase in posted price, an 8 percent increase in income tax rate and production rate increases.

Continental Marathon and Amerada–Hess sign agreement with Libya, but Shell refuses.

Libya prohibits Shell from lifting its share of the Oasis Group's export.

Texaco and Socal agree with Libya.

October. Exxon, Mobil and BP agree with Libya (8 October).

Shell agrees with Libya (16 October).

Forties oil-field discovered in UK sector of North Sea.

OPEC meets in Caracas and maps out its pricing strategy.

1971

Production from offshore fields in Indonsia starts.

Abu Dhabi establishes its Fund for Arab Economic Development (ADFAED).

January. Syria finally allows repair to the Tapline; reopening follows new agreements on transit fees and a lump sum payment to Syria.

Jallud calls oil companies to Tripoli to listen to Libyan demands under the Caracas resolutions.

Libyan Producers' Oil Agreement signed by oil companies in Libya to give stronger opposition to Libya.

Oil and gas production starts in the Norwegian sector of North Sea.

OPEC's six Gulf States raise threat of oil embargos unless the oil companies meet their price demands.

February. Teheren Agreement to increase crude oil prices for the Gulf oil producers announced.

Algeria nationalizes 51 percent of French oil interests.

April. Tripoli Agreement increases crude oil prices for Mediterranean oil producers.

June. Lifting of 21-year embargo on trade with China by USA.

July. Brent oil-field discovered in UK sector of North Sea.

August. USA suspends convertibility of dollars into gold and introduces internal price controls for crude and refined products.

September. Twenty-fifth OPEC Conference in Beirut. OPEC advances compensation claims due to devaluation of the dollar and for participation.

Qatar becomes independent and joins the United Nations and the Arab League.

November. Abu Dhabi National Oil Company (ADNOC) is created.

December. Libya nationalizes BP oil investments in retaliation for UK policy in the Gulf.

United Arab Emirates (UEA) formed, joining the Arab League on 6 December.

1972

United Arab Emirates joins the United Nations.

Oil discovered in the Reforma province of Mexico.

January. Abu Dhabi government acquire an initial stake in Abu Dhabi Petroleum Company (ADPC).

Iraq Petroleum Company cutback liftings of Iraqi crude exports from Mediterranean terminals.

UK miners' stike begins. UK.

Ireland, Norway and Denmark sign EEC treaty.

February. State of emergency in power crisis following UK miners' strike; large-scale power cuts begin.

Military coup d'etat in Ecuador.

March. EEC currency 'Snake' introduced.

June. UK and Ireland leave the 'Snake'.

Iraq Petroleum Company nationalized.

September. Norway votes against joining the EEC.

October. General Agreement on Participation signed between the Gulf States and the oil companies.

November. President Nixon elected for second term of office.

1973

January. Denmark, Ireland and UK join EEC.

Vietnam cease-fire agreement signed.

March. Suharto re-elected as President of the Republic of Indonesia.

April. VAT introduced in UK.

Abolition of import quotas for crude oil and refined products in the USA.

June. Algerian refinery built at Arzew by Japanese consortium.

July. National Iranian Oil Company took over control of oil industry in Iranian Consortium area.

August. Libya nationalizes some of the smaller oil companies' producing interests.

September. Dr Henry Kissinger becomes US Secretary of State.

October. Gulf States under Iranian chairmanship abrogate Teheran and Tripoli Agreements and increase oil prices from $3 to more than $5 per barrel.

Arab–Israeli (Yom Kippur) War starts.

Arab States announce the Arab Oil Embargo which involves cuts in oil supplies until Israel withdraws from occupied territories.

November. Arab Oil Embargo increases production cutback to 25 percent.

Iranians auction oil, selling at over $17 per barrel.

December. UK Prime Minister Edward Heath imposes a three-day week to begin 1 January 1974 in response to disruptions in energy supplies.

OPEC meets in Teheran and agrees to increase the oil price to $11.65.

Kuwait signs agreement with BP and Gulf giving 60 percent participation in the Kuwait Oil Company.

1974

British National Oil Company (BNOC) formed.

January. UK Department of Energy established; Lord Carrington becomes UK Secretary for Energy.

USA places controls on certain oil exports and issues construction permit for trans-Alaska pipeline.

February. USA calls Washington Energy Conference, comprising 13 industrial nations, to discuss cooperation in the energy field.

March. Majority of Arab states decide to lift the oil embargo against USA.

Minority Labour government takes office in the UK.

April. Nigeria takes 55 percent stake in oil production.

June. Biggest explosion in UK since World War II at Flixborough chemical works.

July. UK government announces choice of UK-designed nuclear reactor for third stage of programme in 1980s.

August. President Nixon resigns following the Watergate scandal.

Gerald Ford sworn in as 38th US President.

Norwegians announce discovery of Statfjord field.

State-owned Qatar General Petroleum Corporation (QCPC) set up to operate all of the hydrocarbon sector.

October. General Election in UK; Labour returned with overall majority of three.

November. International Energy Agency (IEA) set up to replace Energy Coordinating Group set up in February.

1975

Algerian National Charter approved.

Dubai government takes control of its own oil industry.

Algeria reveals proposal for OPEC guarantees over oil supply and pricing at Joint Conference of OPEC ministeres in Algeria.

February. EEC Foreign Ministers approve resolution to decrease from 63 percent to 50 percent and if possible 40 percent imported energy content of the total EEC energy consumption.

March. King Faisal of Saudi Arabia assassinated; succeeded by Khaled.

April. First European Nuclear Energy Conference held in Paris.

June. Suez Canal reopened.

OPEC decides to adopt the SDR as unit in which crude oil would be priced due to poor state of the dollar; the recovery of the dollar meant that this change never occurred.

July. Military coup d'etat in Nigeria.

September. International Energy Agency countries agree to raise emergency oil stocks from 60 to 70 days' supply by January 1976.

October. OPEC increases oil price by 10 percent.

Inauguration at Teesside of first pipeline from Norwegian Ekofisk field.

November. Inauguration of North Sea oil from BP Forties field.

Leaders from UK, USA, France, West Germany, Japan and Italy met at

Rambouiller, near Paris for summit conference on economic matters.

Thirteen ministers kidnapped at OPEC conference in Vienna by terrorist gang led by 'Carlos'.

1976

Corporation Estatal Petrolera Ecuatariana (CEPE), Ecuador's state oil company, takes over field operations.

Kuwait sets up Reserve Fund for Future Generations.

January. Petroleos de Venezuela SA (PDVSA) created to enable the nationalization of Venezuela's oil industry.

June. OECD adopts a code of conduct on bribery and disclosure of information for multinational companies.

Leaders from UK, France, West Germany, Italy, Canada, USA and Japan meet in Puerto Rico for second Economic Summit Conference.

September. Mao Tse-Tung dies.

October. Six Arab leaders meet in Riyadh to draw up plans for ending civil war in Lebonan.

November. Jimmy Carter elected President of USA.

December. UK government estimates the UK oil reserves could be worth more than £300 billion at current prices.

OPEC splits over price in Doha. As from 1 January 1977 Saudi Arabia and United Arab Emirates raise price by 5 percent; other OPEC members agree to 10 percent increases.

1977

Gulf sells out Lake Agrio field in Ecuador to CEPE.

Nigerian National Petroleum Company (NNPC) formed.

Abu Dhabi Marine Operating Company (Adma-opco) takes over Abu Dhabi acreage.

January. One of the severest winters of the 20th century in the eastern states of the USA.

February. UK government to spend £6 million during four years on solar energy research and development.

March. Newly established General People's Congress (GPC) proclaims 'Jamahiriya' — the state of the masses — in Libya.

April. President Carter announces reprocessing techniques and fast breeder programme to slow down.

Norwegian Ekofisk Bravo oil-field blowout lasts 181 hours until stopped.

May. Third Economic Summit Conference held in London, attended by Carter (USA), Fukada (Japan), Schmidt (West Germany), Giscard d'Estaing (France), Callaghan (UK), Trudeau (Canada) and Andreotti (Italy).

Ecuador builds refinery at Esmeraldas.

June. Four-day conference on International Economic Cooperation (North–South Dialogue) ends with very limited success.

October. Meeting in Israel between President Sadat of Egypt and Prime Minister Menachem Begin of Israel.

December. OPEC countries freeze oil prices for six months.

1978

January. Second line of Sumed comes into operation.

State of the Union address by President Carter urges the passing of Energy Bill.

March. *Amoco Cadiz* runs aground off the Brittany coast; major pollution of coastal waters results.

Saudi government decides to lift allowable production to 9.5 million barrels per day for first year's quarter.

Suharto re-elected as President of Indonesia.

June. OPEC agrees to peg oil prices until December.

July. President Carter announces new US energy plans including limiting oil imports.

Summit Conference in Bonn of leading industrial nations ends in agreement on strategy to meet economic problems.

September. Martial law declared in Teheran and 11 Iranian cities.

Abu Dhabi Company for Onshore Oil Operations (ADCO) founded.

October. Iranian oil workers strike.

November. Serious riots in Teheran.

December. EEC Council decides to set up European Monetary System.

OPEC announces price rise in four stages in 1979.

1979

January. Saudi Arabi sets temporary production limit at 9.5 million barrels per day to compensate for loss of Iranian crude oil.

Shah leaves Iran.

February. Ayatollah Khomeini returns to Iran; Islamic revolution in Iran.

March. Iran gets new revolutionary government and resumes oil exports at lower levels.

OPEC raises oil price to $14.00 per barrel.

Israel and Egypt sign the Camp David accords in Washington.

Saudi Arabia announces a production cutback as a protest to the Camp David accords.

OPEC announces an increase in oil prices of 9 percent from 1 April.

Radiation leak at Three Mile Island nuclear power plant in Pennsylvania.

Boycott of Egypt agreed by 19 Arab League states.

OPEC guarantees oil supplies to low-income LDCs.

April. Arab petrol embargo on Egypt.

Islamic Republic declared in Iran.

May. Motor spirit shortages in California.

UK General Election; Conservatives win with overall majority of 43. Margaret Thatcher becomes first woman prime minister in UK

June. ULCCs are included in the AFRA ratings for the first time.

President Carter and Brezhnev sign SALT II treaty in Vienna.

July. Nigeria nationalizes BP's production and marketing interests.

UK government announces decision to sell off a substantial part of BNOC.

October. Currency exchange controls, which had been in force since World War II, abolished in UK.

Civilian rule restored in Nigeria.

UK government to sell 5 percent of its holdings in BP.

Gabon's national oil company, Petrogap, formed.

November. Iran imposes pressure on USA to return Shah. Hostages taken in USA Embassy in Teheran.

USA freezes Iranian assets; President Carter orders cessation of all US oil imports from Iran.

December. OPEC meeting at Caracas fails to agree on a unified crude oil price.

1980

Emirates General Petroleum Corporation set up.

January. USSR invades Afghanistan.

Saudi Arabia increases price of Arab light crude by $2 to $26 per barrel.

UK government announces gas price rises of nearly 30 percent.

February. Publication of Brandt report *North–South: a programme for survival*.

March. *Alexander Keilland*, an oil accommodation rig, overturned in North Sea with 124 dead.

April. Windfall profits tax on US domestic crude oil production introduced.

Algerian refinery built at Skikda by Snamprogetti of Italy.

USA breaks diplomatic relations with Iran and embargos exports to Iran; backed by EEC after two weeks.

International Energy Agency emergency meeting held to discuss oil-sharing schemes.

Iran threatens to block all Middle East seaborne exports.

Attempts to free US hostages in Iran unsuccessful.

May. Saudi Arabia increases oil prices by $2 to $28 per barrel for Arabian light.

June. FLN Congress meeting in Algeria sets out objectives of 1980–84 five-year plan.

Economic Summit Conference in Venice attended by seven major industrial nations.

September. Saudi Arabia increases Arabian light price by $2 to $30 per barrel.

Iraq invades Iran.

Iraq attacks Abadan refinery.

October. Syria announces support for Iran.

USA supplies AWACs to Saudi Arabia.

November. Ronald Reagan elected as President of USA. Release of US hostages held in Iran arranged immediately after election.

December. Saudi Arabia at OPEC meeting in Bali raises crude oil prices. OPEC pricing sets official price of Arabian light crude at $32 per barrel and agrees a maximum upward differential of $9.

1981

January. Iran announces go-ahead for Kangan gas treatment plant.

Value of UK fuel exports — mainly North sea oil — exceeds fuel imports — mostly OPEC oil — for the first time in 40 years.

Ronald Reagan installed as US President and US hostages leave Iran.

May. OPEC meeting in Geneva fails to agree on crude prices.

June. Israelis bomb Iraq's nuclear reactor.

UK North Sea crude oil prices cut from $39.25 to $35 per barrel.

July. UK National Coal Board, for the second consecutive year, announces that it expects to double its coal exports.

Unprecedented leap in tanker storage reflects oil glut.

Libyan jets shot down by US aircraft.

August. Du Pont beats Mobil to buy Conoco for about $7.8 billion.

October. First conventional tanker for North Sea crude oil storage successfully tested at the Fulmar field.

OPEC agrees on a unified oil price structure based on a $34 per barrel marker.

President Sadat of Egypt assassinated.

1982

February. *Ocean Ranger* rig sinks off Newfoundland; all 84 on board drowned.

March. OPEC sets production ceiling in an attempt to hold crude oil prices.

Cut in North Sea crude oil announced.

April. Argentina invades Falkland Islands.

June. Khaled of Saudi Arabia dies; King Fahd succeeds.

August. Mexican currency crisis.

September. Non-OPEC WOCA crude oil production tops 20 million barrels per day for first time.

1983

Saudi Arabia creates Norbec, an independent crude sales company.

January. Emergency meeting of OPEC in Geneva fails to reach agreement on prices.

February. BNOC cuts oil prices by 10 percent.

Nigeria cuts oil price by 15 percent.

March. OPEC, after a 12-day meeting in London, agrees on production controls and reduction in Arabian light marker crude oil from $34 to $29 per barrel.

Phillips Petroleum buys General American Oil Company of Texas for $1141 million.

Kuwait Petroleum agrees to purchase Gulf Oil's European interests.

May. China grants licences for offshore exploration. By the end of 1983 27 companies from nine countries sign contracts to explore.

According to the UK government's *Brown Book* rapidly rising output is depleting North Sea oil and gas reserves at a faster rate than they are being replaced.

June. Mr Andropov elected President of the USSR.

July. OPEC nears ceiling of 17.5 million barrels per day as Saudi output swings higher.

UK government announces sales of £500 million assets in BP.

August. OPEC oil output surges past the 17.5 million barrels per day ceiling.

September. BP offers 12.5 percent of Forties field for sale.

November. Iraq threatens to attack Kharg Island and Iran threatens to block the Straits of Hormuz, with the result that Saudi Arabi charters 11 ULCCs to store oil.

December. Military coup d'etat in Nigeria.

1984

January. Texaco bids $9.9 billion for Getty Oil.

BP abandons first oil well drilled in Pearl River Basin of South China Sea.

February. Death of President Andropov of USSR; Konstantin Chernenko succeeds.

March. Socal bids for Gulf Oil.

Mobil bids for Superior Oil.

UK miners' strike starts.

May. UK miners' strike is estimated to be adding as much as 0.5 million barrels per day to current oil demand.

Insurance for tankers using Iranian ports rises.

USA condemns attacks on tankers in Gulf.

June. USA announces increase in military role in Gulf with dispatch of AWACs.

July. OPEC meeting leaves crude oil prices and OPEC oil quotas unchanged.

October. OPEC in Geneva agrees to production cuts from 17.5 to 16 million barrels per day.

November. Ronald Reagan re-elected President of the USA.

December. OPEC in Geneva agrees to new output monitoring scheme to be carried out by a firm of independent auditors.

1985

February. Sterling sinks to an all-time low of $1.037.

March. UK government announces that BNOC is to be abolished.

UK miners vote to end strike.

President Chernenko of the USSR dies; Mikhail Gorbachev succeeds.

May. UK government announces plans to privatize state-owned British Gas Corporation (BGC) in 1986 and to allow BGC to re-enter oil exploration.

July. Shell announces intention to dispose of all ULCCs and VLCCs in its fleet.

OPEC fails to agree action to stabilize oil prices after three-day meeting. After reconvening, OPEC votes 10–3 to reduce heavy crude price by 50 cents per barrel and medium crude price by 20 cents per barrel.

August. Coup d'etat in Nigeria.

UK government sells remaining 49 percent of Britoil stake for £448.8 million.

September. Saudi Arabia announces it is no longer prepared to act as swing producer and moves to netback pricing.

Ecuador announces intention to leave OPEC if request for increased production quota refused.

Saudi Arabia agrees to buy £1 billion of military aircraft (Tornados) from UK with half of the payment in crude oil.

December. OPEC at Geneva unanimously agree to hold market share of crude oil and risk fall in price; its intention is to secure a 'fair share' of the market.

1986

January. USA announces economic sanctions against Libya and freezes Libya's assets in USA.

Crude oil (Brent) traded at $17 per barrel.

February. Forward dealings in Brent crude oil at 12.70 per barrel.

March. OPEC meeting at Geneva breaks up with no agreement on production quotas. Oil prices fall to $11 per barrel.

UK awards 75 licences for oil exploration onshore.

International oil companies announce large cuts in staff and capital expenditure due to low oil price.

April. USA bombs Libya.

OPEC in Geneva fails to agree to on oil production limits.

September. North Sea oil companies to receive an estimated $1.5 billion rebate for overpayment of wellhead revenue taxes.

OPEC cuts output by even more than targetted.

October. Europe's largest onshore oilfield — 230 million barrel Wytch Farm — to be upgraded over next two years.

Sheikh Yamani, Saudi Arabia's oil minister sacked; this signals a change in oil policy.

December. OPEC agrees to Sanoi price target of $18 per barrel.

Energy data conversion tables

These tables have been taken from a U.S. government publication originally titled *Energy Interrelationships*. They are also included in the *Energy Data Conversion Handbook*, published by Macmillan Press; and by McGraw–Hill, Inc. in the United States of America.

Table 1. Mass Equivalents

	Kilograms	Metric Tons	Long Tons	Short Tons	Pounds
1 Kilogram	1.0	0.001	0.000984	0.001102	2.2046
1 Metric Ton	1000	1.0	0.984	1.102	2204.6
1 Long Ton	1016	1.016	1.0	1.120	2240
1 Short Ton	907.2	0.9072	0.893	1.0	2000
1 Pound	0.454	0.000454	0.000446	0.0005	1.0

Table 2. Volume Equivalents

	U.S. Gallons	Imperial Gallons	Barrels		Barrels per Day	
1 U.S. Gallon	1.000	0.8327	0.02381		0.06523	$\times 10^{-3}$
1 Imp. Gallon	1.201	1.000	0.02859		0.07834	$\times 10^{-3}$
1 Barrel	42	34.97	1.000		0.00274	
1 Barrel per day	15300	12764	365		1.000	
1 Kiloliter*	264.2	220	6.289		17.23	$\times 10^{-3}$
1 Hectoliter	26.42	22.0	0.6289		1.723	$\times 10^{-3}$
1 Liter	0.2642	0.220	6.289	$\times 10^{-3}$	17.23	$\times 10^{-6}$

	Kiloliters		Hectoliters		Liters	
1 U.S. Gallon	3.785	$\times 10^{-3}$	3.785	$\times 10^{-2}$	3.785	
1 Imp. Gallon	4.546	$\times 10^{-3}$	4.546	$\times 10^{-2}$	4.546	
1 Barrel	0.159		1.59		159	
1 Barrel per day	58.03		580.3		58.03	$\times 10^{3}$
1 Kiloliter	1.0		10		1000	
1 Hectoliter	0.1		1.0		100	
1 Liter	1	$\times 10^{-3}$	0.01		1.0	

*Or one cubic meter.

Table 3.—Metric System Multiples*

$\times 10$	deka
$\times 100$	hecto
$\times 1000$	kilo
$\times 10^6$	mega
$\times 10^9$	giga
$\times 10^{12}$	tera
$\times 10^{15}$	peta
$\times 10^{18}$	exa

*In the British System the quad (q) is sometimes employed to represent Btu $\times 10^{15}$. This should not be confused with the Q (Btu $\times 10^{18}$) made fashionable some years ago by Palmer Putnam.

Table 4. Volume of Liquids of Different Specific Gravities Contained in One Metric Ton

Specific Gravity	Liters	Kiloliters	Imperial Gallons	U.S. Gallons	Barrels	Barrels per Day	API Gravity*
50	2002	2.002	440	529	12.59	0.0345	—
51	1962	1.962	432	519	12.34	0.0338	—
52	1925	1.925	423	509	12.11	0.0332	—
53	1888	1.888	415	499	11.88	0.0325	—
54	1853	1.853	408	490	11.66	0.0319	—
55	1820	1.820	400	481	11.45	0.0314	—
56	1787	1.787	393	472	11.24	0.0308	—
57	1756	1.756	386	464	11.04	0.0303	—
58	1726	1.726	380	456	10.85	0.0297	—
59	1696	1.696	373	448	10.67	0.0292	—
60	1668	1.668	367	441	10.49	0.0288	—
61	1641	1.641	361	436	10.32	0.0283	—
62	1614	1.614	355	427	10.15	0.0278	96.73
63	1589	1.589	349	420	9.99	0.0274	93.10
64	1564	1.564	344	413	9.84	0.0270	89.59
65	1540	1.540	339	407	9.69	0.0265	86.19
66	1516	1.516	334	401	9.54	0.0261	82.89
67	1494	1.494	329	395	9.40	0.0257	79.69
68	1472	1.472	324	389	9.26	0.0254	76.59
69	1450	1.450	319	383	9.12	0.0250	73.57
70	1430	1.430	315	378	8.99	0.0246	70.64
71	1410	1.410	310	372	8.87	0.0243	67.80
72	1390	1.390	306	367	8.74	0.0240	65.03
73	1371	1.371	302	362	8.62	0.0236	62.34
74	1352	1.352	298	357	8.51	0.0233	59.72
75	1334	1.334	294	353	8.39	0.0230	57.17
76	1317	1.317	290	348	8.28	0.0227	54.68
77	1300	1.300	286	343	8.18	0.0224	52.27
78	1283	1.283	282	339	8.07	0.0221	49.91
79	1267	1.267	279	335	7.97	0.0218	47.61

continued on the following page

Table 4. *continued*

Specific Gravity	Liters	Kiloliters	Imperial Gallons	U.S. Gallons	Barrels	Barrels per Day	API Gravity*
80	1251	1.251	275	331	7.87	0.0216	45.38
81	1236	1.236	272	326	7.77	0.0213	43.19
82	1220	1.220	269	323	7.68	0.0210	41.06
83	1206	1.206	265	319	7.59	0.0208	38.98
84	1191	1.191	262	315	7.50	0.0205	36.95
85	1177	1.177	259	311	7.41	0.0203	34.97
86	1164	1.164	256	308	7.32	0.0201	33.03
87	1150	1.150	253	304	7.24	0.0198	31.14
88	1137	1.137	250	301	7.15	0.0196	29.30
89	1124	1.124	247	297	7.07	0.0194	27.49
90	1112	1.112	245	294	7.00	0.0192	25.72
91	1100	1.100	242	291	6.92	0.0190	23.99
92	1088	1.088	239	287	6.84	0.0189	22.30
93	1076	1.076	237	284	6.77	0.0186	20.65
94	1065	1.065	234	281	6.70	0.0184	19.03
95	1053	1.053	232	278	6.63	0.0182	17.45
96	1043	1.043	229	275	6.56	0.0180	15.90
97	1032	1.032	227	273	6.49	0.0178	14.38
98	1021	1.021	225	270	6.42	0.0176	12.89
99	1011	1.011	222	267	6.36	0.0174	11.43
100	1001	1.001	220	264	6.30	0.0173	10.00
101	991	0.991	218	262	6.23	0.0171	8.60
102	981	0.981	216	259	6.17	0.0169	7.23
103	972	0.972	214	257	6.11	0.0168	5.88
104	962	0.962	212	254	6.05	0.0166	4.56
105	953	0.953	210	252	6.00	0.0164	3.26

*In this table and elsewhere, a single hyphen denotes zero or not applicable; two hyphens denote information is not available.

Table 5. Weights of Liquids of Different Gravities in Kilograms

Specific Gravity	Per Liter	Per Hectoliter	Per Kiloliter	Per Imperial Gallon	Per U.S. Gallon	Per Barel	Per Barrel Per Day (in tons)	API Gravity
50	0.500	50.0	500	2.275	1.894	79.5	29.0	—
51	0.510	51.0	510	2.320	1.932	81.1	29.6	—
52	0.520	52.0	520	2.366	1.970	82.7	30.2	—
53	0.530	53.0	530	2.411	2.008	84.3	30.8	—
54	0.540	54.0	540	2.457	2.046	85.9	31.4	—
55	0.550	55.0	550	2.502	2.083	87.5	31.9	—
56	0.560	56.0	560	2.548	2.121	89.1	32.5	—
57	0.570	57.0	570	2.593	2.159	90.7	33.1	—
58	0.580	58.0	580	2.639	2.197	92.3	33.7	—
59	0.590	59.0	590	2.684	2.235	93.9	34.3	—

continued on the following page

Table 5. *continued*

Specific Gravity	Per Liter	Per Hectoliter	Per Kiloliter	Per Imperial Gallon	Per U.S. Gallon	Per Barel	Per Barrel Per Day	API Gravity
							(in tons)	
60	0.600	60.0	600	2.730	2.273	95.5	34.8	—
61	0.610	61.0	610	2.775	2.311	97.1	35.4	—
62	0.620	62.0	620	2.821	2.349	98.7	36.0	96.73
63	0.631	63.1	631	2.866	2.386	100.2	36.6	93.10
64	0.641	64.1	641	2.912	2.424	101.8	37.2	89.59
65	0.651	65.1	651	2.957	2.462	103.4	37.7	86.19
66	0.661	66.1	661	3.003	2.500	105.0	38.3	82.89
67	0.671	67.1	671	3.048	2.538	106.6	38.9	79.69
68	0.681	68.1	681	3.084	2.576	108.2	39.5	76.59
69	0.691	69.1	691	3.139	2.614	109.8	40.1	73.57
70	0.701	70.1	701	3.185	2.652	111.4	40.7	70.64
71	0.711	71.1	711	3.230	2.689	112.9	41.2	67.80
72	0.721	72.1	721	3.276	2.727	114.5	41.8	65.03
73	0.731	73.1	731	3.321	2.765	116.1	42.4	62.34
74	0.741	74.1	741	3.367	2.803	117.7	43.0	59.72
75	0.751	75.1	751	3.412	2.841	119.3	43.6	57.17
76	0.761	76.1	761	3.458	2.879	120.9	44.1	54.68
77	0.771	77.1	771	3.503	2.917	122.5	44.7	52.27
78	0.781	78.1	781	3.549	2.955	124.1	45.3	49.91
79	0.791	79.1	791	3.594	2.993	125.7	45.9	47.61
80	0.801	80.1	801	3.640	3.030	127.3	46.4	45.38
81	0.811	81.1	811	3.685	3.068	128.9	47.0	43.19
82	0.821	82.1	821	3.731	3.106	130.5	47.6	41.06
83	0.831	83.1	831	3.776	3.144	132.0	48.2	38.98
84	0.841	84.1	841	3.822	3.182	133.6	48.8	36.95
85	0.851	85.1	851	3.867	3.220	135.2	49.4	34.97
86	0.861	86.1	861	3.913	3.258	136.8	49.9	33.03
87	0.871	87.1	871	3.958	3.296	138.4	50.5	31.14
88	0.881	88.1	881	4.004	3.333	140.0	51.1	29.30
89	0.891	89.1	891	4.049	3.371	141.6	51.7	27.49
90	0.901	90.1	901	4.095	3.409	143.2	52.3	25.72
91	0.911	91.1	911	4.140	3.447	144.8	52.8	23.99
92	0.921	92.1	921	4.186	3.485	146.4	53.4	22.30
93	0.931	93.1	931	4.231	3.523	148.0	54.0	20.65
94	0.941	94.1	941	4.277	3.561	149.6	54.6	19.03
95	0.951	95.1	951	4.322	3.599	151.2	55.2	17.45
96	0.961	96.1	961	4.368	3.637	152.8	55.8	15.90
97	0.971	97.1	971	4.413	3.674	154.3	56.3	14.38
98	0.981	98.1	981	4.459	3.712	155.9	56.9	12.89
99	0.991	99.1	991	4.504	3.750	157.5	57.5	11.43
100	1.001	100.1	1001	4.550	3.788	159.1	58.1	10.00
101	1.011	101.1	1011	4.595	3.826	160.7	58.7	8.60
102	1.021	102.1	1021	4.641	3.864	162.3	59.2	7.23
103	1.031	103.1	1031	4.686	3.902	163.9	59.8	5.88
104	1.041	104.1	1041	4.732	3.940	165.5	60.4	4.56
105	1.051	105.1	1051	4.777	3.977	167.0	61.0	3.26

Table 6a. Energy and Work Equivalents: Calorie Base

			IT Calories	Kilocalories	Megaclories	Gigacalories	Teracalories
1 International Steam Table calorie	(IT cal)	=	1.0	10^{-3}	10^{-6}	10^{-9}	10^{-12}
1 Kilocalorie	(kcal)	=	1000	1.0	10^{-3}	10^{-6}	10^{-9}
1 Thermie	(Mcal)	=	1×10^{6}	1×10^{3}	1.0	10^{-3}	10^{-6}
1 Gigacalorie	(Gcal)	=	1×10^{9}	1×10^{6}	1000	1.0	10^{-3}
1 Teracalorie	(Tcal)	=	1×10^{12}	1×10^{9}	1×10^{6}	1000	1.0
1 British thermal unit	(Btu)	=	252	.252	252×10^{-6}	252×10^{-9}	252×10^{-12}
1 Therm	(thm)	=	25.2×10^{6}	25.2×10^{3}	25.2	25.2×10^{-3}	25.2×10^{-6}
1 Joule	(J)	=	.2388	238.8×10^{-6}	238.8×10^{-9}	238.8×10^{-12}	238.8×10^{-15}
1 Kilojoule	(kJ)	=	238.8	.2388	238.8×10^{-6}	238.8×10^{-9}	238.8×10^{-12}
1 Megajoule	(MJ)	=	238.8×10^{3}	238.8	.2388	238.8×10^{-6}	238.8×10^{-9}
1 Gigajoule	(GJ)	=	238.8×10^{6}	238.8×10^{3}	238.8	.2388	238.8×10^{-6}
1 Terajoule	(TJ)	=	238.8×10^{9}	238.8×10^{6}	238.8×10^{3}	238.8	.2388
1 Kilowatt hour	(kWh)	=	860×10^{3}	860	.860	860×10^{-6}	860×10^{-9}
1 Megawatt hour	(MWh)	=	860×10^{6}	860×10^{3}	860	.860	860×10^{-6}
1 Gigawatt hour	(GWh)	=	860×10^{9}	860×10^{6}	860×10^{3}	860	.860
1 Terawatt hour	(TWh)	=	860×10^{12}	860×10^{9}	$860\times^{6}$	860×10^{3}	860
1 Horsepower hour	(hp·h)	=	641.2×10^{3}	641.2	.6412	641.2×10^{-6}	641.2×10^{-9}
1 Cheval vapeur hour	(cv·h)	=	632.4×10^{3}	632.4	.6324	632.4×10^{-9}	632.4×10^{-9}
1 Foot pound	(ft·lb)	=	.3238	323.8×10^{-6}	323.8×10^{-9}	323.8×10^{-12}	323.8×10^{-15}
1 Kilogram-force meter	(kgf·m)	=	2.342	2.342×10^{-3}	2.342×10^{-6}	2.342×10^{-9}	2.342×10^{-12}

Table 6b. Energy and Work Equivalents: Kilowatt Hour Base

			Kilowatt Hours	Megawatt Hours	Gigawatt Hours	Terawatt Hours
1 International Steam Table calorie	(IT cal)	=	1.163×10^{-6}	1.163×10^{-9}	1.163×10^{-12}	1.163×10^{-15}
1 Kilocalorie	(kcal)	=	1.163×10^{-3}	1.163×10^{-6}	1.163×10^{-9}	1.163×10^{-12}
1 Thermie	(Mcal)	=	1.163	1.163×10^{-3}	1.163×10^{-6}	1.163×10^{-9}
1 Gigacalorie	(Gcal)	=	1163	1.163	1.163×10^{-3}	1.163×10^{-6}
1 Teracalorie	(Tcal)	=	1163×10^{3}	1163	1.163	1.163×10^{-3}
1 British thermal unit	(Btu)	=	$.293071\times10^{-3}$	$.293071\times10^{-6}$	$.293071\times10^{-9}$	$.293071\times10^{-12}$
1 Therm	(thm)	=	29.3071	29.3071×10^{-3}	29.3071×10^{-6}	29.3071×10^{-9}
1 Joule	(J)	=	$.2777\times10^{-6}$	$.2777\times10^{-9}$	$.2777\times10^{-12}$	$.2777\times10^{-15}$
1 Kilojoule	(kJ)	=	$.2777\times10^{-3}$	$.2777\times10^{-6}$	$.2777\times10^{-9}$	$.2777\times10^{-9}$
1 Megajoule	(MJ)	=	.2777	$.2777\times10^{-3}$	$.2777\times10^{-6}$	$.2777\times10^{-9}$
1 Gigajoule	(GJ)	=	277.7	.2777	$.2777\times10^{-3}$	$.2777\times10^{-6}$
1 Terajoule	(TJ)	=	277.7×10^{3}	277.7	.2777	$.2777\times10^{-3}$
1 Kilowatt hour	(kWh)	=	1.0	10^{-3}	10^{-6}	10^{-9}
1 Megawatt hour	(MWh)	=	10^{3}	1.0	10^{-3}	10^{-6}
1 Gigawatt hour	(GWh)	=	10^{6}	10^{3}	1.0	10^{-3}
1 Terawatt hour	(TWh)	=	10^{9}	10^{6}	10^{3}	1.0
1 Horsepower hour	(hp·h)	=	.7457	$.7457\times10^{-3}$	$.7457\times10^{-6}$	$.7457\times10^{-9}$
1 Cheval vapeur hour	(cv·h)	=	.7355	$.7355\times10^{-3}$	$.7355\times10^{-6}$	$.7355\times10^{-9}$
1 Foot pound	(ft·lb)	=	$.3766\times10^{-6}$	$.3766\times10^{-9}$	$.3766\times10^{-12}$	$.3766\times10^{-15}$
1 Kilogram-force meter	(kgf·m)	=	2.724×10^{-6}	2.724×10^{-9}	2.724×10^{-12}	2.724×10^{-15}

Table 6c. Energy and Work Equivalents: Btu Base

			Btu	Therms
1 International Steam Table calorie	(IT cal)	=	3.968×10^{-3}	39.68×10^{-9}
1 Kilocalorie	(kcal)	=	3.968	39.68×10^{-6}
1 Thermie	(Mcal)	=	3968	39.68×10^{-3}
1 Gigacalorie	(Gcal)	=	3.968×10^{6}	39.68
1 Teracalorie	(Tcal)	=	3.968×10^{9}	39.68×10^{3}
1 British thermal unit	(Btu)	=	1.0	10^{-5}
1 Therm	(thm)	=	100000	1.0
1 Joule	(J)	=	$.9478 \times 10^{-3}$	9.478×10^{-9}
1 Kilojoule	(kJ)	=	.9478	9.478×10^{-6}
1 Megajoule	(MJ)	=	$.9478 \times 10^{3}$	9.478×10^{-3}
1 Gigajoule	(GJ)	=	$.9478 \times 10^{6}$	9.478
1 Terajoule	(TJ)	=	$.9478 \times 10^{9}$	9478
1 Kilowatt hour	(kWh)	=	3412	34.12×10^{-3}
1 Megawatt hour	(MWh)	=	3412×10^{3}	34.12
1 Gigawatt hour	(GWh)	=	3412×10^{6}	34.12×10^{3}
1 Terawatt hour	(TWh)	=	3412×10^{9}	34.12×10^{6}
1 Horsepower hour	(hp·h)	=	2544.43	25.4443×10^{-3}
1 Cheval vapeur hour	(cv·h)	=	2509.62	25.0962×10^{-3}
1 Foot pound	(ft·lb)	=	1.2851×10^{-3}	12.851×10^{-9}
1 Kilogram-force meter	(kgf·m)	=	9.2949×10^{-3}	92.949×10^{-9}

Table 6d. Energy and Work Equivalents: Joule Base

			Joules	Kilojoules	Megajoules	Gigajoules	Terajoules
1 International Steam Table calorie	(IT cal)	=	4.1868	4.1868×10^{-3}	4.1868×10^{-6}	4.1868×10^{-9}	4.1868×10^{-12}
1 Kilocalorie	(kcal)	=	4186.8	4.1868	4.1868×10^{-3}	4.1868×10^{-6}	4.1868×10^{-9}
1 Thermie	(Mcal)	=	4186.8×10^{3}	4186.8	4.1868	4.1868×10^{-3}	4.1868×10^{-6}
1 Gigacalorie	(Gcal)	=	4186.8×10^{6}	4186.8×10^{3}	4186.8	4.1868	4.1868×10^{-3}
1 Teracalorie	(Tcal)	=	4186.8×10^{9}	4186.8×10^{6}	4186.8×10^{3}	4186.8	4.1868
1 British thermal unit	(Btu)	=	1055.1	1055.1×10^{-3}	1055.1×10^{-6}	1055.1×10^{-9}	1055.1×10^{-12}
1 Therm	(thm)	=	105.51×10^{6}	105.51×10^{3}	105.51	105.51×10^{-3}	105.51×10^{-6}
1 Joule	(J)	=	1.0	10^{-3}	10^{-6}	10^{-9}	10^{-12}
1 Kilojoule	(kJ)	=	10^{3}	1.0	10^{-3}	10^{-6}	10^{-9}
1 Megajoule	(MJ)	=	10^{6}	10^{3}	1.0	10^{-3}	10^{-6}
1 Gigajoule	(GJ)	=	10^{9}	10^{6}	10^{3}	1.0	10^{-3}
1 Terajoule	(TJ)	=	10^{12}	10^{9}	10^{6}	10^{3}	1.0
1 Kilowatt hour	(kWh)	=	3.6×10^{6}	3.6×10^{3}	3.6	3.6×10^{-3}	3.6×10^{-6}
1 Megawatt hour	(MWh)	=	3.6×10^{9}	3.6×10^{6}	3.6×10^{3}	3.6	3.6×10^{-3}
1 Gigawatt hour	(GWh)	=	3.6×10^{12}	3.6×10^{9}	3.6×10^{6}	3.6×10^{3}	3.6
1 Terawatt hour	(TWh)	=	3.6×10^{15}	3.6×10^{12}	3.6×10^{9}	3.6×10^{6}	3.6×10^{3}
1 Horsepower hour	(hp·h)	=	2684.52×10^{3}	2684.52	2.68453	2.68452×10^{-3}	2.68452×10^{-6}
1 Cheval vapeur hour	(cv·h)	=	2647.8×10^{3}	2647.8	2.6478	2.6478×10^{-3}	2.6478×10^{-6}
1 Foot pound	(ft·lb)	=	1.3558	1.3558×10^{-3}	1.3558×10^{-6}	1.3558×10^{-9}	1.3558×10^{-12}
1 Kilogram-force meter	(kgf·m)	=	9.80665	9.80665×10^{-3}	9.80665×10^{-6}	9.80665×10^{-9}	9.80665×10^{-12}

Table 6e. Energy and Work Equivalents: Work Base

			Horsepower Hours	Cheval vapeur Hours	Foot Pounds	Kilogram-force Meters
1 International Steam Table calorie	(IT cal)	=	1.5596×10^{-6}	1.5812×10^{-6}	3.088	.4269
1 Kilocalorie	(kcal)	=	1.5596×10^{-3}	1.5812×10^{-3}	3.088×10^{3}	426.9
1 Thermie	(Mcal)	=	1.5596	1.5812	3.088×10^{6}	426.9×10^{3}
1 Gigacalorie	(Gcal)	=	1559.6	1581.2	3.088×10^{9}	426.9×10^{6}
1 Teracalorie	(Tcal)	=	1559.6×10^{3}	1581.2×10^{3}	3.088×10^{12}	426.9×10^{9}
1 British thermal unit	(Btu)	=	$.3930\times10^{-3}$	$.39847\times10^{-3}$	778.17	107.6
1 Therm	(thm)	=	39.3	39.847	77.817×10^{6}	10.759×10^{6}
1 Joule	(J)	=	$.3725\times10^{-6}$	$.3777\times10^{-6}$	.7376	.10197
1 Kilojoule	(kJ)	=	$.3725\times10^{-3}$	$.3777\times10^{-3}$	737.6	$.10197\times10^{3}$
1 Megajoule	(MJ)	=	.3725	.3777	737.6×10^{3}	$.10197\times10^{6}$
1 Gigajoule	(GJ)	=	372.5	377.7	737.6×10^{6}	$.10197\times10^{9}$
1 Terajoule	(TJ)	=	372.5×10^{3}	377.7×10^{3}	737.6×10^{9}	$.10197\times10^{12}$
1 Kilowatt hour	(kWh)	=	1.341	1.360	2.655×10^{6}	$.3671\times10^{6}$
1 Megawatt hour	(MWh)	=	1341	1360	2.655×10^{9}	$.3671\times10^{9}$
1 Gigawatt hour	(GWh)	=	1341×10^{3}	1360×10^{3}	2655×10^{9}	367.1×10^{9}
1 Terawatt hour	(TWh)	=	1341×10^{6}	1360×10^{6}	1655×10^{12}	367.1×10^{12}
1 Horsepower hour	(hp·h)	=	1.0	1.01387	1.98×10^{6}	$.2737\times10^{6}$
1 Cheval vapeur hour	(cv·h)	=	.98632	1.0	1.953×10^{6}	$.27\times10^{6}$
1 Foot pound	(ft·lb)	=	$.50505\times10^{-6}$	$.51206\times10^{-6}$	1.0	.13825
1 Kilogram-force meter	(kgf·m)	=	3.653×10^{-6}	3.7037×10^{-6}	7.233	1.0

Table 7. Power Equivalents

		ft·lb/s	kgf·m/s	kW	hp	cv*
1 Foot pound per second (ft·lb/s)	=	1.0	0.1383	1.355×10^{-3}	1.818×10^{-3}	1.843×10^{-3}
1 Kilogram-force meter per second (kgf·m/s)	=	7.233	1.0	9.803×10^{-3}	0.01315	0.01333
1 Kilowatt (kW)	=	738.0	102.0	1.0	1.341	1.360
1 Horsepower (hp)	=	550	76.04	0.7457	1.0	1.014
1 Metric horsepower (cv)	=	542.6	75.0	0.7353	0.9862	1.0

*Cheval vapeur.

Table 8. Heat Content of Oils of Different Gravities* (Disregarding Water, Ash, and Sulphur Content)

		Net Heat Content	
A.P.I. Gravity, 60° F.	Specific Gravity, 60°/60° F.	Specific Megacalories per Metric Ton	Thousand Btu per Barrel
10	1.0000	9740	6140
11	.9930	9770	6115
12	.9861	9790	6086
13	.9792	9810	6056
14	.9725	9840	6031
15	.9659	9860	6002
16	.9593	9880	5977
17	.9529	9900	5947
18	.9465	9920	5918
19	.9402	9940	5893
20	.9340	9960	5863
21	.9279	9980	5838
22	.9218	10000	5809
23	.9159	10020	5783
24	.9100	10040	5758
25	.9042	10050	5729
26	.8984	10070	5704
27	.8927	10090	5678
28	.8871	10110	5653
29	.8816	10120	5624
30	.8762	10140	5599
31	.8708	10150	5573
32	.8654	10170	5548
33	.8602	10180	5523
34	.8550	10200	5498
35	.8498	10210	5473
36	.8448	10230	5447
37	.8398	10240	5422
38	.8348	10260	5397
39	.8299	10270	5372
40	.8251	10280	5347
41	.8203	10300	5321
42	.8155	10310	5300
43	.8109	10320	5275
44	.8063	10330	5250
45	.8017	10340	5225
46	.7972	10360	5204
47	.7927	10370	5179
48	.7883	10380	5158
49	.7839	10390	5132
50	.7796	10400	5111
51	.7753	10410	5086
52	.7711	10420	5065
53	.7669	10430	5044
54	.7628	10440	5019
55	.7587	10450	4998
56	.7547	10460	4977
57	.7507	10470	4956
58	.7467	10480	4935
59	.7428	10490	4910
60	.7389	10500	4889
61	.7351	10510	4868
62	.7313	10520	4847
63	.7275	10530	4826
64	.7238	10540	4805
65	.7201	10540	4784
66	.7165	10550	4763
67	.7128	10560	4742
68	.7093	10570	4725
69	.7057	10580	4704
70	.7022	10580	4683
72	.6953	10600	4645
74	.6886	10610	4607
76	.6819	10630	4569
78	.6754	10640	4532
80	.6690	10650	4494
82	.6628	10670	4456
84	.6566	10680	4418
86	.6506	10690	4385
88	.6446	10700	4347
90	.6388	10710	4313
92	.6331	10720	4280
94	.6275	10740	4246
96	.6220	10750	4212
98	.6166	10760	4179
100	.6112	10770	4145
105	.5983	10790	4066
110	.5859	10810	3994
115	.5740	10830	3919
120	.5626	10850	3847
125	.5517	10860	3780
130	.5411	10880	3713
135	.5310	10900	3646
140	.5212	10910	3583
145	.5118	10920	3524

*Based on data given in *Thermal Properties of Petroleum Products*. Misc. Publication No. 97. U.S. Department of Commerce, Washington, D.C.: National Bureau of Standards. 1933.

Table 9. Gravities and Energy Values of Selected Liquid Fuels

				Energy Content					
				Megacalories per Kilogram		Megacalories per Liter		Btu per Barell	
Commodity	Specific Gravity Range	"Norm"	API Gravity	Gross	Net	Gross	Net	Gross	Net
LPG/LRG	.50– .58	.54	- -	11.8	10.8	6.4	5.8	4.0	3.7
Propane	.50– .51	.51	- -	11.9	10.9	6.1	5.6	3.8	3.5
Butane	.56– .61	.58	- -	11.7	10.7	6.8	6.2	4.3	3.9
Natural gasoline	.62– .64	.63	93	11.6	10.6	7.3	6.7	4.6	4.2
Mfd. gasoline	.70– .75	.74	60	11.3	10.5	8.4	7.8	5.3	4.9
Aviation gasoline	.71	.71	68	11.5	10.5	8.2	7.5	5.2	4.7
Motor gasoline	.73– .75	.74	60	11.4	10.5	8.4	7.8	5.3	4.9
Gasoline-type jet fuel	.76	.76	54	11.3	10.4	8.6	7.9	5.4	5.0
Naphthas	.72– .79	.74	60	11.3	10.5	8.4	7.8	5.3	4.9
Petrochemical feedstocks	.72– .79	.74	60	11.3	10.5	8.4	7.8	5.3	4.9
White spirit	.78	.78	50	11.4	10.6	8.9	8.3	5.5	5.1
Jet fuel	.79– .83	.82	41	11.3	10.4	9.3	8.5	5.9	5.4
Kerosine	.78– .83	.81	43	11.3	10.3	9.1	8.3	5.7	5.2
Kerosine-type jet fuel	.79– .83	.82	41	11.3	10.4	9.3	8.5	5.9	5.4
Other	.81	.81	43	11.3	10.3	9.1	8.3	5.7	5.2
Distillate fuel oil	.82– .90	.86	33	11.1	10.2	9.5	8.8	6.0	5.5
Heating oils	.82– .85	.83	39	11.1	10.2	9.2	8.5	5.8	5.4
Highway diesel	.82– .85	.84	36	11.1	10.2	9.3	8.6	5.9	5.4
Indust. diesel	.86– .90	.88	29	11.0	10.1	9.7	8.9	6.1	5.6
Residual fuel oil	.91– .99	.94	17	10.6	9.8	10.0	9.2	6.3	5.8
Light	.92– .94	.93	21	10.7	9.9	9.9	9.1	6.2	5.7
Heavy	.94– .98	.96	16	10.4	9.6	10.0	9.2	6.3	5.8
Lubes	.85– .95	.88	29	10.8	9.9	9.5	8.7	6.0	5.5
Asphalt	1.0– 1.1	1.05	- -	10.2	10.0	10.7	10.5	6.7	6.6
Petroleum coke	1.28– 1.42	1.35	- -	8.4	8.3	11.3	11.2	7.1	7.1
Wax	.87– .91	.89	27	9.8	9.0	8.7	8.0	5.5	5.0
Crude (not further defined)	.80– .87	.86	33	11.1	10.2	9.5	8.8	6.0	5.5
Liquefied natural gas	- -	.42	- -	13.8	12.6	5.8	5.3	3.7	3.3
Unidentified products	- -	.86	33	11.1	10.2	9.5	8.8	6.0	5.5
Ethyl alcohol (100%)	- -	.79	- -	7.3	6.6	5.8	5.2	3.7	3.3
Methyl alcohol (100%)	- -	.80	- -	5.7	5.0	4.6	4.0	2.9	2.5
Tar	1.20	1.20	- -	9.2	- -	11.0	- -	6.9	- -

Table 10. Gravities and Approximate Energy Values of Particular Crude Oils (Disregarding Water, Ash, and Sulphur Content)

		Low Heat Value		
Country	Gravity Specific	Mega-calories per Kilogram	Thousand Btu per Pound	Thousand Btu per Barrel
Albania	.94	9.94	17.9	5893
Algeria	.82	10.30	18.5	5321
Angola	.89	10.10	18.2	5666
Argentina	.88	10.13	18.2	5616
Australia	.85	10.21	18.4	5473
Austria	.90	10.06	18.1	5711
Bahrain	.86	10.18	18.3	5523
Bolivia	.80	10.35	18.6	5217
Brazil	.82	10.30	18.5	5321
Brunei	.84	10.24	18.4	5422
Bulgaria	.86	10.18	18.3	5523
Burma	.89	10.10	18.2	5676
Canada	.85	10.21	18.4	5475
Chile	.82	10.30	18.5	5321
China	.86	10.18	18.3	5523
Colombia	.89	10.10	18.2	5666
Congo	.84	10.24	18.4	5422
Cuba	.95	9.91	17.8	5934
Czechoslavakia	.93	9.97	17.9	5876
Ecuador	.88	10.13	18.2	5616
Egypt	.88	10.13	18.2	5616
France	.86	10.18	18.3	5523
Gabon	.87	10.15	18.3	5573
Germany (W)	.89	10.10	18.2	5666
Hungary	.94	9.94	17.9	5893
India	.83	10.27	18.5	5372
Indonesia	.85	10.21	18.4	5473
Iran	.86	10.18	18.3	5523
Iraq	.85	10.21	18.4	5474
Israel	.87	10.15	18.3	5573
Italy	.92	10.01	18.0	5801
Japan	.86	10.18	18.3	5523

		Low Heat Value		
Country	Gravity Specific	Mega-calories per Kilogram	Thousand Btu per Pound	Thousand Btu per Barrel
Kuwait	.87	10.15	18.3	5573
Libya	.83	10.27	18.5	5372
Malaysia	.82	10.30	18.5	5321
Mexico	.88	10.13	18.2	5616
Mongolia	.86	10.18	18.3	5523
Morocco	.83	10.27	18.5	5372
Netherlands	.92	10.01	18.0	5801
Neutral Zone	.87	10.15	18.3	5573
New Zealand	.78	10.40	18.7	5111
Nigeria	.86	10.18	18.3	5523
Norway	.84	10.24	18.4	5422
Oman	.86	10.18	18.3	5523
Pakistan	.86	10.18	18.3	5523
Perua	.85	10.21	18.4	5473
Poland	.85	10.21	18.4	5473
Qatar	.83	10.27	18.5	5371
Romania	.84	10.24	18.4	5422
Sarawak	.84	10.24	18.4	5422
Saudia Arabia	.86	10.18	18.3	5523
Spain	.84	10.24	18.4	5422
Sweden	.97	9.85	17.7	6017
Syria	.91	10.04	18.1	5758
Taiwan	.86	10.18	18.3	5523
Thailand	.86	10.18	18.3	5523
Trinidad	.89	10.10	18.2	5666
Trucial Oman	.84	10.24	18.4	5422
Turkey	.88	10.13	18.2	5616
USSR	.86	10.18	18.3	5523
UAE Abu Dhabi	.83	10.27	18.5	5372
UAE Dubai	.86	10.18	18.3	5523
United Kingdom	.86	10.18	18.3	5523
United States	.85	10.21	18.4	5473
Venezuela	.90	10.06	18.1	5711
Yugoslavia	.85	10.21	18.4	5473

Table 11. Geographical Variations in the Energy Values of Selected Natural Gases

	Megacalories per Cubic Meter			Btu per Cubic Foot		
Country	Gross	*	Net	Gross	*	Net
Afghanistan	- -		- -	- -		- -
Algeria	- -		- -	- -		- -
Angola	- -		- -	- -		- -
Argentina		9.3			1045	
Australia	8.9		**(8.0)	1000		(899)
Austria	9.8		8.9	1102		1000
Bangladesh (Titas)		9.3			1045	
Bahrain	- -		- -	- -		- -
Barbados	- -		- -	- -		- -
Belgium	8.4		7.6	944		854
Bolivia	- -		- -	- -		- -
Brazil	- -		- -	- -		- -
Brunei	(9.9)		9.0	(1113)		1012
Bulgaria		8.4			944	
Burma	- -		- -	- -		- -
Canada		9.1			1023	
Chile	- -		- -	- -		- -
China	- -		- -	- -		- -
Colombia		10.7			1203	
Congo	- -		- -	- -		- -
Czechoslovakia	(8.7)		7.8	(978)		877
Denmark	- -		- -	- -		- -
Ecuador	- -		- -	- -		- -
Egypt	- -		- -	- -		- -
France	9.1		8.2	1023		912
Finland (Imported)	8.3		(7.5)	933		(843)
Gabon	- -		- -	- -		- -
Germany (E)	- -		3.1	- -		- -
Germany (W)	8.5		7.6	955		854
Hungary	(9.8)		8.8	(1102)		989
India	- -		- -	- -		- -
Indonesia	- -		- -	- -		- -
Iran	**(10.4)		9.4	(1169)		1057
Iraq	- -		- -	- -		- -
Israel	(10.2)		9.2	(1147)		1034
Italy	9.2		8.2	1023		922
Japan	9.8		8.5	1102		956
Kuwait	- -		- -	- -		- -
Libya	- -		- -	- -		- -
Luxembourg	8.4		7.6	944		854
Mexico		8.7			978	
Morocco	- -		- -	- -		- -
Netherlands	8.5		(7.6)	955		(854)
New Zealand		***10.1			***1135	
Nigeria	- -		- -	- -		- -
Norway	10.1		(9.1)	1135		(1023)

continued on the following page

Table 11. *continued*

	Megacalories per Cubic Meter			Btu per Cubic Foot		
Country	Gross	*	Net	Gross	*	Net
Oman	- -		- -	- -		- -
Pakistan (Sui field)		8.3			933	
Peru		8.9			1000	
Poland	8.0		(7.2)	899		(809)
Qatar	- -		- -	- -		- -
Romania	10.3		9.2	1158		1034
Rwanda	- -		- -	- -		- -
Saudia Arabia	- -		- -	- -		- -
Switzerland	9.6		8.6	1079		967
Spain	10.1		(9.1)	1135		(1023)
Trinidad		8.9			1000	
Tunisia		11.0			1237	
Turkey	- -		- -	- -		- -
USSR	(9.2)		8.3	(1034)		933
United Arab Emirates	- -		- -			- -
United Kingdom	10.0		(9.0)	1124		(1012)
United States	9.2		(8.3)	1031		(928)
Venezuela		10.3			1158	
Yugoslavia	9.9		(8.9)	1113		(1000)

*Not identified as gross or not.
**Numbers in parentheses are estimated from companion (i.e., gross or net) data.
***After removal of CO_2.

Table 12. Average Energy Values of Gaseous Fuels

	Megacalories per Cubic Meter		Thousand Btu per Cubic Foot	
	Gross	Nett	Gross	Net
Natural gas	9.2	8.3	1.0	0.9
Mine gas	9.0	8.0	1.0	0.9
Refinery gas	12.0	11.0	1.3	1.2
Methane	9.0	8.0	1.0	0.9
Ethane	15.9	14.2	1.7	1.6
Propane	22.4	20.5	2.5	2.3
Isobutane	28.1	25.8	3.2	2.9
Butane	29.0	26.7	3.3	3.0
Pentane	34.7	32.0	3.9	3.6
Oven gas	4.6	4.2	0.5	0.45
City gas	4.3	4.0	0.5	0.45
Producer gas	1.5	1.4	0.17	0.16
Blast furnace gas	0.9	0.9	0.10	0.10

Table 13. Approximate Efficiencies of Energy Utilization, by Commodity and Sector

	Air, Highway, Agriculture	Railways	Inland Waterways	International Bunkers	Pipelines	Industry	Domestic Sector
Coal	–	.05	.15	.15	–	.60	.50
Lignite	–	.04	.10	–	–	.60	.50
Coal briquettes	–	.05	.15	.15	–	.60	.50
Lignite briquettes	–	.05	.15	.15	–	.60	.50
Coke	–	–	–	–	–	.70	.50
Coke breeze	–	–	–	–	–	.60	–
Crude oil	–	–	–	–	–	.70	–
LPG/LRG	.30	–	–	–	–	.80	.70
Aviation gasoline	.20	–	–	–	–	–	–
Motor gasoline	.20	.20	.20	–	–	.20	–
Jet fuel	.20	–	–	–	–	.20	–
Kerosine	.20	.15	–	–	–	.15	.30
Distillate fuel oil	.25	.30	.30	.30	.30	.30	.60
Residual fuel oil	.50	.06	.20	.20	–	.70	.60
Refinery fuel	–	–	–	–	–	.70	–
Petroleum coke	–	–	–	–	–	.70	–
Refinery gas	–	–	–	–	–	.80	–
Natural and mf'd. gases	.25	–	–	–	.35	.80	.70
Noncommercial fuel	–	.03	–	–	–	.40	.20
Electricity	1.00	1.00	–	–	1.00	1.00	.90

Table 14. "Normal" Fuel Values and Equivalents*

		Coal kt	Oil kt	Oil ML	Oil kB	Oil b/d	Nat. Gas GL	Electricity GWh
Coal: 1000 metric tons (kt)	=	1.0	0.70	0.78	4.9	13.4	0.84	8.14
Oil:								
1000 metric tons (kt)	=	1.43	1.0	1.11	7.0	19.1	1.20	11.63
1000 kiloliters (ML)	=	1.29	0.9	1.0	6.3	17.2	1.08	10.47
1000 barrels (kB)	=	0.204	0.143	0.159	1.0	2.74	0.172	1.66
1 barrel per day (b/d)	=	0.075	0.052	0.058	0.365	1.0	0.063	0.61
Natural Gas: 1 Million m^3 (GL)	=	1.19	0.83	0.92	5.80	15.9	1.0	9.65
Electricity: 1 Gigawatt (GWh)	=	0.123	0.086	0.0955	0.60	1.65	0.104	1.0

*Based on the following unit values: Coal, 7.0 teracalories per 1000 tons; oil, 10.0 teracalories (net) per 1000 tons, or 9.0 teracalories (net) per 1000 ML: natural gas, 8.3 teracalories (net) per GL; electricity, 860 teracalories per TWh.

Table 15a. Calorie Equivalents of Solid Fuels

	Teracalories (net)		
Fuels	Per Thousand Metric Tons	Per Thousand Long Tons	Per Thousand Short Tons
Bituminous coal			
Imported or exported	7.0	7.1	6.4
Consumed elsewhere	7.0	7.1	6.4
Source and use not known	6.0	6.1	5.4
Anthracite			
United States	7.0	7.1	6.4
Other countries	7.5	7.6	6.8
Coal equivalent			
Coal and other fuels	7.0	7.1	6.4
Peat	3.5	3.6	3.2
Coal briquettes, patent fuel	7.0	7.1	6.4
Lignite briquettes	4.8	4.9	4.4
Coke briquettes	5.7	5.8	5.2
Peat briquettes	5.2	5.3	4.7
Gas coke	6.8	6.9	6.2
Oven coke	6.8	6.9	6.2
Soft coke (India)	6.0	6.1	5.4
Brown coal coke	4.8	4.9	4.4
Semicoke (char)	6.8	6.9	6.2
Coke breeze	5.5	5.6	5.0
Charcoal	6.9	7.0	6.3
Petroleum coke	8.4	8.5	7.6

Table 15b. Calorie Equivalents of Liquid Fuels

	Teracalories (net)					
Fuels	Per Thousand Metric Tons	Per Thousand Long Tons	Thousand Kiloliters (cubic meters)	Per Thousand Barrels	Per Million Imperial Gallons	Per Million U.S. Gallons
LPG/LRG	10.8	11.0	5.8	.93	26.5	22.1
Propane	10.9	11.1	5.6	.88	25.3	21.0
Butane	10.7	10.9	6.2	.99	28.2	23.5
Natural gasoline	10.6	10.8	6.7	1.06	30.3	25.3
Gasoline	10.5	10.7	7.8	1.24	35.3	29.4
Aviation	10.5	10.7	7.5	1.19	33.9	28.2
Motor	10.5	10.7	7.8	1.24	35.3	29.4
Jet fuel (gas type)	10.4	10.6	7.9	1.26	35.9	29.9
Naphtha	10.5	10.7	7.8	1.24	35.3	29.4
Petrochemical feedstocks	10.5	10.7	7.8	1.24	35.3	29.4

continued on following page

Table 15b. *continued*

Fuels	Teracalories (net) Per Thousand Metric Tons	Per Thousand Long Tons	Thousand Kiloliters (cubic meters)	Per Thousand Barrels	Per Million Imperial Gallons	Per Million U.S. Gallons
White spirit	10.4	10.6	8.1	1.29	36.9	30.7
Jet fuel (n.f.d.)*	10.4	10.6	8.5	1.36	38.7	32.3
Kerosine	10.3	10.5	8.3	1.33	37.9	31.6
Jet fuel (kero type)	10.4	10.6	8.5	1.36	38.7	32.3
Other	10.3	10.5	8.3	1.33	37.9	31.6
Distillate fuel oil	10.2	10.4	8.8	1.39	39.8	33.2
Heating oil	10.2	10.4	8.5	1.35	38.5	32.0
Highway diesel oil	10.2	10.4	8.6	1.36	38.9	32.4
Indust. diesel oil	10.1	10.3	8.9	1.41	40.4	33.6
Fuel oil (n.f.d.)*	10.0	10.2	9.0	1.43	40.9	34.1
Fuel oil equivalent						
Fuels other than oil	10.0	10.2	9.0	1.43	40.9	34.1
Residual fuel oil	9.8	10.0	9.2	1.46	41.8	34.9
Light	9.9	10.1	9.2	1.46	41.8	34.9
Heavy	9.6	9.8	9.2	1.47	41.9	34.9
Lubes	9.9	10.1	8.7	1.39	39.6	33.0
Asphalt/bitumen	10.0	10.2	10.5	1.67	47.7	39.8
Petroleum coke	8.3	8.4	11.2	1.78	50.9	42.4
Wax	9.0	9.1	8.0	1.27	36.4	30.3
Crude oil (n.f.d.)*	10.2	10.4	8.8	1.39	39.8	33.2
Identified crudes			*(See Table 10)*			
Petroleum products (n.f.d.)*	10.2	10.4	8.8	1.39	39.8	33.2
Liquefied natural gas	12.6	12.8	5.3	0.84	24.1	20.1
Ethyl alcohol	6.6	6.7	5.2	0.83	23.6	19.7
Methyl alcohol	5.0	5.1	4.0	0.64	18.2	15.1
Tar	9.2	9.3	11.0	1.75	50.0	41.6

*Not further defined.

Table 15c. Calorie Equivalents of Gaseous Fuels and Electricity

Gaseous Fuels		Teracalories (net)			
		Per Teracalories	Per Thousand Million Btu	Per Million Meters³	Per Million Feet³
Natural gas in particular countries	gross	0.9	.227	*(See Table 11)*	
	net	1.0	.252		
Natural gas (n.f.d.)*					
Wet	gross	0.9	.227		
	net	1.0	.252	8.9	.252
Dry	gross	0.9	.227		
	net	1.0	.252	8.3	.235
Oven gas	gross	0.9	.227		
	net	1.0	.252	4.2	.119
City gas	gross	0.9	.227		
	net	1.0	.252	4.0	.113
Producer gas	gross	0.9	.227		
	net	1.0	.252	1.4	.040
Blast furnace gas	gross	0.9	.252	0.9	.026
	net	0.9	.252	0.9	.026
Refinery gas	gross	0.9	.227		
	net	1.0	.252	11.0	.311
Methane	gross	0.9	.227		
	net	1.0	.252	8.0	.227
Ethane	gross	0.9	.227		
	net	1.0	.252	14.2	.402
Propane	gross	0.9	.227		
	net	1.0	.252	20.5	.580
Isobutane	gross	0.9	.227		
	net	1.0	.252	25.8	.731
Butane	gross	0.9	.227		
	net	1.0	.252	26.7	.756
Pentane	gross	0.9	.227		
	net	1.0	.252	32.0	.905

Electricity	Teracalories per Gigawatt Hour	
	@ 100% efficiency	0.860
	@ 40% efficiency	2.150
	@ 35% efficiency	2.457
	@ 30% efficiency	2.867
	@ 25% efficiency	3.440
	@ 20% efficiency	4.330

*Not further defined.

Table 16a. Joule Equivalents of Liquid Fuels

Fuels	Terajoules (net) Per Thousand Metric Tons	Per Thousand Long Tons	Per Thousand Kiloliters (cubic meters)	Per Thousand Barrels	Per Million Imperial Gallons	Per Million U.S. Gallons
LPG/LRG	45.2	45.9	24.4	3.9	110.9	92.4
Propane	45.6	46.4	23.3	3.7	105.7	88.1
Butane	44.8	45.5	26.0	4.1	118.0	98.4
Natural gasoline	44.4	45.1	28.0	4.4	127.0	105.8
Gasoline	44.0	44.7	32.5	5.2	147.8	123.1
Aviation	44.0	44.7	31.2	5.0	141.8	118.1
Motor	44.0	44.7	32.5	5.2	147.8	123.1
Jet fuel (gas type)	43.5	44.2	33.1	5.3	150.3	125.3
Naphtha	44.0	44.7	32.5	5.2	147.8	123.1
Petrochemical feedstocks	44.0	44.7	32.5	5.2	147.8	123.1
White spirit	43.5	44.2	34.0	5.4	154.3	128.6
Jet fuel (n.f.d)*	43.5	44.2	35.7	5.7	162.2	135.2
Kerosine	43.1	43.8	34.9	5.6	158.7	132.2
Jet fuel (kero type)	43.5	44.2	35.7	5.7	162.2	135.2
Other	43.1	43.8	34.9	5.6	158.7	132.2
Distillate fuel oil	42.7	43.4	36.7	5.8	166.8	139.0
Heating oil	42.7	43.4	35.4	5.6	161.0	134.2
Highway diesel oil	42.7	43.4	35.9	5.7	163.0	135.8
Indust. diesel oil	42.3	43.0	37.2	5.9	169.0	140.9
Fuel oil (n.f.d.)*	41.9	42.5	37.7	6.0	171.2	142.6
Fuel oil equivalent						
Fuels other than oil	41.9	42.5	37.7	6.0	171.2	142.6
Residual fuel oil	41.0	41.7	38.6	6.1	175.2	146.0
Light	41.4	42.1	38.5	6.1	175.1	145.9
Heavy	40.2	40.8	38.6	6.1	175.3	146.1
Lubes	41.4	42.1	36.5	5.8	165.7	138.1
Asphalt/bitumen	41.9	42.5	44.0	7 0	199.7	166.4
Petroleum coke	34.8	35.3	46.9	7.5	213.1	177.6
Wax	37.7	38.3	33.5	5.3	152.3	157.0
Crude oil (n.f.d.)*	42.7	43.4	36.7	5.8	166.8	139.0
Identified crudes			*(See Table 10)*			
Petroleum products (n.f.d.)*	42.7	43.4	36.7	5.8	166.8	139.0
Liquefied natural gas	52.8	53.6	22.2	3.5	100.8	84.0
Ethyl alcohol	27.6	28.1	21.8	3.5	98.9	82.4
Methyl alcohol	20.9	21.3	16.7	2.7	76.1	63.4
Tar	38.5	39.1	46.1	7.3	209.2	174.3

*Not further defined.

Table 16b. Joule Equivalents of Gaseous Fuels and Electricity

Gaseous Fuels		Terajoules (net)			
		Per Teracalorie	Per Thousand Million Btu	Per Million Meters³	Per Million Feet³
Natural gas in particular countries	gross	3.7681	.950	*(See Table 11)*	
	net	4.1868	1.055	*(See Table 11)*	
Natural gas (n.f.d.)*					
Wet	gross	3.7681	.950		
	net	4.1868	1.055	37.3	1.055
Dry	gross	3.7681	.950		
	net	4.1868	1.055	34.8	.985
Oven gas	gross	3.7681	.950		
	net	4.1868	1.055	17.6	.498
City gas	gross	3.7681	.950		
	net	4.1868	1.055	16.7	.473
Producer gas	gross	3.7681	.950		
	net	4.1868	1.055	5.9	.167
Blast furnace gas	gross	4.1868	1.055	3.8	.107
	net	4.1868	1.055	3.8	.107
Refinery gas	gross	3.7681	.905		
	net	4.1868	1.055	46.1	1.305
Methane	gross	3.7681	.950		
	net	4.1868	1.055	33.5	.949
Ethane	gross	3.7681	.950		
	net	4.1868	1.055	59.5	1.684
Propane	gross	3.7681	.950		
	net	4.1868	1.055	85.8	2.428
Isobutane	gross	3.7681	.950		
	net	4.1868	1.055	108.0	3.056
Butane	gross	3.7681	.950		
	net	4.1868	1.055	111.8	3.164
Pentane	gross	3.7681	.950		
	net	4.1868	1.055	134.0	3.792

Electricity	Terajoules per Terawatt Hour	
	@ 100% efficiency	3600
	@ 40% efficiency	9000
	@ 35% efficiency	10285
	@ 30% efficiency	12000
	@ 25% efficiency	14400
	@ 20% efficiency	18000

*Not further defined.

Table 17a. Net Btu Equivalents of Liquid Fuels

Fuels	Thousand Million Btu					
	Per Thousand Metric Tons	Per Thousand Long Tons	Thousand Kiloliters (cubic meters)	Per Thousand Barrels	Per Million Imperial Gallons	Per Million U.S. Gallons
LPG/LRG	42.9	43.5	23.1	3.7	105.1	87.6
Propane	43.3	43.9	22.1	3.5	100.2	83.5
Butane	42.5	43.1	24.6	3.9	111.9	93.2
Natural gasoline	42.1	42.7	26.5	4.2	120.4	100.3
Gasoline	41.7	42.3	30.8	4.9	140.1	116.7
Aviation	41.7	42.3	29.6	4.7	134.4	112.0
Motor	41.7	42.3	30.8	4.9	140.1	116.7
Jet fuel (gas type)	41.3	41.9	31.4	5.0	142.5	118.7
Naphtha	41.7	42.3	30.8	4.1	140.1	116.7
Petrochemical feedstocks	41.7	42.3	30.8	4.9	140.1	116.7
White spirit	41.3	41.9	32.2	5.1	146.2	121.8
Jet fuel (n.f.d.)**	41.3	41.9	33.8	5.4	153.7	128.2
Kerosine	40.9	41.5	33.1	5.3	150.4	125.3
Jet fuel (kero type)	41.3	41.9	33.8	5.4	153.7	128.1
Other	40.9	41.5	33.1	5.3	150.4	125.3
Distillate fuel oil	40.5	41.1	34.8	5.5	158.1	131.8
Heating oil	40.5	41.1	33.6	5.3	152.6	127.2
Highway diesel oil	40.5	41.1	34.0	5.4	154.4	128.7
Indust. diesel oil	40.1	40.7	35.3	5.6	160.2	133.5
Fuel oil (n.f.d.)**	39.7	40.3	35.7	5.7	162.2	135.2
Fuel oil equivalent						
Fuels other than oil	39.7	40.3	35.7	5.7	162.2	135.2
Residual fuel oil	38.9	39.5	36.6	5.8	166.1	138.4
Light	39.3	39.9	36.6	5.8	166.0	138.3
Heavy	38.1	38.7	36.6	5.8	166.1	138.4
Lubes	39.3	39.9	34.6	5.5	157.0	130.9
Asphalt/bitumen	39.7	40.3	41.7	6.6	189.3	157.7
Petroleum coke	32.9	33.5	44.5	7.1	202.0	168.3
Wax	35.7	36.3	31.8	5.1	144.4	120.3
Crude oil (n.f.d.)**	40.5	41.1	34.8	5.5	158.1	131.8
Identified crudes			*(See Table 10)*			
Petroleum products (n.f.d.)**	40.5	41.1	34.8	5.5	158.1	131.8
Liquefied natural gas	50.0	50.8	21.0	3.3	95.5	79.6
Ethyl alcohol	26.2	26.6	20.6	3.3	93.7	78.1
Methyl alcohol	19.8	20.2	15.9	2.5	72.1	60.1
Tar	36.5	37.1	43.7	6.9	199.3	165.2

*For gross Btu values, see Table 31b.
**Not further defined.

Table 17b. Net Btu Equivalents of Gaseous Fuels and Electricity*

Gaseous Fuels		Thousand Million Btu			
		Per Teracalorie	Per Thousand Million Btu	Per Million Meters^3	Per Million Feet^3
Natural gas in particular countries	gross	3.571	0.9	*(See Table 11)*	
	net	3.968	1.0	*(See Table 11)*	
Natural gas					
Wet	gross	3.571	0.9		
	net	3.968	1.0	35.3	1.000
Dry	gross	3.571	0.9		
	net	3.968	1.0	32.9	.932
Oven gas	gross	3.571	0.9		
	net	3.968	1.0	16.7	.472
City gas	gross	3.571	0.9		
	net	3.968	1.0	15.9	.450
Producer gas	gross	3.571	0.9		
	net	3.968	1.0	5.6	.157
Blast furnace gas	gross	3.968	1.0	3.6	.102
	net	3.968	1.0	3.6	.102
Refinery gas	gross	3.571	0.9		
	net	3.968	1.0	43.7	1.236
Methane	gross	3.571	0.9		
	net	3.968	1.0	31.7	.899
Ethane	gross	3.571	0.9		
	net	3.968	1.0	56.3	1.596
Propane	gross	3.571	0.9		
	net	3.968	1.0	81.3	2.304
Isobutane	gross	3.571	0.9		
	net	3.968	1.0	102	2.899
Butane	gross	3.571	0.9		
	net	3.968	1.0	106	3.002
Pentane	gross	3.571	0.9		
	net	3.968	1.0	127	3.596

Electricity	Thousand Million Btu per Gigawatt Hour
@ 100% efficiency	3.412
@ 40% efficiency	8.530
@ 35% efficiency	9.748
@ 30% efficiency	11.373
@ 25% efficiency	13.648
@ 20% efficiency	17.060

Table 18a. Metric Ton Oil Equivalents of Liquid Fuels

	Thousand Metric Tons Actual Measure					
Fuels	Per Thousand Metric Tons	Per Thousand Long Tons	Per Thousand Kiloliters (cubic meters)	Per Thousand Barrels	Per Million Imperial Gallons	Per Million U.S. Gallons
LPG/LRG	1.0	1.016	0.54	0.086	2.45	2.04
Propane	1.0	1.016	0.51	0.081	2.32	1.93
Butane	1.0	1.016	0.58	0.092	2.63	1.96
Natural gasoline	1.0	1.016	0.63	0.100	2.86	2.38
Gasoline	1.0	1.016	0.74	0.118	3.36	2.80
Aviation	1.0	1.016	0.71	0.113	3.23	2.69
Motor	1.0	1.016	0.74	0.118	3.36	2.80
Jet fuel (gas type)	1.0	1.016	0.76	0.121	3.45	2.88
Naphtha	1.0	1.016	0.74	0.118	3.36	2.80
Petrochemical feedstocks	1.0	1.016	0.74	0.118	3.36	2.80
White spirit	1.0	1.016	0.78	0.124	3.54	2.95
Jet fuel (n.f.d.)*	1.0	1.016	0.82	0.130	3.72	3.10
Kerosine	1.0	1.016	0.81	0.129	3.68	3.07
Jet fuel (kero type)	1.0	1.016	0.82	0.130	3.72	3.10
Other	1.0	1.016	0.81	0.129	3.68	3.07
Distillate fuel oil	1.0	1.016	0.86	0.137	3.91	3.26
Heating oil	1.0	1.016	0.83	0.132	3.77	3.14
Highway diesel oil	1.0	1.016	0.84	0.134	3.82	3.18
Indust. diesel oil	1.0	1.016	0.88	0.140	4.00	3.33
Fuel oil (n.f.d.)*	1.0	1.016	0.90	0.143	4.09	3.41
Fuel oil equivalent						
Fuels other than oil	1.0	1.016	0.90	0.143	4.09	3.41
Residual fuel oil	1.0	1.016	0.94	0.149	4.27	3.56
Light	1.0	1.016	0.93	0.148	4.22	3.52
Heavy	1.0	1.016	0.96	0.153	4.36	3.32
Lubes	1.0	1.016	0.88	0.140	4.00	3.33
Asphalt/bitument	1.0	1.016	1.05	0.167	4.77	3.97
Petroleum coke	1.0	1.016	1.35	0.215	6.13	5.11
Wax	1.0	1.016	0.89	0.142	4.04	3.37
Crude oil (n.f.d.)*	1.0	1.016	0.86	0.137	3.91	3.26
Identified crudes	1.0	1.016		*(See Table 10)*		
Petroleum products (n.f.d.)*	1.0	1.016	0.86	0.137	3.91	3.26
Nonpetroleum Products	Thousand Metric Tons Oil Equivalent					
Liquefield natural gas	1.26	1.28	0.53	0.084	2.41	2.01
Ethyl alcohol	0.66	0.67	0.52	0.083	2.36	1.97
Methyl alcohol	0.50	0.51	0.40	0.064	1.82	1.51
Tar	0.92	0.93	1.10	0.175	5.00	4.16

*Not further defined.

Table 18b. Metric Ton Oil Equivalents of Gaseous Fuels and Electricity

		Thousand Metric Tons			
Gaseous Fuels		Per Teracalorie	Per Thousand Million Btu	Per Million Meters³	Per Million Feet³
Natural gas in particular countries	gross	0.09	.0227	*(See Table 11)*	
	net	0.10	.0252	*(See Table 11)*	
Natural gas (n.f.d.)*					
Wet	gross	0.09	.0227		
	net	0.10	.0252	0.89	.0252
Dry	gross	0.09	.0227		
	net	0.10	.0252	0.83	.0235
Oven gas	gross	0.09	.0227		
	net	0.10	.0252	0.42	.0119
City gas	gross	0.09	.0227		
	net	0.10	.0252	0.40	.0113
Producer gas	gross	0.09	.0227		
	net	0.10	.0252	0.14	.0040
Blast furnace gas	gross	0.10	.0252	0.09	.0025
	net	0.10	.0252	0.09	.0025
Refinery gas	gross	0.09	.0227		
	net	0.10	.0252	1.1	.0311
Methane	gross	0.09	.0227		
	net	0.10	.0252	0.80	.0227
Ethane	gross	0.09	.0227		
	net	0.10	.0252	1.42	.0402
Propane	gross	0.09	.0227		
	net	0.10	.0252	2.05	.0581
Isobutane	gross	0.09	.0227		
	net	0.10	.0252	2.58	.0731
Butane	gross	0.09	.0227		
	net	0.10	.0252	2.67	.0756
Pentane	gross	0.09	.0227		
	net	0.10	.0252	3.20	.0906

Electricity	Metric Tons per Gigawatt Hour
@ 100% efficiency	86
@ 40% efficiency	215
@ 35% efficiency	246
@ 30% efficiency	287
@ 25% efficiency	344
@ 20% efficiency	430

*Not further defined.

Table 19a. Kiloliter Oil Equivalents of Liquid Fuels

Fuels	Thousand Kiloliters Actual Measure					
	Per Thousand Metric Tons	Per Thousand Long Tons	Per Thousand Kiloliters (cubic meters)	Per Thousand Barrels	Per Million Imperial Gallons	Per Million U.S. Gallons
LPG/LRG	1.85	1.88	1.0	0.159	4.546	3.785
Propane	1.96	1.99	1.0	0.159	4.546	3.785
Butane	1.72	1.75	1.0	0.159	4.546	3.785
Natural gasoline	1.59	1.61	1.0	0.159	4.546	3.785
Gasoline	1.35	1.37	1.0	0.159	4.546	3.785
Aviation	1.41	1.43	1.0	0.159	4.546	3.785
Motor	1.35	1.37	1.0	0.159	4.546	3.785
Jet fuel (gas type)	1.32	1.34	1.0	0.159	4.546	3.785
Naphtha	1.35	1.37	1.0	0.159	4.546	3.785
Petrochemical feedstocks	1.35	1.37	1.0	0.159	4.546	3.785
White spirit	1.28	1.30	1.0	0.159	4.546	3.785
Jet fuel (n.f.d.)*	1.22	1.24	1.0	0.159	4.546	3.785
Kerosine	1.23	1.25	1.0	0.159	4.546	3.785
Jet fuel (kero type)	1.22	1.24	1.0	0.159	4.546	3.785
Other	1.23	1.25	1.0	0.159	4.546	3.785
Distillate fuel oil	1.16	1.18	1.0	0.159	4.546	3.785
Heating oil	1.20	1.22	1.0	0.159	4.546	3.785
Highway diesel oil	1.19	1.21	1.0	0.159	4.546	3.785
Indust. diesel oil	1.14	1.16	1.0	0.159	4.546	3.785
Fuel oil (n.f.d.)*	1.11	1.13	1.0	0.159	4.546	3.785
Fuel oil equivalent						
Fuels other than oil	1.11	1.13	1.0	0.159	4.546	3.785
Residual fuel oil	1.06	1.08	1.0	0.159	4.546	3.785
Light	1.08	1.09	1.0	0.159	4.546	3.785
Heavy	1.04	1.06	1.0	0.159	4.546	3.785
Lubes	1.14	1.15	1.0	0.159	4.546	3.785
Asphalt/bitumen	0.95	0.97	1.0	0.159	4.546	3.785
Petroleum coke	0.74	0.75	1.0	0.159	4.546	3.785
Wax	1.12	1.14	1.0	0.159	4.546	3.785
Crude oil (n.f.d.)*	1.16	1.18	1.0	0.159	4.546	3.785
Identified crudes		*(See Table 10)*		0.159	4.546	3.785
Petroleum products (n.f.d.)*	1.16	1.18	1.0	0.159	4.546	3.785
Nonpetroleum Products	Thousand Kiloliters Oil Equivalent					
Liquefield natural gas	1.40	1.42	0.59	0.094	2.68	2.23
Ethyl alcohol	0.73	0.75	0.58	0.092	2.62	2.19
Methyl alcohol	0.56	0.56	0.44	0.071	2.02	1.68
Tar	1.02	1.04	1.22	0.194	5.55	4.63

*Not further defined.

Table 19b. Kiloliter Oil Equivalents of Gaseous Fuels and Electricity

Gaseous Fuels		Thousand Kiloliters Per Teracalorie	Per Thousand Million Btu	Per Million Meters^3	Per Million Feet^3
Natural gas in particular countries	gross	.100	.0252	*(See Table 11)*	
	net	.111	.0280	*(See Table 11)*	
Natural gas (n.f.d.)*					
Wet	gross	.100	.0252		
	net	.111	.0280	0.989	.2080
Dry	gross	.100	.0252		
	net	.111	.0280	0.92	.0273
Oven gas	gross	.100	.0252		
	net	.111	.0280	0.47	.0138
City gas	gross	.100	.0252		
	net	.111	.0280	0.44	.0132
Producer gas	gross	.100	.0252		
	net	.111	.0280	0.16	.0046
Blast furnace gas	gross	.111	.0280	0.10	.0030
	net	.111	.0280	0.10	.0030
Refinery gas	gross	.100	.0252		
	net	.111	.0280	1.22	.0362
Methane	gross	.100	.0252		
	net	.111	.0280	0.89	.0252
Ethane	gross	.100	.0252		
	net	.111	.0280	1.58	.0447
Propane	gross	.100	.0252		
	net	.111	.0280	2.28	.0646
Isobutane	gross	.100	.0252		
	net	.111	.0280	2.87	.0812
Butane	gross	.100	.0252		
	net	.111	.0280	2.97	.0840
Pentane	gross	.100	.0252		
	net	.111	.0280	3.56	.1007

Electricity	Kiloliters per Gigawatt Hour
@ 100% efficiency	96
@ 40% efficiency	239
@ 35% efficiency	273
@ 30% efficiency	319
@ 25% efficiency	382
@ 20% efficiency	478

*Not further defined.

Table 20a. Barrel Oil Equivalents of Liquid Fuels

Fuels	Thousand Barrels Oil Actual Measure					
	Per Thousand Metric Tons	Per Thousand Long Tons	Per Thousand Kiloliters (cubic meters)	Per Thousand Barrels	Per Million Imperial Gallons	Per Million U.S. Gallons
LPG/LRG	11.6	11.8	6.29	1.0	28.6	23.8
Propane	12.3	12.5	6.29	1.0	28.6	23.8
Butane	10.8	11.0	6.29	1.0	28.6	23.8
Natural gasoline	10.0	10.1	6.29	1.0	28.6	23.8
Gasoline	8.5	8.6	6.29	1.0	28.6	23.8
Aviation	8.9	9.0	6.29	1.0	28.6	23.8
Motor	8.5	8.6	6.29	1.0	28.6	23.8
Jet fuel (gas type)	8.3	8.4	6.29	1.0	28.6	23.8
Naphtha	8.5	8.6	6.29	1.0	28.6	23.8
Petrochemical feedstocks	8.5	8.6	6.29	1.0	28.6	23.8
White spirit	8.1	8.2	6.29	1.0	28.6	23.8
Jet fuel (n.f.d.)*	7.7	7.8	6.20	1.0	28.6	23.8
Kerosine	7.8	7.9	6.29	1.0	28.6	23.8
Jet fuel (kero type)	7.7	7.8	6.20	1.0	28.6	23.8
Other	7.8	7.9	6.29	1.0	28.6	23.8
Distillate fuel oil	7.3	7.4	6.29	1.0	28.6	23.8
Heating oil	7.6	7.7	6.29	1.0	28.6	23.8
Highway diesel oil	7.5	7.6	6.29	1.0	28.6	23.8
Indust. diesel oil	7.1	7.3	6.29	1.0	28.6	23.8
Fuel oil (n.f.d.)*	7.0	7.1	6.29	1.0	28.6	23.8
Fuel oil equivalent						
Fuels other than oil	7.0	7.1	6.29	1.0	28.6	23.8
Residual fuel oil	6.7	6.8	6.29	1.0	28.6	23.8
Light	6.8	6.9	6.29	1.0	28.6	23.8
Heavy	6.6	6.7	6.29	1.0	28.6	23.8
Lubes	7.1	7.2	6.29	1.0	28.6	23.8
Asphalt/bitumen	6.0	6.1	6.29	1.0	28.6	23.8
Petroleum coke	4.7	4.8	6.29	1.0	28.6	23.8
Wax	7.1	7.2	6.29	1.0	28.6	23.8
Crude oil (n.f.d.)*	7.3	7.4	6.29	1.0	28.6	23.8
Identified crudes	*(See Table 10)*		6.29	1.0	28.6	23.8
Petroleum products (n.f.d.)*	7.3	7.4	6.29	1.0	28.6	23.8
Nonpetroleum Products	Thousand Barrels Oil Equivalent					
Liquefied natural gas	8.8	8.9	3.70	0.59	16.8	14.0
Ethyl alcohol	4.6	4.7	3.63	0.58	16.5	13.8
Methyl alcohol	3.5	3.6	2.80	0.44	12.7	10.6
Tar	6.4	6.5	7.69	1.22	34.9	29.1

*Not further defined.

Table 20b. Barrel Oil Equivalents of Gaseous Fuels and Electricity

Gaseous Fuels		Thousand Barrels Per Teracalorie	Per Thousand Million Btu	Per Million Meters^3	Per Million Feet^3
Natural Gas in particular countries	gross	.629	.158	*(See Table 11)*	
	net	.699	.176	*(See Table 11)*	
Natural gas (n.f.d.)*					
Wet	gross	.629	.158		
	net	.699	.176	6.22	0.1761
Dry	gross	.629	.158		
	net	.699	.176	5.80	0.1642
Oven gas	gross	.629	.158		
	net	.699	.176	2.94	0.0831
City gas	gross	.629	.158		
	net	.699	.176	2.80	0.0792
Producer gas	gross	.629	.158		
	net	.699	.176	0.978	0.0277
Blast furnace gas	gross	.699	.176	0.629	0.0178
	net	.699	.176	0.629	0.0178
Refinery gas	gross	.629	.158		
	net	.699	.176	7.688	0.2177
Methane	gross	.629	.158		
	net	.699	.176	5.591	0.1583
Ethane	gross	.629	.158		
	net	.699	.176	9.924	0.2810
Propane	gross	.629	.158		
	net	.699	.176	14.33	0.4057
Isobutane	gross	.629	.158		
	net	.699	.176	18.031	0.5106
Butane	gross	.629	.158		
	net	.699	.176	18.6	0.5284
Pentane	gross	.629	.158		
	net	.699	.176	22.36	0.63327

Electricity	Barrels per gigawatt Hour
@ 100% efficiency	601
@ 40% efficiency	1530
@ 35% efficiency	1717
@ 30% efficiency	2003
@ 25% efficiency	2404
@ 20% efficiency	3005

*Not further defined.

Table 21a. Barrel per Day Oil Equivalents of Solid Fuels

Fuels	Barrels per Day		
	Per Thousand Metric Tons	Per Thousand Long Tons	Per Thousand Short Tons
Bituminous coal			
Imported or exported	13.4	13.6	12.2
Consumed elsewhere		13.4	
Source and use not known	11.5	11.7	10.4
Anthracite			
United States	13.4	13.6	12.2
Other countries	14.4	14.6	13.0
Coal equivalent			
Coal and other fuels	13.4	13.6	12.2
Peat		6.7	
Coal briquettes, patent fuel		13.4	
Lignite briquettes	9.2	9.3	8.3
Coke briquettes	10.9	11.1	9.9
Peat briquettes	10.0	10.1	9.0
Gas coke	13.0	13.2	11.8
Oven coke	13.0	13.2	11.8
Soft coke (India)	11.5	11.7	10.4
Brown coal coke	9.2	9.3	8.3
Semicoke (char)	13.0	13.2	11.8
Coke breeze	10.5	10.7	9.6
Charcoal	13.2	13.4	12.0
Petroleum coke	16.1	16.3	14.6

Table 21b. Barrel per Day Oil Equivalents of Liquid Fuels

Fuels	Barrel per Day Actual Measure					
	Per Thousand Metric Tons	Per Thousand Long Tons	Per Thousand Kiloliters (cubic meters)	Per Thousand Barrels	Per Million Imperial Gallons	Per Million U.S. Gallons
LPG/LRG	31.9	32.4	17.2	2.74	78.3	65.2
Propane	33.8	34.3	17.2	2.74	78.3	65.2
Butane	29.7	30.2	17.2	2.74	78.3	65.2
Natural gasoline	27.4	27.8	17.2	2.74	78.3	65.2
Gasoline	23.3	23.7	17.2	2.74	78.3	65.2
Aviation	24.3	24.7	17.2	2.74	78.3	65.2
Motor	23.3	23.7	17.2	2.74	78.3	65.2
Jet fuel (gas type)	22.7	23.0	17.2	2.74	78.3	65.2
Naphtha	23.3	23.7	17.2	2.74	78.3	65.2
Petrochemical feedstocks	23.3	23.7	17.2	2.74	78.3	65.2
White spirit	22.1	22.4	17.2	2.74	78.3	65.2
Jet fuel (n.f.d.)*	21.0	21.4	17.2	2.74	78.3	65.2
Kerosine	21.3	21.6	17.2	2.74	78.3	65.2
Jet fuel (kero type)	21.0	21.4	17.2	2.74	78.3	65.2
Other	21.3	21.6	17.2	2.74	78.3	65.2
Distillate fuel oil	20.0	20.4	17.2	2.74	78.3	65.2
Heating oil	20.8	21.1	17.2	2.74	78.3	65.2
Highway diesel oil	20.5	20.8	17.2	2.74	78.3	65.2
Indust. diesel oil	19.6	19.9	17.2	2.74	78.3	65.2
Fuel oil (n.f.d.)*	19.1	19.5	17.2	2.74	78.3	65.2
Fuel oil equivalent						
Fuels other than oil	19.1	19.5	17.2	2.74	78.3	65.2
Residual fuel oil	18.3	18.6	17.2	2.74	78.3	65.2
Light	18.5	18.8	17.2	2.74	78.3	65.2
Heavy	17.9	18.2	17.2	2.74	78.3	65.2
Lubes	19.6	19.9	17.2	2.74	78.3	65.2
Asphalt/bitumen	16.4	16.7	17.2	2.74	78.3	65.2
Petroleum coke	12.8	13.0	17.2	2.74	78.3	65.2
Wax	19.4	19.7	17.2	2.74	78.3	65.2
Crude oil (n.f.d.)*	20.0	20.4	17.2	2.74	78.3	65.2
Identified crudes	*(See Table 10)*		17.2	2.74	78.3	65.2
Petroleum products (n.f.d.)*	20.0	20.4	17.2	2.74	78.3	65.2
Nonpetroleum Products	Barrel per Day Oil Equivalent					
Liquefied natural gas	24.1	24.5	10.1	1.61	46.1	38.4
Ethyl alcohol	12.6	12.8	10.0	1.58	45.2	37.7
Methyl alcohol	9.6	9.7	7.7	1.22	34.8	29.0
Tar	17.6	17.9	21.1	3.35	95.7	79.7

*Not further defined.

Table 21c. Barrel per Day Oil Equivalents of Gaseous Fuels and Electricity

Gaseous Fuels		Barrels per Day			
		Per Teracalorie	Per Thousand Million Btu	Per Million Meters³	Per Million Feet³
Natural gas in particular countries	gross	1.7	.43	*(See Table 11)*	
	net	1.9	.48	*(See Table 11)*	
Natural gas (n.f.d.)*					
Wet	gross	1.7	.43		
	net	1.9	.48	17.04	.4825
Dry	gross	1.7	.43		
	net	1.9	.48	15.89	.450
Oven gas	gross	1.7	.43		
	net	1.9	.48	8.04	.229
City gas	gross	1.7	.43		
	net	1.9	.48	7.66	.217
Producer gas	gross	1.7	.43		
	net	1.9	.48	2.68	.076
Blast furnace gas	gross	1.9	.48	1.72	.049
	net	1.9	.48	1.72	.049
Refinery gas	gross	1.7	.43		
	net	1.9	.48	21.0	.596
Methane	gross	1.7	.43		
	net	1.9	.48	15.3	.434
Ethane	gross	1.7	.43		
	net	1.9	.48	27.19	.770
Propane	gross	1.7	.43		
	net	1.9	.48	39.25	1.11
Isobutane	gross	1.7	.43		
	net	1.9	.48	49.40	1.40
Butane	gross	1.7	.43		
	net	1.9	.48	51.12	1.44
Pentane	gross	1.7	.43		
	net	1.9	.48	61.27	1.735

Electricity	Barrels per Day per Gigawatt Hour
@ 100% efficiency	1.65
@ 40% efficiency	4.12
@ 35% efficiency	4.70
@ 30% efficiency	5.49
@ 25% efficiency	6.59
@ 20% efficiency	8.23

*Not further defined.

Table 22a. Gross Btu Equivalents of Liquid Fuels

Fuels	Thousand Million Btu					
	Per Thousand Metric Tons	Per Thousand Long Tons	Per Thousand Kiloliters (cubic meters)	Per Thousand Barrels	Per Million Imperial Gallons	Per Million U.S. Gallons
LPG/LRG	46.8	47.6	25.4	4.0	115.4	96.1
Propane	47.2	48.0	24.2	3.8	110.0	91.6
Butane	46.4	47.1	27.0	4.3	122.7	102.2
Natural gasoline	41.0	41.7	29.0	4.6	131.8	109.8
Gasoline	44.8	45.5	33.3	5.3	151.4	126.0
Aviation	45.6	46.3	32.5	5.2	147.7	123.0
Motor	45.2	45.9	33.3	5.3	151.4	126.0
Jet fuel (gas type)	44.8	45.5	34.1	5.4	155.0	129.1
Naphthas	44.8	45.5	33.3	5.3	151.4	126.0
Petrochemical feedstocks	44.8	45.5	33.3	5.3	151.4	126.0
White spirit	45.2	45.9	35.3	5.6	160.5	133.6
Jet fuel (n.f.d.)*	44.8	45.5	36.9	5.9	167.7	139.7
Kerosine	44.8	45.5	36.1	5.7	164.1	136.6
Jet fuel (kero type)	44.8	45.5	36.9	5.9	167.7	139.7
Other	44.8	45.5	36.1	5.7	164.1	136.6
Distillate fuel oil	44.0	44.7	37.7	6.0	171.4	142.7
Heating	44.0	44.7	36.5	5.8	165.9	138.2
Highway disel oil	44.0	44.7	36.9	5.9	167.7	139.7
Indust. diesel oil	43.7	44.4	38.5	6.1	175.0	145.7
Fuel oil (n.f.d.)*	43.0	43.7	38.0	6.0	172.7	143.8
Fuel oil equivalent						
Fuels other than oil	43.0	43.7	38.0	6.0	172.7	143.8
Residual fuel oil	42.1	42.8	39.7	6.3	180.5	150.3
Light	42.5	43.2	39.3	6.2	178.7	148.8
Heavy	41.3	42.0	39.7	6.3	180.5	150.3
Lubes	42.9	43.6	37.7	6.0	171.4	142.7
Asphalt/bitumen	40.5	41.1	42.5	6.8	193.2	160.9
Petroleum coke	33.3	33.8	44.8	7.1	203.7	169.6
Wax	38.9	39.5	34.5	5.5	156.8	130.6
Crude oil (n.f.d.)*	44.0	44.7	37.7	6.0	171.4	142.7
Identified crudes			*(See Table 10)*			
Petroleum products (n.f.d.)*	44.0	44.7	37.7	6.0	171.4	142.7
Liquefied natural gas	54.8	55.7	23.0	3.7	104.6	87.1
Ethyl alcohol	29.0	29.5	23.0	3.7	104.6	87.1
Methyl alcohol	22.6	23.0	18.3	2.9	83.2	69.3
Tar	36.5	37.1	43.7	6.9	198.7	165.4

*Not further defined.

Table 22b. Gross Btu Equivalents of Gaseous Fuels and Electricity

Gaseous Fuels		Thousand Million Btu (gross)			
		Per Teracalorie	Per Thousand Million Btu	Per Million Meters³	Per Million Feet³
Natural gas in particular countries	gross	3.968	1.0	*(See Table 11)*	
	net	4.408	1.11	*(See Table 11)*	
Natural gas					
Wet	gross	3.968	1.0	39.2	1.1
	net	4.408	1.11		
Dry	gross	3.968	1.0	36.5	1.0
	net	4.408	1.11		
Oven gas	gross	3.968	1.0	18.3	0.5
	net	4.408	1.11		
City gas	gross	3.968	1.0	17.1	0.5
	net	4.408	1.11		
Producer gas	gross	3.968	1.0	6.0	0.17
	net	3.968	1.0		
Blast furnace gas	gross	3.968	1.0	3.6	0.102
	net	3.968	1.0	3.6	0.102
Refiner gas	gross	3.968	1.0	47.6	1.3
	net	4.408	1.11		
Methane	gross	3.968	1.0	35.7	1.0
	net	4.408	1.11		
Ethane	gross	3.968	1.0	63.1	1.8
	net	4.408	1.11		
Propane	gross	3.968	1.0	88.9	2.5
	net	4.408	1.11		
Isobutane	gross	3.968	1.0	111.5	3.2
	net	4.408	1.11		
Butane	gross	3.968	1.0	115.1	3.3
	net	4.408	1.11		
Pentane	gross	3.968	1.0	137.7	3.9
	net	4.408	1.11		

Electricity	Thousand Million Btu per Gigawatt Hour
@ 100% efficiency	3.412
@ 40% efficiency	8.530
@ 35% efficiency	9,748
@ 30% efficiency	11.373
@ 25% efficiency	13.648
@ 20% efficiency	17.060

*Not further defined.

Table 22c. Calorie Equivalents of Gaseous Fuels and Electricity

Gaseous Fuels		Teracalories (net)			
		Per Teracalories	Per Thousand Million Btu	Per Million Meters³	Per Million Feet³
Natural gas in particular countries	gross	0.9	.227	*(See Table 17)*	
	net	1.0	.252		
Natural gas (n.f.d.)*					
Wet	gross	0.9	.227		
	net	1.0	.252	8.9	.252
Dry	gross	0.9	.227		
	net	1.0	.252	8.3	.235
Oven gas	gross	0.9	.227		
	net	1.0	.252	4.2	.119
City gas	gross	0.9	.227		
	net	1.0	.252	4.0	.113
Producer gas	gross	0.9	.227		
	net	1.0	.252	1.4	.040
Blast furnace gas	gross	0.9	.252	0.9	.026
	net	0.9	.252	0.9	.026
Refinery gas	gross	0.9	.227		
	net	1.0	.252	11.0	.311
Methane	gross	0.9	.227		
	net	1.0	.252	8.0	.227
Ethane	gross	0.9	.227		
	net	1.0	.252	14.2	.402
Propane	gross	0.9	.227		
	net	1.0	.252	20.5	.580
Isobutane	gross	0.9	.227		
	net	1.0	.252	25.8	.731
Butane	gross	0.9	.227		
	net	1.0	.252	26.7	.756
Pentane	gross	0.9	.227	32.0	.905
	net	1.0	.252	32.0	.905

Electricity	Teracalories per Gigawatt Hour	
	@ 100% efficiency	.860
	@ 40% efficiency	.150
	@ 35% efficiency	.457
	@ 30% efficiency	.867
	@ 25% efficiency	.440
	@ 20% efficiency	.330

*Not further defined.

SPECIFIC GRAVITY AND API GRAVITY RELATED TO PARTICULAR PETROLEUM PRODUCTS AT 60°F

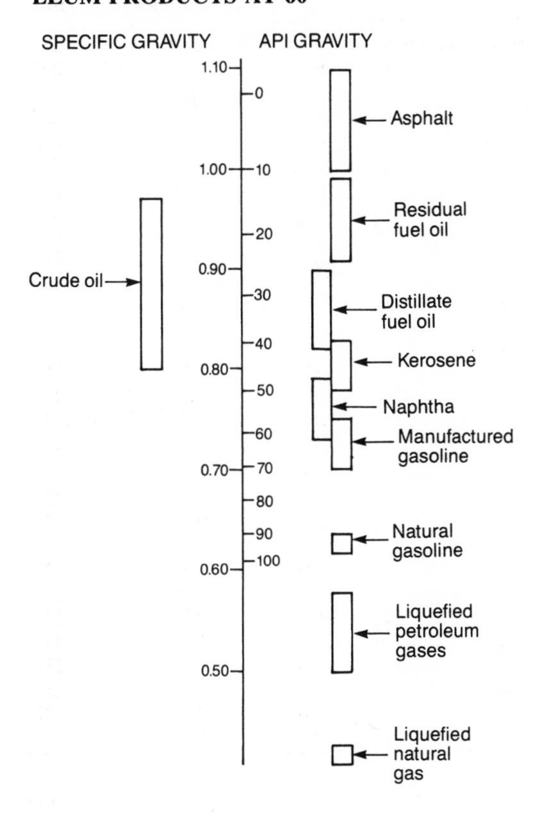

HIGH AND LOW HEAT VALUES OF PETROLEUM PRODUCTS RELATED TO API AND SPECIFIC GRAVITIES

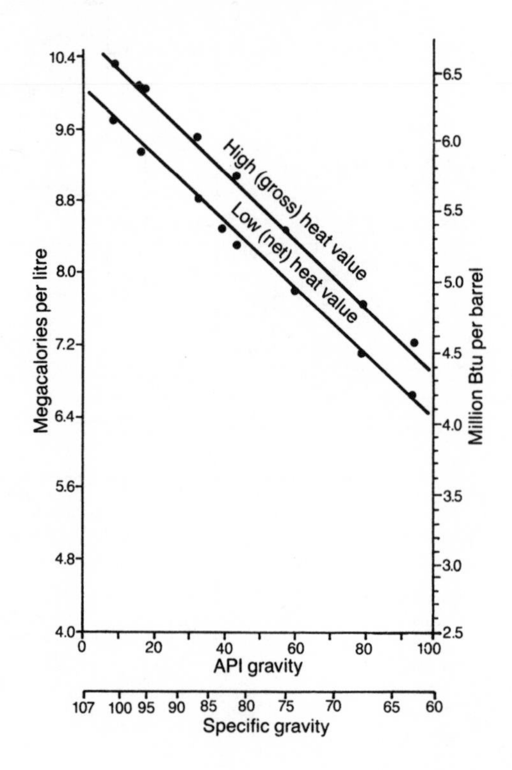